W9-ADZ-313
WITHDRAWN
L. R. COLLEGE LIBRARY

CHROMOSOME STRUCTURE AND FUNCTION

Impact of New Concepts

STADLER GENETICS SYMPOSIA SERIES

CHROMOSOME STRUCTURE AND FUNCTION
Impact of New Concepts
Edited by J. Perry Gustafson and R. Appels

GENE MANIPULATION IN PLANT IMPROVEMENT
Edited by J. Perry Gustafson

GENETICS, DEVELOPMENT, AND EVOLUTION
Edited by J. Perry Gustafson, G. Ledyard Stebbins, and Francisco J. Ayala

CHROMOSOME STRUCTURE AND FUNCTION

Impact of New Concepts

Edited by

J. Perry Gustafson

University of Missouri
Columbia, Missouri

and

R. Appels

CSIRO
Canberra, Australia

PLENUM PRESS • NEW YORK AND LONDON

Library of Congress Cataloging in Publication Data

Stadler Genetics Symposium (18th: 1987: University of Missouri–Columbia)
Chromosome structure and function: impact of new concepts / edited by J. Perry Gustafson and R. Appels.
p. cm.–(Stadler genetics symposia series)
"18th Stadler Genetics Symposium"–Cover.
Bibliography: p.
Includes index.
ISBN 0-306-42933-0
1. Chromosomes–Congresses. I. Gustafson, J. P. II. Appels, R. III. Title. IV. Series.
QH600.S73 1988 88-17814
574.87′322–dc19 CIP

A Division of Plenum Publishing Corporation
233 Spring Street, New York, N.Y. 10013

Printed in the United States of America

Drs. Ernie and Lotti Sears have devoted many decades of their lives toward the study of cereal genetics and chromosome manipulation. We would like to dedicate this volume to them in recognition of their lifelong commitment to science, and to the generous sharing of their knowledge and germplasm. Their doors have always been open to students and scientists from around the world. We are sure their future research will continue to be an inspiration to us all. They will always remain our best of friends.

FOREWORD

A Historical Perspective on the Study of Chromosome Structure and Function

R. Appels

Division of Plant Industry

CSIRO

P.O. Box 1600

A.C.T. AUSTRALIA

"Modern physical science gives us no model to explain the reduplication of the gene-string in each cell generation, or to explain the production of effective quantities of specific enzymes or other agents by specific genes. The precise pairing and interchange of segments by homologous gene-strings at meiosis also suggest novel physical properties of this form of matter".

Stadler (1954)

The very strong influence of reductionism in the history of understanding chromosome structure and function is evident in the above quotation from Stadler's 1954 paper, "The gene". Early observations on the constancy of the cytological appearance of chromosomes and their regular behaviour in cell division led to speculation on their biological importance. As genetics became more refined in the early decades of the 20th century the genes-on-a-string model of chromosomes developed and greater emphasis was placed on the further dissection of these structures. As a result, in the 1980's the reductionist approach is reaching a crest as extensive regions of the genetic material are being sequenced. However, although high levels of structural resolution have been attained, several aspects of the statement by Stadler remain unsolved. Problems such as the nature of the chromosome pairing at meiosis serve as a reminder that a process of building up a concept of the chromosome as a whole, from the parts generated by the genetic and the molecular/biochemical analyses, remains an important challenge.

The present symposium covers a broad spectrum of studies on chromosome structure and function and it is useful to view progress in this area with an appreciation of the history on which many of the studies are built; the following is a brief survey of the historical developments in our understanding of chromosome structure and function.

Era	Chromosome feature	Impact
1900 and before	Chromosome cytology. Primary and secondary constrictions as well as nucleolus organiser regions are clearly recognised. General chemical nature of RNA and DNA are understood.	Constancy of appearance and behaviour at cell division suggest chromosomes are of basic importance in cell biology.
1900-1930's	The chromosome as the bearer of genes: changes in chromosome structure (cytological) are correlated with genetic and phenotypic changes. Linkage of genes and crossing-over are found to occur in chromosomes.	Understanding of aneuploidy, polyploidy, translocations, inversions, and deficiencies. Specific phenotypic effects are associated with specific chromosome segments.
	Mitotic and meiotic (pachytene) chromosomes define euchromatin and heterochromatin in plants and animals. Salivary gland chromosomes of Drosophila, in particular, reveal fine detail in euchromatin.	The combination of salivary gland cytology and genetics in Drosophila with the pachytene cytology and genetics in corn provide a new level of appreciation of genetic material. Heterochromatin introduces the new concept of apparently genetically inert material which can influence gene expression by its position in the chromosome.
1930's-1940's	Misdivision of centromeres is described.	Provides a rationale for one level of change in chromosome structure (translocations involving

		whole chromosome arms) observed in later experiments (1950-1970's).
	Microscope spectrophotometry measures the chemical composition of cell components directly. Biochemical studies show nucleoprotein to be composed of deoxyribonucleic acid and histone protein. The chemical structure of the components making up DNA are defined.	Provide a starting point for studies on how DNA is organised into the interphase nucleuc and chromosomes.
1940's	Genetic instability in maize is demonstrated and reinforces earlier observations of genetic instability in other systems.	Concept of a unit in the chromosome which can move to create mutations. Concepts of regulatory and structural elements.
1950's	The overall structure of DNA and its importance as the genetic material is generally accepted. Semiconservative replication is demonstrated cytogenetically and biochemically.	New appreciation of the transmission of genetic information and the understanding of recombination and mutations.
	Polytene chromosomes in *Chironomus* correlate chromosome puffs and gene activity.	Concept of structural change in the chromosome related to gene activity.
	Successful manipulation of cereal chromosomes to introduce leaf rust into wheat. Homoeologous pairing mutation is described in wheat.	One product of chromosome engineering finds its way into a commercial wheat variety.
	Operon, and repressor protein, interpretation of bacterial gene expression.	Concept of specific regions of DNA to accomplish control of gene expression.
1960's	Development of preparatory techniques for examining unsectioned chromosomes by electron microscopy.	Concept of a chromosome comprised of folded or looped chromatin fibers develops.

	Definition, by amino acid sequencing, of the conserved character of the histones.	Concept of a DNA-protein combination providing the basic building blocks of a chromosome devlops.
	Ribosomal DNA genes are shown to be located in the nucleolus organiser region.	The first detailed understanding of a classic chromosome landmark.
	High content of repetitive sequences in eukaryotic chromosomes is demonstrated by Cot curves.	New questions are raised regarding the function of DNA and how genes are arranged in chromosomes.
1960's-1970's	The structure of highly purified 18S-26S ribosomal DNA and 5S DNA is determined. Ribosomal DNA transcription is visualised by electron microscopy.	Gene-spacer-gene-spacer etc., structure provides a detailed insight into eukaryotic genome structure.
	Banding techniques on mitotic chromosomes are developed.	New era of diagnostic work using mitotic chromosomes is initiated. In the analysis of human chromosomes this is particularly significant.
1970's	Specific classes of highly repetitive DNA sequences are shown to be located in heterochromatin. Genetic analyses demonstrate a low density of genes in heterochromatin.	An explanation of the apparent genetic inertness of heterochromatin is revealed.
	The repeated folding structure of chromatin is discovered.	The nucleosome concept develops and progress is made toward understanding how the DNA is arranged in chromosomes.
	A segment of 18S-26S ribosomal DNA from <u>Xenopus</u> is cloned into bacteria.	A new era of the high resolution analysis of gene structure is started.
	Demonstration that each chromosome of <u>Drosophila</u> contains only a single	The composition of chromosomes is clarified.

	continuous DNA strand, and that the length of it may increase by translocations but is unaffected by inversions.	
1980's	Major progress in the linear analysis of DNA.	Understanding the structure of transposable elements and how they cause genetic instability. Promoters, enhancers, and other controlling elements are identified for many genes. The exon/intron structures of many genes are determined.
	Specific eukaryote chromosomal proteins are identified. X-ray crystallography of DNA-protein complexes.	Details of how DNA interacts with protein molecules are revealed.
	Transformation systems in eukaryotes are developed.	Factors affecting the structure and function of chromosomes can be analysed with a great degree of precision.
	Cloning of yeast centromeres.	Reconstruction of a small chromosome to determine its essential features.
	Use of DNA markers are combined with earlier work (from 1960's) on isozyme markers in constructing genetic maps.	Major expansion of genetic maps from many organisms.

Acknowledgements

The author is indebted to Dr. A. J. Hilliker for his critical comments on an early version of this foreword.

The following symposium participants provided valuable remarks on omissions and errors in the summary, and their comments are gratefully acknowledged: G. D. Burkholder, S.C.R. Elgin, B. S. Gill, M. Maguire, G. P. Redei.

REFERENCES

Discussions with Drs. E. Coe, M. Green, J. P. Gustafson, B. John, W. J. Peacock and E. R. Sears. These researchers cannot be held responsible for deficiencies in the historical survey presented.

de Robertis, E.D.P., Nowinski, W. W., and Saez, F. A., 1949, General Cytology, published by W.B.S. Saunders, Philadelphia and London.

Redei, G. P., 1974, Steps in the evolution of genetic concepts, Biol. Zbl., 93:385-424.

Stadler, L. J., 1954, The gene, Science, 120:811-819.

Strickberger, M. W., 1976, Genetics, published by MacMillan, New York, 3rd edition.

Sturtevant, A. H., 1965, A history of genetics, published by Harper and Row, New York.

Wilson, E. B., 1925, The cell in development and heredity, published by the MacMillian Co., New York, thrid edition.

Wyckoff, R.W.G., 1959, Optical methods in cytology, In the Cell, ed. J. Brachet and A. E. Mirsky, 1:1-20.

ACKNOWLEDGEMENT

The editors gratefully acknowledge the generous support of the following contributors: Agracetus; College of Agriculture, University of Missouri; Division of Biological Sciences, University of Missouri; Calgene, Inc.; CIBA-GEIGY Corporation; DeKalb Pfizer Genetics; Del Monte Company; E. I. du Pont de Nemours & Company; Garst Seed Company; Illinois Foundation Seeds, Inc.; School of Medicine, University of Missouri; Molecular Genetics, Inc.; Monsanto Company; Northrup King Company; Philip Morris Inc. and Pioneer Hi-Bred International, Inc. who made the 18th Stadler Genetics Symposium a success.

The speakers, who spent a tremendous amount of time preparing their manuscripts and lectures are gratefully acknowledged. Without their expertise and dedication the Symposium could not have taken place.

I wish to thank the local chairpersons for their effort to see that everyone in the respective sessions were well taken care of during the Symposium.

The behind-the-scene and on-site preparation was excellently handled by Joanne Fredmeyer and Joy Gasparovic from Conferences and Specialized Courses, University of Missouri, who tirelessly handled all of my peculiar requirements and made sure everything was extremely well organized.

Many thanks are due to Joyce Reinbott, University of Missouri, for her excellent secretarial help in handling all the correspondence and typing. Thanks to R. J. Kaufman for helping to arrange financial support. A special thanks goes to Kathleen Ross for keeping the lab running.

J. P. Gustafson
R. Appels

April 1, 1988
Columbia, Missouri

CONTENTS

THE ANALYSIS OF CHROMOSOME ORGANIZATION BY EXPERIMENTAL MANIPULATION

Gary D. Burkholder

Department of Anatomy
College of Medicine
University of Saskatchewan
Saskatoon, Saskatchewan S7N OWO
Canada

INTRODUCTION

Chromosomes have occupied a pivotal position in genetics ever since their involvement in heredity became firmly established in the early years of the twentieth century. There has consequently been a longstanding and fervent desire to understand how chromosomes are organized, and in particular, to ascertain how the hereditary material is arranged and regulated at the molecular level.

Historically, the evolution of concepts concerning chromosome organization has often occurred in a step-wise fashion, with major advances and shifts in knowledge following closely upon the heels of new technological advances. As in many other areas of genetics, the rapid development of innovative technology during the past few years has provided the opportunity for increasingly sophisticated approaches to the study of chromosome organization. The application of an armamentarium of new cytogenetic, morphological, and molecular biological techniques has led to radical and sometimes controversial changes in our views about the way in which chromosomes are constructed.

The purpose of this review is to highlight emerging concepts of chromosome organization. More specifically, two fundamental but interrelated aspects of this organization will be addressed: 1. the higher-order arrangement of chromatin in mitotic chromosomes, and 2. the underlying molecular and functional organization revealed by the chromosome banding techniques. In order to place recent research in its proper perspective, the problems and pitfalls of chromosome research are considered, and alternative possibilities of chromosome organization are discussed. References have been chosen to illustrate specific points as opposed to providing a comprehensive literature review. For the most part, the emphasis will be on the organization of mammalian chromosomes since these have been studied in the greatest detail.

QUESTIONS, APPROACHES AND PROBLEMS IN CHROMOSOME RESEARCH

Major questions relating to chromosome organization include: 1. How is DNA arranged in the chromosome? 2. How are the other molecular components (e.g. proteins) of chromosomes involved in this organization? 3. How is the structural organization of the chromosome related to its functional organization?

In spite of intensive study, the answers to these questions have been elusive. The reason for this is apparent from the following considerations. A human diploid nucleus contains about 6.4 pg of DNA which corresponds to a combined total length of about 1.74 meters. This DNA is subdivided amongst the 46 chromosomes and, based on studies in other organisms (Laird, 1971; Kavenoff and Zimm, 1973; Molitor et al., 1974), exists as a single uninterrupted molecule within each chromatid. At metaphase, a chromosome only a few µm long may contain anywhere from 1.4 to 7.3 cm of DNA packaged into each sister chromatid. This translates into DNA and histone concentrations of approximately 25 mg/ml each in a condensed chromosome while the nonhistone protein concentration may be as high as 50-75 mg/ml (Okada and Comings, 1980). Such incredibly high concentrations mean that the chromosomes are chock-full of highly-condensed chromatin, effectively hindering any and all attempts by morphological approaches to examine the manner in which constituent chromatin fibers are organized in native metaphase chromosomes.

The vast majority of morphological studies on mammalian metaphase chromosomes have utilized light or electron microscopy. For light microscopy, used in most cytogenetic studies, the mitotic cells are exposed to a hypotonic salt solution (usually 0.075 M KCl), fixed in a mixture of acetic acid and methanol (1:3), and dried on slides. This is the only method known that gives good chromosome morphology and adequate spreading. Additional treatments are usually required to produce chromosome banding. It is obvious that an understanding of the effects of these preparatory procedures and banding treatments on chromosome constituents is required in order to fully comprehend what the observed morphological effects mean in terms of normal chromosome organization.

Thin-section electron microscopy, which has been extremely informative in demonstrating the intricate structure of other cellular organelles, has been notably unproductive in revealing chromatin organization in mitotic cells. The chromosomes have a homogeneous, electron-dense appearance in thin-sections (DuPraw, 1970). Short segments or cut ends of fibers can be seen, suggesting a densely-packed, tangled network of chromatin, but there is no evidence of any kind of regular organization or internal structure. In fact, the only revealing aspect of chromosome structure seen in thin-sections is the kinetochore (Comings and Okada, 1971; Roos, 1973). This is the plate-like structure, located in the centromeric region on the superficial surface of each sister chromatid, to which the spindle microtubules are attached.

The dirth of information obtained from thin-section electron microscopy has necessitated the development of alternate ultrastructural methods for examining chromatin organization in chromosomes. Without exception, these methods all involve some kind of experimental manipulation designed to loosen or disaggregate the typical condensed chromosome structure so that internal detail can be observed microscopically. In some instances, this involves hypotonic shocks and spreading the chromosomes on aqueous hypophases; in others, it involves the isolation of mitotic chromosomes, followed by their exposure to a variety of chemical treatments. Unfortunately, these approaches are problematic because they can unknowingly modify the chromosomes so that the observed morphology does not accurately reflect the organization that exists *in vivo*. The fibrous nature of the chromosomes renders them highly susceptible to the formation of artifacts by any kind of manipulative procedure. This appears to be reflected in the diversity of results, opinions, and models that have appeared and disappeared over the years.

In view of these inherent problems, it is imperative to constantly question the effect of the experimental manipulations on the results obtained. Caution must also be exercised in extrapolating the implications of the results to the *in vivo* organization of chromosomes until the conclusions can be verified by other independent approaches. Surprisingly, in spite of the problems, an integrated concept of chromosome organization is slowly emerging.

ORGANIZATION OF THE CHROMATIN FIBER

Although many of the fine details have yet to be worked out, it has been convincingly established that nuclear DNA is organized into several hierarchical levels within chromatin (reviewed by Felsenfeld, 1978; McGhee and Felsenfeld, 1980; Igo-Kemenes et al., 1982). The fundamental unit of chromatin organization is the nucleosome, consisting of a well-protected core and a nuclease-sensitive linker region which varies in length but is approximately 60 base-pairs long. The core consists of 146 base-pairs of DNA wrapped around the outside of a histone octamer comprised of two molecules each of histones H2A, H2B, H3, and H4. One molecule of histone H1 is located at the site where DNA enters and exits the core particle, effectively closing two full turns of DNA around the histone octamer. The linker DNA interconnects adjacent nucleosomal DNA, producing a nucleosome chain containing a single continuous DNA molecule. As seen by electron microscopy of chromatin prepared under conditions of low ionic strength, this chain has a typical beads-on-a-string appearance.

In vivo, the elemental chromatin fiber consists of a linear chain of closely-packed nucleosomes without any apparent internucleosomal space, forming a 10-11 nm filament. The next level of order is the 25-30 nm chromatin fiber characteristically seen in most electron microscopy studies. This thicker fiber appears to form through a continuous coiling of the 10 nm fiber, giving rise to a solenoid with 6-7 nucleosomes per turn. It is thought that histone H1 may be involved in stabilizing the thick fiber, perhaps by crosslinking adjacent turns of the coil (Felsenfeld, 1978). Moderate ionic strengths are required to preserve the stability of this fiber, and it can be reversibly converted to the 10 nm fiber by successively lowering and raising the ionic strength.

The manner in which nonhistones may associate with the 10 or 30 nm fibers is largely unknown. However, the high mobility group (HMG) proteins 14 and 17, which appear to be preferentially associated with transcriptionally-active chromatin, have been shown to reversibly bind to mononucleosomes (reviewed by Igo-Kemenes et al., 1982). Additional details of how these and other nonhistones bind to chromatin, and the implications of these associations, must await further study.

CONSIDERATIONS FOR THE ORGANIZATION OF THE MITOTIC CHROMOSOME

During the transition from interphase to mitosis, there is no change in the basic structure of nucleosomes or their organization into chromatin fibers; the 10 and 30 nm fibers of interphase chromatin are also found in mitotic chromosomes (Rattner and Hamkalo, 1978a, 1978b; Labhart et al., 1982). On the other hand, there is a major rearrangement of the 30 nm fibers during prophase to form the compact, condensed structure of the mitotic chromosomes. This higher-order organization of chromatin has been remarkably refractory to elucidation.

Out of the myriad of chromosome models that have appeared over the years, variations on three major themes are predominant: 1. an organization based on chromatin folding; 2. an organization based on higher levels of helical coiling; or 3. an organization based on a central core that dictates chromatin arrangement. Each of these models is supported by experimental data, and this has led to both confusion and controversy.

In considering any model of chromosome structure, it is important to bear in mind the following: 1. DNA is responsible for maintaining the linear continuity of the chromosome. If the DNA is broken, this damage manifests itself as a break in the mitotic chromosome. 2. The organization of the chromosome is ordered. The genetic and cytogenetic evidence clearly indicates that the genes are arranged in a linear order along the chromosome. This reflects the order of genes along the DNA molecule. Both the chromomere patterns in meiotic pachytene chromosomes and the banding patterns in mitotic chromosomes are constant, further implying an ordered, as opposed to random organization. 3. The existence of chromosome banding imposes additional complications on models of chromosome organization. This has often been ignored. Bands appear to represent the highest level of chromosome organization but their origins are probably determined at a grass-roots molecular level. An understanding of what the bands represent in terms of underlying molecular arrangements and the implications of these arrangements for the structural and functional organization of chromosomes will be essential for a complete understanding of how chromosomes are put together.

EVIDENCE SUPPORTING A FOLDED-FIBER ORGANIZATION OF CHROMOSOMES

Since thin-sectioning reveals very little of the organization of the mitotic chromosome, alternative preparative techniques have been developed. One of the more successful of these involves spreading mitotic cells on an air-liquid (usually water) interface (DuPraw, 1965). The hypotonic shock plus the surface-tension forces existing at the interface cause the cells to lyse, scattering the chromosomes, some of which remain suspended from the interface in the liquid hypophase. When a specimen grid is touched to the surface, the spread chromosomes attach to the support film and can then be processed by staining, dehydration and critical-point drying. This is commonly called the whole mount-method of electron microscopy because it allows for the observation of whole, presumably intact chromosomes.

This method clearly visualizes the fibrous nature of the chromosomes, but the degree of preservation of chromosome morphology is variable, presumably depending on the surface-tension forces. When the morphology of the chromosomes is well-preserved, the chromatin fibers are so densely-packed that only those emerging from the periphery can be clearly observed. With greater degrees of spreading, there is either a tangle of fibers or a high degree of fiber stretching accompanied by a loss of chromosome morphology. Nonetheless, in moderately-dispersed chromosomes, the chromosome is composed of a jumbled mass of irregular, bumpy fibers with an average diameter of 25-30 nm (DuPraw, 1965, 1966; Comings and Okada, 1972). These fibers correspond to the solenoidal fiber.

On the basis of observations made on chromosomes prepared by the whole-mount technique, DuPraw (1965, 1966) proposed a folded-fiber model in which a single 25 nm chromatin fiber is multiply-folded both transversely and longitudinally to form the highly-condensed structure of a chromatid. Comings (1972a) modified this model by suggesting that the chromatin fibers have a net-like arrangement in each chromatid, with the folding occurring over relatively short distances. The existence of fibers running longitudinally throughout the chromatid would be inconsistent with the established correlation between the genetic and cytological maps as well as with the occurrence of reciprocal translocations. The centromere was viewed as a region in which the folded chromatin fibers formed interdigitated loops between the two sister chromatids (Comings and Okada, 1970).

In the folded-fiber model, the DNA itself maintains the linear continuity of the chromosome. If the DNA is cut somewhere along its length, the break if unrepaired would manifest itself as a chromosome or chromatid break at the subsequent mitosis. This is consistent with the cytogenetic data. It is assumed that the chromatin fiber inherently contains the mechanism responsible for its folding, presumably as a result of the strategic distribution along the fiber of proteins involved in this function.

The advantage of the whole-mount method is that it permits a direct examination of the chromosomes immediately after lysing and spreading the mitotic cells. The surface-tension forces alone are responsible for loosening and dispersing the chromatin fibers in order to reveal the internal structure. Compared to other methods which rely on chemical treatments to disperse the chromosomes, this method may provide a reasonably representative overview of a native chromosome. On the other hand, the spreading forces can lead to extensive stretching of the chromatin fibers, distorting normal relationships.

EVIDENCE SUPPORTING A CHROMOSOME ORGANIZATION BASED ON SUPERCOILING

Since the organization of the chromatin fiber is based on coiling, it seems a natural extension to suggest that the higher-levels of chromosome organization may depend on further levels of supercoiling. In fact, there is incontrovertible evidence that supercoiling is at least partially involved in chromosome organization.

When human mitotic cells are exposed to a hypotonic solution composed of a special mixture of salts, a spiral organization of the mitotic chromatids is observed by light microscopy (Ohnuki, 1968). The number of gyres appears to be constant for the same chromosome from different cells, but the direction of coiling is random, i.e. the gyre direction is not specific to any chromosome region. In these preparations, the centromere is uncoiled and is frequently elongated. Chromatid spiralization has been observed in many different cell types from a variety of species, and is also apparent in the prophase chromosomes of living Haemanthus endosperm (Bajer, 1957).

A coiled configuration of the chromatids can also be seen by whole-mount electron microscopy. This so-called quaternary coiling appears to be superimposed on a folded-fiber organization (Dupraw, 1966; Abuelo and Moore, 1969). Scanning electron microscopy (SEM) of prematurely-condensed chromosomes has demonstrated coiling in G_1 chromatids, and in G_2 chromatids during the transition into mitosis (Hanks et al., 1983). This spiralization probably represents the final condensation of chromatin into the compact metaphase chromosome.

In some cell types, spiralization of the chromosomes occurs as a natural phenomenon (Rattner and Lin, 1985). When chromosomes are released from these cells at physiological ionic strength and examined by electron microscopy, the chromatids are composed of a helically-coiled fiber having a uniform diameter of 200-300 nm. Within these chromatid coils are 25-30 nm chromatin fibers having a looped configuration. An identical configuration can also be demonstrated in cells that do not normally show chromosome spiralization. Growth of such cells in the presence of 33258-Hoechst or 5-azacytidine produces a marked increase in chromosome length and the appearance of a coiled organization of the chromatid. SEM studies showed that successive coils of the 200-300 nm fiber are normally in very close apposition giving the condensed chromosome a smooth topology. This is probably why the coiled configuration is not usually apparent.

These data provide clear evidence that the chromatids of condensed metaphase chromosomes are constructed by quaternary coiling. Is there any evidence of a level of coiling between the solenoid (secondary coiling) and the the quaternary coils? Based on a variety of microscopical evidence, Sedat and Manuelidis (1978) proposed that the solenoidal fiber is supercoiled in an orderly fashion into a 200 nm tube which may ultimately undergo quaternary coiling during chromosome condensation. A similar hierarchy of helices was suggested by Bak et al. (1977).

In biochemically-isolated chromosomes studied by SEM, the surface topology of the chromosomes supports the view that the 30 nm solenoidal fibers are extensively supercoiled (Daskal et al., 1976) but apparently not in a very orderly fashion. Microconvules about 52 nm in diameter cover the surface of the chromosomes and appear to consist of several orders of coiling. Other SEM data also suggest a superhelical arrangement of the 30 nm fiber (Taniguchi and Takayama, 1986).

THE PROS AND CONS OF CHROMOSOME CORES

Chromosome models based on cores are conceptually attractive because, unlike other models, they offer a clearly-defined means to produce chromatin order out of apparent chaos; the core provides a framework to which the chromatin is attached in a consistent and reproducible fashion. Historically, core models have involved either a proteinaceous (Taylor, 1958) or deoxyribonucleoprotein (Stubblefield and Wray, 1971) chromatid axis, to which the chromosomal DNA was attached as epichromatin. These early models became the focus of strong controversy and ultimately fell into disrepute (Comings, 1972a). Current evidence favoring a core organization of the chromosome comes from two sources: observations made on histone-depleted chromosomes and on chromosomes stained with silver.

When isolated chromosomes are exposed to 2M NaCl or a mixed dextran sulfate-heparin solution, virtually all of the histones and many nonhistones are extracted (Adolph et al., 1977a). The residual chromosome consists of an axial fibrous network, called a scaffold, that is surrounded by a halo of DNA loops (Paulson and Laemmli, 1977). Each loop of DNA is between 10 and 30 μm in length, corresponding to 0.5-2.0 μm of 30 nm chromatin fiber, and its base is anchored within the scaffold. These results suggest a radial-loop organization in which a central scaffold arranges the constituent chromatin fiber of a chromatid into a series of peripherally-radiating loops (Laemmli et al., 1978).

The scaffold can be isolated as an independent structure (Adolph et al., 1977b), indicating that it is structurally distinct from the chromatin fibers. It has the shape and size of an untreated chromosome (Adolph et al., 1977b; Laemmli et al., 1978). Recent data indicates that the scaffold contains residual kinetochores (Earnshaw et al.,1984), axial elements, and a diffuse peripheral network extending throughout the chromosome (Earnshaw and Laemmli, 1983). Depending on the Mg^{2+} concentration, the scaffold can apparently expand or contract, suggesting that it might be a dynamic structure involved in chromosome condensation. Scaffolds are stabilized by metalloprotein interactions involving Cu^{2+} and possibly Ca^{2+}, whereas treatment with 2-mercaptoethanol or metal chelators causes them to fall apart (Lewis and Laemmli, 1982).

Biochemically, the scaffold is virtually devoid of histones but contains a subset of approximately 30 nonhistone proteins

(Adolph et al.,1977a, 1977b). Highly purified scaffolds, containing 3-4% of the total chromosomal proteins and 1-3% of the DNA, have only 2 major proteins, Sc1 and Sc2, with molecular weights of 170 and 135 kD, respectively (Lewis and Laemmli, 1982; Earnshaw and Laemmli, 1983). Recent work utilizing polyclonal antibodies has identified the Sc1 protein as topoisomerase II (topo II) (Earnshaw et al., 1985; Gasser et al., 1986). This enzyme alters the topology of DNA by allowing a double-stranded DNA segment to pass through a transient break made in a second double-stranded segment (Wang, 1985). It can apparently relax closed supercoiled loops of DNA and may function in DNA replication, repair, and transcription (Mattern et al., 1982; Wang, 1985).

In experimentally-expanded chromosomes, immunolocalization studies involving fluorescence and electron microscopy have shown that topo II is concentrated around the center of each chromatid where the axial elements of the scaffold are expected to be (Earnshaw and Heck, 1985; Gasser et al., 1986). These studies imply that topo II may be bound to the bases of the radial loop domains of DNA. The pattern of antibody binding suggests that there may be a series of discrete foci containing topo II within the scaffold, possibly representing clusters of looped DNA domains associated with topo II at their bases (Earnshaw and Heck, 1985; Gasser et al., 1986). Unfortunately, the localization of topo II has apparently not been possible in normal condensed chromosomes (Gasser et al., 1986). Thus, some caution is required in utilizing these data from expanded chromosomes to predict the intrachromosomal localization of topo II *in vivo*.

These studies have led to speculation that, in addition to an enzymatic role, topo II may have a structural function within the scaffold, stabilizing the repeated loops of DNA. Alternately, if topo II only has an enzymatic function, it may associate with other scaffold proteins that are principally structural in function. In this regard, no function has yet been assigned to Sc2.

Assuming that the chromosomal DNA is organized in loops through binding to the scaffold, topo II might function to control the topology of these looped domains, particularly if it was directly responsible for closing the loops at their bases. There is circumstantial evidence that torsional stress may be involved in regulating the transcription of active genes (Weintraub, 1985). If topo II regulates superhelical stress in looped domains through its association with the bases of the loops, it may possibly influence gene activity within the loop domain. Another possibility is that topo II plays a role in regulating the segregation of looped domains at the termination of DNA replication, thereby allowing the disentangling of daughter DNA molecules and the segregation of sister chromatids at the ensuing mitosis. The enzyme appears to have such a function in prokaryotes (Steck and Drlica, 1984) and yeast

(*Saccharomyces cerevisiae*) (DiNardo et al., 1984; Uemura and Yanagida, 1984). The strategic localization of topo II at the base of the DNA loops would also facilitate this possible function. At the moment, however, there is no experimental proof that topo II is directly involved in closing the loop domains. In fact, the *in vivo* function of this enzyme in eukaryotic cells remains uncertain (DiNardo et al., 1984; Earnshaw and Heck, 1985), although a recent report (Nelson et al., 1986) indicates that topo II in mammalian cells is associated with nascent fragments of newly-replicated DNA, suggesting a function in DNA fork progression. A further clarification of the *in vivo* function and location of this enzyme will be required to assess the validity of the proposals concerning its role in the maintenance, regulation, and segregation of chromosomal looped domains.

The occurrence of looped chromatin domains as a characteristic feature of chromosome organization was first suggested by the DNA loops in dehistonized chromosomes (Paulson and Laemmli, 1977), but the act of dehistonization itself is likely to have major distortive and disruptive effects on chromosomes, making a looped arrangement in untreated chromosomes tentative. However, other evidence based on less disruptive manipulations convincingly indicates the reality of chromatin loops. Metaphase chromosomes, exposed to hypotonic conditions or to chemicals that chelate divalent cations, swell and expand outwards from their periphery. When these chromosomes are examined by thin-section electron microscopy, their constituent chromatin fibers are clearly organized in loops that radiate outward in all directions (Marsden and Laemmli, 1979; Adolph 1980a, 1980b, 1981). The bases of the loops merge into the longitudinal central axis of the chromatid, which is characteristically condensed and densely-stained. Due to the compactness of this axial region, it was not possible to ascertain how the bases of the loops are anchored and interconnected, but it is generally assumed that a scaffold is involved (Marsden and Laemmli, 1979; Adolph, 1980b).

The results from a variety of experiments suggest that the chromatin fibers in interphase nuclei are also topologically constrained in loop configurations (Benyajati and Worcel, 1976; Cook and Brazell, 1978; Igo-Kemenes and Zachau, 1978; Vogelstein et al., 1980; Lebkowski and Laemmli, 1982). These interphase chromatin loops are about the same size as those found in metaphase chromosomes. This suggests that the looped domain organization may be maintained throughout the cell cycle.

Recent data suggest that there are specific sequences involved in the attachment of DNA to the nuclear matrix (Mirkovitch et al., 1984). In the HSP70 gene and the histone gene repeats of *Drosophila*, the attachment sites are within nontranscribed DNA segments and have been mapped to regions of a few hundred base-pairs. They may be comprised of unique sequences (Mirkovitch et al., 1986). The looped domains delimited by the attachment sites are heterogeneous in size, and contain from one to eight unrelated genes, more than one of which may be transcriptionally-active (Mirkovitch et al., 1986). In the mouse (Mus musculus), the transcription unit of the immunoglobulin kappa gene contains a nuclear matrix attachment region that has two topo II sites and is adjacent to the tissue-specific enhancer (Cockerill and Garrard, 1986). This association of topo II with the matrix attachment region is a further indication that this enzyme may regulate torsional stress in the looped domains and play a role in their functional organization.

In addition to the scaffold structure observed in dehistonized chromosomes, a second line of evidence suggests that chromosomes have organizational cores. Howell and Hsu (1979) found that when chromosomes were exposed to a prolonged hypotonic treatment during harvesting, cores could be observed after staining the preparations with ammoniacal silver. This procedure also visualizes the active nucleolar organizer regions (NORs) of the chromosomes (Goodpasture and Bloom, 1975; Miller et al., 1976).

By both light and electron microscopy, the core elements are intensely-stained with silver, extend linearly throughout the axial region of each chromatid, and are surrounded by a halo of palely-stained, dispersed chromatin (Kaiserman and Burkholder, 1980; Burkholder and Kaiserman, 1982). The morphology of the silver cores is variable, and they may be continuous, discontinuous or spiralized. Within the centromeric region, the cores frequently appear thicker and more intensely-stained than elsewhere in the chromosome, and the silver-stained NORs are attached to the core elements.

Cytochemical studies suggest that the silver reacts with nonhistones (Howell and Hsu, 1979; Howell et al., 1975), and particularly with protein sulfhydryl groups (Buys and Osinga, 1980). Howell and Hsu (1979) proposed that the silver cores are real components of metaphase chromosome structure, representing axially-located nonhistones which have a higher affinity for silver than those of the hypotonically-dispersed peripheral chromatin. In normal chromosomes, cores would not be apparent because they are covered with chromatin that contains proteins which have a low affinity for silver.

This hypothesis suggests that there may be a relationship between the silver-stained core and the scaffold observed after histone depletion; the silver could potentially stain the protein components of the scaffold. When chromosomes are histone-depleted with ammonium acetate, the resulting scaffold has a high affinity for silver under conditions where core staining is observed in intact chromosomes (Burkholder, 1983). Isolated scaffolds also stain with silver (Earnshaw and Laemmli, 1984). These results suggest that the silver core in ordinary chromosomes represents the nonhistone protein scaffold visualized by histone depletion. On the other hand, the peripherally-dispersed DNA radiating from the scaffold is also silver-reactive (Burkholder, 1983), indicating that cytological silver staining is not simply a matter of staining nonhistone protein but may also involve other components of the chromosome.

The discovery of the scaffold in histone-depleted chromosomes and the appearance of a core structure in silver-stained chromosomes has seemingly produced wide-spread acceptance of a core model of chromosome structure. This may be premature. Before accepting such a model, it is imperative to convincingly ascertain whether the cores represent a real component of chromosome structure or whether they are artifacts of preparation or staining. It is also essential to consider whether a core model is compatible with traditional genetic and cytogenetic data.

Sister chromatid exchanges can be clearly conceptualized in terms of DNA pairing and recombination (Holliday, 1964), but become substantially more complex to explain when the exchange also involves a core structure. In fact, it is difficult to imagine a mechanism by which a bulky fibrous scaffold could participate in chromatid exchanges. This problem would only be alleviated if the cores were assembled in G_2 after sister chromatid exchange had occurred. However, if the scaffold is a modified version of the nuclear matrix (Lewis and Laemmli, 1982), the chromosomal DNA may be constantly constrained throughout the cell cycle. This would create enormous logistical problems which have not been addressed. For instance, a break in a DNA molecule during interphase would rarely, if ever, manifest itself as a chromatid or chromosome break unless there was a complete breakdown of the scaffold sometime prior to chromosome condensation. Similarly, it is hard to imagine how chromosome rearrangements would occur. If a scaffold is involved in organizing the chromosome, the existence of a normal chromosome morphology in G_1 and G_2 prematurely-condensed chromosomes (Sperling and Rao, 1974; Hanks et al., 1983) indicates that the scaffold must exist at other stages of the cell cycle or be rapidly assembled during premature chromosome condensation.

Although the data provide strong support for a radial loop model of chromatin organization within chromosomes, the actual mechanism by which the loops are held in place is uncertain. A scaffold provides one possible means of organizing the loops, but whether or not it exists as a discrete structure within chromosomes _in vivo_ is an open question. This element of uncertainty stems from the extensive manipulations required in order to visualize the scaffold. A very real possibility is that the massive, wholesale extraction of histones does not preserve the normal arrangements of the residual nonhistones, regardless of whether dehistonization is accomplished under high, low, or isotonic ionic conditions. There may be specific rearrangements of the residual nonhistones and/or alterations of protein/DNA interactions, leading to the formation of a precipitated protein network having a defined protein composition.

A variety of evidence suggests that the scaffold structure in dehistonized chromosomes may be induced by the experimental manipulation and does not exist *in vivo*. 1. Chromosomes with core-like structures have only been observed in experimentally-manipulated chromosomes. A scaffold has never been identified in normal condensed chromosomes *in situ*, nor have they ever been seen in untreated chromosomes examined by thin-section, whole-mount, stereo, or freeze-fracture electron microscopy (reviewed by Okada and Comings, 1980; Lepault et al., 1980; Laughlin et al., 1982). Even in chromosomes digested with DNase, there is no evidence of a protein core (Rattner et al., 1978; Hadlaczky et al., 1981b). The failure to demonstrate scaffolds in these preparations has been blamed on their similarity to chromatin fibers in appearance and staining properties (Paulson and Laemmli, 1977), their diffuse nature, and their small mass (5-10%) relative to the intact chromosomes (Earnshaw and Laemmli, 1983).

2. Assuming that the scaffold is involved in organizing the reproducible shape and size of each chromosome, it is logical to assume that the scaffold would have a consistent organization. It does not. There is an incredible variability in the morphology of the scaffold amongst those prepared in the same or different ways (Paulson and Laemmli, 1977; Earnshaw and Laemmli, 1983). The protein extraction and spreading techniques have major effects on the appearance of the scaffold (Hadlaczky et al., 1981a) and demonstrate that no particular appearance of the scaffold can be considered as representative of the *in vivo* state. If the scaffold is a real structure, this variability in morphology is anomalous compared to other biological structures which invariably have a consistent morphological appearance.

3. No cores are seen in chromosomes stained with protein-specific stains (Okada and Comings, 1980). Neither the EDTA-regressive stain nor the phosphotungstic acid stain demonstrate any internal proteinaceous network even when the chromosomes are partially dispersed.

4. If the chromosome is partially dispersed before histone depletion, no scaffold is observed (Okada and Comings, 1980).

5. If protein aggregation is inhibited during histone depletion, no scaffold forms (Hadlaczky et al., 1981b). Extraction of histones with 2M NaCl in the presence of sucrose inhibits protein

aggregation and results in a loose structure lacking a scaffold. These data suggest that protein aggregation is required for the visualization of a scaffold.

6. If the spatial relationships within chromosomes are preserved during dehistonization, only a network of fine fibers and the condensed kinetochore, without a scaffold, is observed (Hadlaczky et al., 1981b).

7. If the nonhistone proteins, including the major scaffold proteins Sc1 (topo II) and Sc2, are selectively and quantitatively extracted from chromosomes with the zwitterionic detergent CHAPS, the residual histone-DNA complex has a typical chromosome morphology (Matsui et al., 1985). This implies that chromosome structure is preserved in the absence of nonhistones.

8. In plant chromosomes such as those from wheat (*Triticum monococcum* L.), a 2M NaCl-resistant scaffold does not exist (Hadlaczky et al., 1982). It was found that 0.6M NaCl is the highest salt concentration that preserves chromosome shape. At higher concentrations, there was an immediate disintegration of chromosome structure.

9. Chelation of divalent cations coupled with the extraction of histone H1 produces extensive swelling of isolated chromosomes and the conversion of the 30 nm chromatin fibers into 10 nm nucleosome chains (Nasedkina and Slesinger, 1982). Subsequent digestion with nuclease led to the disappearance of the chromatin fibers, leaving behind a residual structure consisting of randomly-scattered aggregates that are not organized into a scaffold. Although it could be argued that the scaffold disintegrated under these conditions, the methods of preparation and treatment are thought to minimize artifactual associations compared to the usual dehistonization methods.

10. Labhart et al. (1982) found that exposure of chromosomes to 2M NaCl caused a total loss of morphology, in direct contradiction to the residual scaffold observed by Paulson and Laemmli (1977) at this ionic strength. According to Labhart and coworkers (1982), chromosome morphology is ion strength dependent; at NaCl concentrations above 0.5M, chromosomes completely disintegrate but at lower concentrations, chromosome morphology is preserved. The loss of chromosome morphology above 0.5 M NaCl correlates with the extraction of histone H1 and a concomitant destabilization of the chromatin fiber. These data indicate that intact chromatin fibers with H1-containing nucleosomes are necessary for the preservation of metaphase chromosome morphology. They further imply that fiber-fiber contacts, rather than a scaffold, are required to organize the chromosome.

11. When chromosomes are prepared under conditions of low ionic strength and then treated with 2M NaCl, no scaffold is found (Wunderli et al., 1983). Instead, a network composed of many small complexes interconnected with DNA filaments is observed. These chromosomes were isolated under metal-depleting conditions which are supposed to cause the scaffold and thus chromosome structure to disintegrate (Lewis and Laemmli, 1982). In fact, the metal-depleted chromosomes, although swollen, were still morphologically-intact and could be recondensed into a typical metaphase configuration by the addition of Mg^{2+} (Wunderli et al., 1983) or 100 mM NaCl (Labhart et al., 1982). This indicates that a scaffold is not required to maintain chromosome morphology.

Collectively, these data create substantial doubt that the scaffold seen following dehistonization is of structural importance to the intact chromosome. As proposed (Paulson and Laemmli, 1977; Laemmli et al., 1978), the view of the scaffold as an independent core element forming a framework along the axis of each sister chromatid does not seem realistic. Although it has been argued that specific conditions are required for the visualization of the scaffold, the conditions which allow the scaffold to be seen are precisely those which would be expected to lead to large-scale molecular rearrangements, aberrant interactions, and structural distortion within the chromosomes. Conditions that minimize the distortion of normal relationships within the chromosome have never demonstrated a scaffold.

The discovery that topo II is a major component of the scaffold is not proof that the scaffold exists as a structural entity. In fact, there is no solid evidence that this enzyme performs any structural function. During dehistonization, topo II may precipitate, thus becoming a major contributor to the scaffold. Nonetheless, although topo II may contribute to the artifactual formation of the scaffold, it could still be involved in closing the looped chromatin domains and may reside to a major extent within the central region of the chromatid.

Other evidence favoring a core organization of chromosomes is the presence of silver-stained chromatid axes after staining with ammoniacal silver. It was originally proposed that nonhistones in the axial region of the chromatids have a higher affinity for silver than peripheral nonhistones (Howell and Hsu, 1979). However, electron microscopy has revealed that within the same stained preparation, some chromosomes demonstrate cores, while many others are totally covered with silver, completely masking the differentially-staining regions (cores, centromeres, NORs) (Burkholder and Kaiserman, 1982). This indicates that the whole chromosome is capable of reacting strongly with silver, not just some constituents in the core region. It thus seems unlikely that a higher affinity of silver for core nonhistones, relative to other chromosomal nonhistones, accounts for the visualization of cores.

Another pertinent observation is the dense aggregate of silver found in the centromeric region of silver-stained chromosomes (Howell and Hsu, 1979; Kaiserman and Burkholder, 1980). This was initially thought to represent the kinetochore (Howell and Hsu, 1979), but its size and shape indicates that it is actually the centromeric constitutive heterochromatin (Kaiserman and Burkholder, 1980; Zheng and Burkholder, 1982). This was confirmed in studies of mouse and human cells where non-nucleolar secondary constrictions, known to contain highly-condensed constitutive heterochromatin but no kinetochores, were positively-stained with silver (Zheng and Burkholder, 1982).

These results suggested that the intensity of the silver deposit may simply reflect the concentration of chromatin in a particular region of the chromosome. To test this hypothesis, variations in chromatin density along the chromosomes were induced by G-banding with trypsin. The chromosomes were then stained with silver and examined by light and electron microscopy (Zheng and Burkholder, 1982). The condensed aggregates of chromatin within G-band regions were more heavily-stained with silver than the dispersed interband (R-band) chromatin. It was subsequently shown that there are many binding sites for silver on chromatin fibers and silver can even bind to protein-free DNA in staining reactions that are identical to those used to produce silver cores (Burkholder, 1983). Since single chromatin fibers and exposed DNA react with silver, the degree of silver staining in any region of the chromosome may be contingent upon the concentration of chromatin or DNA in that region.

This provides an explanation for the appearance of the silver-stained cores after extensive hypotonic treatment of chromosomes. In these preparations, there is likely a preferential dispersion of the chromatin at the chromosome periphery, leaving the central region of each chromatid relatively unaffected. The difference in the concentration of chromatin between the axial region of the chromatids and the periphery may result in differential silver staining of these regions, producing a silver core. Thus, the silver core may have nothing to do with specific axially-located nonhistones.

CHROMOSOME BANDING: EVIDENCE FOR A REGIONAL ORGANIZATION

The discovery of the diverse methods of chromosome banding ranks as one of the most significant developments in the field of cytogenetics, and has had a major impact on all aspects of this field. These methods revolutionized the approaches to chromosome analysis in humans and other mammalian species and found immediate application in human clinical diagnosis. They have become important tools in genetic studies involving human and animal gene mapping, and have facilitated studies on the chromosomal evolution of species. Last, but not least, they have stimulated intense research activity on the mechanisms underlying the production of bands, and have significantly altered our conceptual view of chromosome organization.

Banding has disclosed an aspect of chromosome organization of which we previously had no concept and has raised the question of what the bands represent in terms of the organization of chromosomes. Although many hypotheses have been advanced to explain the mechanisms of banding, we still do not fully understand the exact role of the underlying structural and molecular organization of the chromosome in the determination of bands, nor the precise effects of the banding treatments on these aspects of chromosome organization. This is of obvious importance since the organization of mammalian chromosomes into bands must be taken into account in any comprehensive model of chromosome structure.

The common methods of chromosome banding include Q-, G-, R-, and C-banding (reviewed by Hsu, 1974; Dutrillaux and Lejeune, 1975; Latt, 1976). Q-, G-, and R-banding represent the same substructure of the chromosome. For the most part, Q- and G-banding patterns are identical to one another while R-banding creates the reverse staining patterns, i.e. positively-stained G- and Q-bands correspond to negatively-stained R-bands, and vice versa. C-bands represent constitutive heterochromatin (Arrighi and Hsu, 1971). The banding patterns are consistent and reproducible features of chromosomes; they do not vary from tissue to tissue or change during the course of development (Burkholder and Comings, 1972). This emphasizes that the inherent organization responsible for banding is not random but highly ordered.

The evidence indicates that the bands reflect a structural, molecular, and functional organization within chromosomes. In general, G- and Q-bands replicate their DNA late in the S phase while R-bands are early-replicating (Ganner and Evans, 1971; Epplen et al., 1975). This is further emphasized by studies involving BrdU substitution for thymine during DNA replication followed by staining with 33258 Hoechst (Latt, 1973). The fluorescence of Hoechst is quenched by BrdU-substituted DNA, so if DNA is substituted with BrdU during only part of the S phase, chromosome regions containing substituted and unsubstituted DNA can be identified at the subsequent mitosis. These studies have directly correlated late-replicating regions with G-bands.

Other data suggest that the late-replicating DNA is enriched in AT base-pairs while early-replicating DNA is GC-enriched (Tobia et al., 1970; Bostock and Prescott, 1971). Direct evidence on this point was obtained by Holmquist et al. (1982), who were able to isolate DNA from G-bands and R-bands. The G-band DNA was 3.2% richer in AT base-pairs than the R-band DNA. The S phase appeared to be bimodal; R-band DNA replicated in the first half of S and G-band DNA in the last half, with a short lull in between the two periods of DNA synthetic activity.

A recent intriguing idea, related to the timing of DNA replication, is that R-bands are intimately involved in meiotic chromosome pairing (Chandley, 1986). During DNA replication in R-bands early in the premeiotic S phase, proteins required for the subsequent initiation of pairing may bind to these chromosome regions. This may establish a homolog-specific pattern along the chromosome which ultimately forms the basis for homologous chromosome recognition and pairing at zygotene. In this model, the initiation of homologous chromosome pairing and synaptonemal complex formation is postulated to occur in R-bands followed by progressive pairing in G- and C-band regions. Stringent pairing involving DNA sequence complementarity may only be achieved in the R-band regions; thus, crossing over may be restricted to these segments of the paired chromosomes and may not occur in G- or C-bands. Although it is hard to judge the validity of this model due to the difficulty of investigating the molecular events in chromosome pairing and recombination, it does provide an explanation for these phenomena based on the unique banding organization of individual homologous chromosomes.

In terms of gene activity, it has been suggested that active genes are principally located in R-bands and inactive genes in G-bands. In fact, the G-bands have been viewed as intercalary heterochromatin (Comings, 1972a). Supporting these views is the following circumstantial evidence: 1. Most of the human trisomies that are compatible with live birth involve chromosomes that have a preponderance of G-bands and few R-bands (Ganner and Evans, 1971; Korenberg et al., 1978). The assumption is that these chromosomes would have few active genes and would therefore produce fewer deleterious effects than an extra chromosome containing many genetically-active R-bands. 2. If the amino acid sequence of proteins is translated back to the DNA base composition, it appears that active, essential genes are relatively GC-rich, thereby locating them in R-bands (Comings, 1972b). 3. _In situ_ nick translation studies have shown that active genes are sensitive to DNase I digestion in mitotic chromosomes and generally map to R-bands (Gazit et al., 1982; Kerem et al., 1984; Adolph and Hameister, 1985). 4. An investigation of the replication timing of specific genes during the S phase led Goldman et al. (1984) to conclude that the early-replicating R-bands contain housekeeping genes active in all cells as well as potentially-expressed, tissue-specific genes; the late-replicating G-bands contain genes that are inactive in a given tissue. This suggests the existence of two functionally-distinct genomes within the nucleus and chromosomes.

More direct data on the question of gene activity relative to chromosome bands was obtained by Yunis et al. (1977). Labelled complementary DNA to total cytoplasmic poly-A RNA was hybridized to chromosomes in situ and the labeling patterns compared to the location of bands. The results indicate that human repetitive DNA sequences producing mRNA are largely localized in R-bands. On the other hand, Holmquist et al. (1982) found that Chinese hamster (Cricetulus griseus) total cellular poly-A RNA hybridized equally to isolated G- and R-band DNA. In addition, G- and R-band chromatin fractions were equally sensitive to DNase I, which preferentially digests active chromatin. These results suggest that the DNA from G- and R-band regions is equally transcriptionally-active. In view of the conflicting data, it is difficult to draw firm conclusions concerning the relationship of bands to transcriptionally-active genes.

Although the distribution of active genes in G- and R-bands is uncertain, some gene arrangements appear to have been conserved during evolution and this is reflected in the preservation of the banding patterns. There is a close similarity of the human chromosome banding patterns to those seen in related primates (Dutrillaux et al., 1975; Yunis and Prakash, 1982), and some banding segments may be similar even in more evolutionarily distant species (Nash and O'Brien, 1982). High-resolution banding has shown that regions of human and mouse chromosomes which are known to contain conserved gene assignments have similar banding patterns (Sawyer and Hozier, 1986). Thus, at the cytogenetic level, chromosome organization in these regions has been conserved during an estimated 60 million years of species divergence.

G- and Q-banding patterns on mitotic chromosomes correspond very closely to the chromomere patterns of meiotic chromosome bivalents at pachytene (Okada and Comings, 1974). Chromomeres appear to represent foci of chromatin condensation along the chromosome, and may be the sites where chromatin condensation is initiated (Sumner, 1982). If this is true, G- and Q-bands may be the corresponding initiation sites for chromosome condensation during mitosis. Alternatively, chromosome regions which are more condensed during interphase may remain so during both meiosis and mitosis (Okada and Comings, 1974). The reason why the chromomeres are naturally apparent in meiotic but not mitotic chromosomes probably relates to the relative degree of condensation of the chromosomes. Meiotic chromosomes are very extended and the homologs are paired, enhancing the chromomere patterns. In mitotic chromosomes, which have a much greater overall degree of condensation, the chromomeres are usually masked and require banding pretreatments to reveal them.

In mitotic prophase chromosomes, up to two thousand or more bands can be observed (Yunis et al., 1978; Yunis, 1981). The bands in these high-resolution chromosomes also correspond to the meiotic chromomeres (Jhanwar et al., 1982). As chromosomes condense during the prophase to metaphase transition, there is a progressive coalescence of bands such that each band and its component sub-bands retain the same relative location and staining intensity in the chromosome. Therefore, the bands observed at metaphase are actually comprised of a collection of finer sub-bands.

It is curious that although the pachytene chromosomes of both plant and animal meiotic cells demonstrate chromomeres, only the chromosomes of mammals demonstrate the corresponding G-bands at mitosis. In plant mitotic chromosomes, the equivalent of C-banding but no G-banding has been produced. It has been suggested that there are much greater quantities of DNA per unit length in plant mitotic chromosomes compared to those in mammals (Greilhuber, 1977); thus, the chromomeres may simply be too closely-packed together to be resolved by G-banding. However, this view has been challenged since other data imply that plant and animal chromosomes are compacted to roughly the same extent (Anderson et al., 1982). Recent work suggests that plant chromosomes can be G-banded if appropriate conditions of chromosome preparation and treatment are utilized (Wang, 1987).

THE EFFECTS OF PREPARATORY TECHNIQUES ON CHROMOSOMES

The starting material for all methods of chromosome banding is standard cytogenetic preparations of chromosomes obtained by exposing cells to hypotonic solution, fixation in acetic-methanol, and air-drying. It is certain that these preparatory techniques produce alterations in both the structure and biochemical composition of the chromosomes. As a result, fixed chromosomes can not be considered to be representative of chromosomes *in vivo*. It is therefore essential to understand the nature of fixed chromosomes in order to fully appreciate the effects of the various banding techniques and how the bands are produced.

Fixation does not appear to extract any significant amount of DNA from the chromosomes; however, it does result in the extraction of both histone and nonhistone proteins. The amount and types of histone extracted is variable (cf. Sumner et al., 1973; Sivak and Wolman, 1974; Comings and Avelino, 1974; Brody, 1974; Pothier et al., 1975; Bustin et al., 1976; Retief and Ruchel, 1977), suggesting that the fixative may not always have a quantitative, reproducible effect on these proteins.

The effect of the hypotonic solution and acetic-methanol fixative on the proteins of isolated nuclei, chromatin, and chromosomes has been investigated by electrophoresis (Burkholder and Duczek, 1980a, 1982a, 1982b). The hypotonic solution has little or no effect on the histone and nonhistone proteins. Fixation removes some of each of the five types of histones, but significant quantities of each remain in the post-fixed chromosomes. Histone H1 appears to be more sensitive to extraction than the core histones. A few nonhistones are also partially solubilized, but these have not been characterized. In spite of these specific changes, the overall protein composition of unfixed and fixed chromosomes is surprisingly similar.

G- and C-banding have recently been obtained by digesting unfixed, isolated chromosomes with certain restriction endonucleases (Burkholder and Schmidt, 1986; Burkholder, 1987). Interestingly, C-bands can be directly visualized after the enzyme digestion by phase-contrast microscopy, whereas G-bands are observed only after the post-digested chromosomes are fixed with acetic-methanol. This suggests that a fixation step is required to visualize G-bands. In this regard, Hancock and Sumner (1982) have indicated that some protein extraction during fixation is essential for the subsequent production of G- and R-bands by normal methods (Hancock and Sumner, 1982).

IMPLICATIONS AND MECHANISMS OF Q-BANDING

Q-banding is obtained by staining chromosome preparations with quinacrine or quinacrine mustard (Caspersson et al., 1970). It is technically the simplest of all the banding methods. *In vitro* experiments utilizing defined polynucleotide sequences demonstrated that quinacrine fluorescence is enhanced by AT base-pairs and quenched by GC base-pairs, with the degree of quenching being related to the GC content of the DNA (Weisblum and deHaseth, 1972; Pachmann and Rigler, 1972). This suggests that regional variations in base composition along the chromosome are the primary basis for Q-banding.

Several observations support this view: 1. Regions of *Samoaia leonensis* chromosomes containing AT-rich satellite DNA are intensely-fluorescent (Ellison and Barr, 1972). 2. Staining chromosomes with fluorescently-labeled anti-adenosine antibodies produces banding patterns similar to Q-bands (Dev et al., 1972). 3. Late-replicating DNA, known to be located in Q-bands is enriched in sequences containing AT base-pairs (Tobia et al., 1970; Bostock and Prescott, 1971; Holmquist et al., 1982). 4. When GC-specific DNA-binding antibiotics, such as olivomycin, chromomycin A3, and mithramycin, are used to stain chromosomes, R-bands are produced (Schweizer, 1976; van de Sande et al., 1977). The ability of these GC-specific DNA-binding agents to produce bands that are the reverse of those seen with quinacrine or other AT-specific fluorochromes strongly implies that a nonuniform arrangement of nucleotides of differing base composition along the chromosomes is a major factor in the production of fluorescent bands.

On the other hand, there is evidence that factors other than base composition may also be involved in determining quinacrine fluorescence. In the mouse, the centromeric constitutive heterochromatin contains an AT-rich satellite DNA, but fluoresces poorly with quinacrine (Rowley and Bodmer, 1971). A similar situation occurs in the secondary constrictions of human chromosomes 1, 9, and 16, which contain an AT-rich satellite II DNA (Corneo et al., 1970; Jones and Corneo, 1971) but fluoresce poorly (Caspersson et al., 1970). It was originally suggested that guanine bases appropriately interspersed in these satellite sequences could quench fluorescence even though the total DNA was AT-rich (Weisblum and deHaseth, 1973). However, this interpretation is unlikely since mouse satellite DNA is now known to contain tracts consisting solely of several AT base-pairs (Horz and Altenburger, 1981).

Other evidence suggests that protein interactions with DNA may play a role in quenching the fluorescence of these regions. Gottesfeld et al. (1974) found that when rat (*Rattus rattus*) or *Drosophila* chromatin was separated into extended and condensed fractions, the former fraction quenched quinacrine fluorescence more than the latter. When the proteins were extracted from the two fractions and the purified DNA was retested, these differences were abolished. This result implies that proteins may have a significant influence on quinacrine fluorescence, at least in some regions of the chromosome.

The data are conflicting concerning the exact mechanism by which quinacrine binds to DNA. X-ray microanalysis of the distribution of quinacrine on chromosomes suggests that this dye binds uniformly to the chromosome (Sumner, 1981), possibly by external ionic binding along the DNA (Sumner, 1982). Banding appears to result from a differential excitation or quenching of the dye rather than by differential binding to the chromosome. Alternately, Comings et al.(1975) have observed that quinacrine binds to DNA by intercalation between base-pairs, with little external side-stacking. Molecules which covered the small groove of DNA inhibited intercalation, suggesting that pale-staining regions may result from a decreased binding of quinacrine, possibly due to nonhistones located in the small groove of the DNA.

The problem of aberrant quinacrine fluorescence of mouse centromeric heterochromatin has been reinvestigated relative to the binding of quinacrine. X-ray microanalysis has shown that the lack of fluorescence in these regions of AT-rich DNA appears to be a complex phenomenon involving a decrease in the amount of quinacrine bound and a lower efficiency of fluorescence excitation (Sumner,

1985). When mouse chromatin was fractionated, spectrophotometric titration and equilibrium dialysis experiments showed that the fraction containing presumptive C-band heterochromatin had a decreased number of binding sites for quinacrine compared to other chromatin fractions. This fraction was enriched in nonhistone proteins (Comings et al., 1975). These data suggest that the decreased fluorescence of mouse centromeric heterochromatin is caused by specific protein-DNA interactions which inhibit dye binding.

The question of whether or not differences in base composition along the arms of the chromosome are sufficient to account for differences in quinacrine fluorescence has been addressed by Comings and Drets (1976). By quantitating the effect of base composition on fluorescence using isolated calf (Bos taurus) DNA fractions differing in their AT- or GC-content, it was found that a change in base composition of 6% in AT-richness results in a 50% change in relative quinacrine fluorescence. These results suggest that the differences in base composition existing along the chromosome are sufficient to account for differences in quinacrine fluorescence. In support of this view, more recent studies have found significant variation in base ratios in segments of human chromosomes (Korenberg and Engels, 1978), and a sequence heterogeneity that is more than adequate to account for Q-banding (Mayfield and McKenna, 1978).

Collectively, the available data suggest that regional variations in base composition along the chromosome play a major role in the determination of fluorescent bands. However, in some chromosome regions such as mouse centromeres, proteins interacting with specific nucleotide sequences may modulate the fluorescence anticipated on the basis of base composition by limiting the accessibility of dye binding sites. At the same time, it cannot be excluded that chromatin condensation or other inherent features of chromosome structure may also affect the distribution of available quinacrine binding sites along the chromosomal DNA.

IMPLICATIONS AND MECHANISMS OF G-BANDING

G-banding usually involves a pretreatment of the chromosomes, followed by staining with Giemsa. Many different pretreatments have been described, the most frequently used of which are a warm sodium chloride/sodium citrate (SSC) incubation (Sumner et al., 1971), or mild trypsin digestion (Wang and Fedoroff, 1972). The common denominator in the diverse G-banding treatments is that they all involve proteolytic enzymes or protein denaturants; thus, it seems reasonable to believe that the protein components of chromosomes may be implicated in G-banding mechanisms.

Several questions are pertinent: 1. What, if any, structural rearrangements or alterations occur in the chromosomes during the banding pretreatments? 2. What specific effects do the pretreatments have on the chromosomal proteins? 3. Which proteins are involved? 4. How does the Giemsa dye stain chromosomes? 5. How do the proteins in different regions of the chromosome modulate Giemsa staining to produce banding? Although many aspects of the mechanism of G-banding are still in doubt, partial answers to some of these questions are now known.

As a result of the normal condensation of chromatin to form the compact metaphase chromosome and the additional contraction induced by the standard use of colcemid as a mitotic blocking agent, acid-fixed chromosomes are almost always uniformly-stained by Giemsa. However, the organization that underscores banding appears to be tucked away inside these condensed chromosomes. This is suggested by several observations. Occasionally, G-banding can be observed in acid-fixed but otherwise untreated chromosomes stained with Giemsa or Feulgen reagent (McKay, 1973; Yunis and Sanchez, 1973). The latter is quantitative for DNA. In addition, the mass distribution of chromatin along the arms of untreated chromosomes correlates closely with the banding patterns (Golomb and Bahr, 1974). When DNA-binding or cross-linking reagents are added to cells growing *in vivo* (Kitchin and Loudenslager, 1976) or *in vitro* (Hsu et al., 1973), spontaneous G-banding occurs in the subsequently-prepared chromosomes. These bands correspond to regions of condensed chromatin (Coming and Okada, 1975), and the reagents apparently act by interfering with the normal condensation of the R-band regions. Finally, the naturally-occurring meiotic chromomere patterns correspond to the mitotic G-banding patterns, as discussed above.

Together, these results indicate that the organization responsible for G-bands preexists in the chromosome. The banding treatments induce the alterations required to disclose the inherent regional condensation patterns along the chromosome arms. How does this occur?

When examined by electron microscopy, acid-fixed chromosomes are uniformly electron-dense, but after trypsinization, bands containing highly-packed chromatin fibers are seen to be separated by regions containing less densely-packed chromatin (Burkholder, 1974; 1975). Although chromatin dispersion was not a prerequisite for banding, the patterns of dispersion observed after increased trypsinization indicate that the G-bands are relatively more resistant to chromatin dispersion than the R-bands. These results suggest that trypsin may induce a reorganization of the chromatin. Some chromatin may pull away from the R-bands into the G-bands and there may be some extraction or dispersion of the R-band chromatin, particularly in overdigested chromosomes. Since Giemsa was not used to stain these chromosomes, this dye is not essential for visualizing the bands. However, at the light microscope level, Giemsa does enhance the relative difference in intensity between G- and R-band regions (Comings, 1978).

A reorganization of chromatin during G-banding is also suggested by other approaches. In an examination of the surface topography of chromosomes during banding, it was found that the SSC treatment causes the chromosomes to collapse (Gormley and Ross, 1972; 1976; Ross and Gormley, 1973). Subsequent staining with Giemsa results in the formation of ridges corresponding to the G-bands. These ridges may be a consequence of swelling due to the accumulation of stain. In support of this view, SEM has shown that the G-band regions are raised proportionally to the intensity of Giemsa staining while R-bands are flat or depressed (Utsumi, 1982).

Another SEM study suggests that there may be a relationship between G-banding and the quaternary coiling of the chromosomes (Harrison et al., 1981, 1982, 1983). Unbanded chromosomes had a smooth surface morphology, while trypsinized chromosomes became relaxed, revealing circumferential grooves lying perpendicular to the long axis of the chromatids. These grooves apparently demarcate the individual gyres of the quaternary coils and correspond to G-bands. At the light microscope level, Takayama (1976) observed that quaternary-coiled chromosomes could be progressively converted into G-banded structures by several rounds of trypsinization. Successive reexamination of the same chromosomes demonstrated configurational changes involving the loss of helical structure, thickening of the chromatids, and apparent intrachromosomal shifts in the distribution of chromatin leading to the emergence of bands. These observations provide clear evidence that chromatin rearrangement can occur during banding.

From these various investigations, it is apparent that the G-banding treatments induce specific structural alterations in chromosomes which contribute to the disclosure of the inherent chromomere organization and the ultimate visualization of banding patterns after Giemsa-staining. At the molecular level, banding implies that there is regional specificity between the DNA and chromosomal proteins that dictates how a region will behave during the banding treatment and subsequent staining.

Either the histones or the nonhistones or perhaps both could potentially be involved in G-banding. Although the role of the histones has been a matter of some debate, and evidence both supporting (Brown et al., 1975; Retief and Ruchel, 1977) and disclaiming (Sivak and Wolman, 1974; Comings and Avelino, 1974; Holmquist and Comings, 1976) their involvement has been presented, it now seems most likely that G-banding occurs independently of the histones.

When the histones remaining in fixed chromosomes are totally solubilized by a lengthy acid extraction, G-bands can still be produced (Comings and Avelino, 1974). Therefore, by a process of elimination, the nonhistones are presumed to be involved in G-banding. Other data also implicate the nonhistones and their interactions with DNA, in band formation. Histochemical (Kato and Moriwaki, 1972; Vogel et al., 1974), dye binding (Sumner and Evans, 1973), equilibrium dialysis (Comings and Avelino, 1975), dansyl chloride-fluorescence (Matsukuma and Utakoji, 1976), and DNase digestion (Burkholder and Weaver, 1977) studies have all pointed a finger at the nonhistones; however, exactly how these proteins are involved in banding, or which ones are important, is not clear.

Proteins appear to be uniformly distributed in unbanded chromosomes as judged by the uniformity of staining with dansyl chloride, a protein-specific fluorochrome (Matsukuma and Utakoji, 1976), and by immune techniques for nonhistone proteins (Bosman and Nakane, 1978). Kato and Moriwaki (1972) suggested that the pretreatments may solubilize some nonhistones, and the distribution of dansyl chloride fluorescence in G-banded chromosomes implies that there is an uneven extraction of nonhistones, apparently affecting R-bands more than G-bands (Matsukuma and Utakoji, 1976). On the other hand, the work of Comings et al. (1973), using radioisotope analyses of ^{3}H-leucine-labeled chromosomes, indicated that very little protein was removed by G-band procedures.

The question of protein alterations occurring during G-band pretreatments has been investigated by SDS gel electrophoresis (Burkholder and Duczek, 1980a, 1982a, 1982b). Isolated and acetic-methanol-fixed nuclei, chromatin, or chromosomes were treated with SSC, trypsin, urea, or NaCl-urea under conditions which produced G-banding, followed by an analysis of the extracted and residual chromosomal proteins. Each of the G-banding treatments extracts a characteristic subset of proteins (including a variety of nonhistones) which are similar irrespective of the type of treatment. The residual proteins remaining in the chromosomes are also similar, indicating that different G-band treatments have common effects on the chromosomal proteins. These effects were markedly different from those produced by R- or C-banding procedures, which in turn differed from one another. This implys a specificity of effect on the chromosomal proteins depending on the type of banding produced.

Unfortunately, the origin of the extracted proteins within G- or R-bands cannot be ascertained from the foregoing experiments, and the role of protein changes in altering the regional dye-binding properties of the chromosome is not clear. If the extracted proteins are mainly from the R-band regions, the disruption of normal DNA-protein interactions may somehow alter the staining properties of these regions. Banding can be inhibited by protein-

protein or protein-DNA crosslinking (Hancock and Sumner, 1982), suggesting that either the extraction of specific proteins or a rearrangement of DNA and protein constituents during the pretreatment is a prerequisite for banding to occur. It has also been found that reagents which break protein disulfide bonds, or those which cross-link protein sulfhydryls, inhibit subsequent banding and affect the overall staining properties of the chromosomes (Sumner, 1974). The results suggest that the intensely-stained G-bands may contain many disulfide bonds which hold the chromatin in a compact state, while the palely-stained R-bands contain a preponderance of protein sulfhydryl groups. Whether or not these differences exist in untreated chromosomes or arise during banding is not apparent. In spite of these uncertainties, the available evidence emphasizes that the conformation of proteins and their interrelationships with DNA are important factors in the production of banding patterns.

A satisfactory explanation of G-banding requires an understanding of how the Giemsa interacts with the chromosomal components. Giemsa is a Romanowsky-type dye mixture consisting of various thiazin dyes (methylene blue, Azure B, Azure A, Azure C, and thionin) and eosin (Comings, 1975). These dye components bind solely to DNA and not to chromosomal proteins (Sumner and Evans, 1973; Comings, 1975; Comings and Avelino, 1975). The thiazins are positively-charged planar molecules that interact ionically with the phosphate groups of DNA and side-stack along the molecule (Comings, 1975). Sumner and Evans (1973) found that staining involves the ionic binding of two thiazin molecules to a pair of DNA phosphates to produce a configuration that can then bind one eosin molecule. The formation of the thiazin-eosin complex is apparently favored in a hydrophobic environment, and once formed, the complex precipitates, releasing the DNA phosphates to take part in the process again (Sumner, 1980). In this proposal, the DNA serves as a catalyst for the formation of precipitate and the level of stain can build up to a high density relatively independently of the quantity or quality of the DNA.

Comings (1975) obtained evidence that eosin is not required for banding; methylene blue or any of the Azures alone can effectively stain banded chromosomes. However, superior staining is obtained in the presence of eosin (Comings, 1975, 1978), suggesting that the ability of this molecule to crosslink thiazins and produce a dye precipitate increases the resolution of the banding patterns.

In spite of all the data relating to the manifold aspects of G-banding, it has proved difficult to integrate the diverse structural and biochemical data into a comprehensive explanation of the mechanism involved. Nonetheless, several possible mechanisms have been proposed. Comings (1978) has proposed that several related changes occur within the chromosome during G-banding: 1. There may be an enhancement of the existing but normally-hidden chromomere patterns by chromatin fiber rearrangement, possibly involving an aggregation of R-band fibers onto existing chromomeres. 2. There might be some extraction of R-band DNA. 3. The phosphate groups in R-band DNA may become less accessible to the thiazin components of Giemsa as a result of the covering of this DNA with denatured non-histones. This is suggested by a study of the binding of methylene blue to DNA in chromatin (Comings and Avelino, 1975). Unfixed chromatin binds only about 0.6 times as much dye as free DNA, and fixed chromatin binds about 0.8 times as much dye. When fixed chromatin was exposed to the SSC used in G-banding, it again only bound about 0.5 times as much dye as the DNA. These results indicate that although fixation increases the number of dye binding sites relative to native chromatin, G-banding techniques reduce the number. This reduction in dye binding during banding may be due to the denaturation of nonhistone proteins. For example, amino groups in the denatured proteins may become available to interact with the DNA phosphates, thus decreasing the number of binding sites for the thiazins. 4. There may be an enhancement of the G-bands by the ability of the thiazin dyes to side-stack on the available DNA within these regions. This same basic mechanism may also explain R-banding if the proteins in G- and R-band regions are selectively denatured under different conditions of pH, temperature, and salt concentrations.

An alternative mechanism of G-banding has been proposed by Sumner (1982), based on the view that the formation of the thiazin-eosin precipitate is favored in chromosome regions enriched in hydrophobic proteins. This proposal implies that there are regional variations in the hydrophobic-hydrophilic character along the chromosome. Relatively hydrophobic sites include AT-enriched DNA and disulfide-rich proteins (Curtis and Horobin, 1982), both of which are characteristic of G-band regions (Holmquist et al., 1982; Sumner, 1974). Sumner (1982) suggests that G-band treatments may specifically extract hydrophobic proteins from R-bands or may change their conformation to a more hydrophilic state, possibly by breaking the disulfide bonds involved in maintaining the hydrophobic configuration. In G-band regions, the disulfide bonds may remain unaltered during banding; thus, the resident hydrophobic proteins will not be removed or changed and will subsequently provide a suitable environment for the formation of the dye precipitate. G-band treatments could also involve a selective removal of hydrophilic proteins (such as histones) from G-bands, creating a relative increase in the hydrophobic nature of these regions (Curtis and Horobin, 1982).

The mechanisms proposed by Comings (1978) and Sumner (1982) do not appear to be mutually exclusive. Inherent in both these proposals is the assumption that the G- and R-band regions of the chromosome contain different spectra of nonhistone proteins which behave differentially during the G-band treatments. Although this seems likely, there is currently no definitive evidence indicating that this is so. A means of analyzing regionally-localized nonhistones along the chromosome is required to establish the validity of these proposals. Through the use of antibodies to specific chromosomal proteins, this may eventually be possible.

Recently, a role for nucleosomes in the mechanism of G-banding has been proposed by van Duijn et al (1985). It is suggested that Giemsa staining of fixed but unbanded chromosomes may involve the swelling of nucleosomal DNA and the insertion of a hydrophobic Giemsa dye complex between swollen DNA and the denatured histone cores. During G-banding with SSC, the arginine-rich segments of the core histones may bind tightly to GC-rich but not AT-rich DNA. The preferential interaction of histones and GC-rich DNA may be responsible for G-banding by inhibiting the swelling of DNA around the histone cores in GC-rich chromosome regions, thereby preventing the insertion of the Giemsa complex and resulting in negatively-stained R-bands. Since no changes occur in DNA-histone relationships within AT-rich regions, DNA swelling and dye insertion proceed normally, giving rise to positively-stained G-bands.

Unlike the other hypotheses of G-banding, which rely on unidentified band-specific nonhistones (Comings, 1978, Sumner, 1982), this hypothesis is entirely dependent on regional differences in DNA base composition and the binding strength of the histones to explain the production of bands. The retention of the histones and preservation of the fundamental aspects of nucleosome structure in fixed chromosomes are obviously essential for this hypothesis to be correct. Many histones are retained in fixed and even in banded chromosomes (Burkholder and Duczek, 1982b), but whether or not the nucleosome organization is preserved relatively intact throughout the fixation and banding procedures is not known. The data of Comings and Avelino (1974), indicating that chromosomes can still be banded after the histones are completely extracted with HCl, argues strongly against an involvement of these proteins in banding. In that study, the extraction of histones with HCl was measured in nuclei and extrapolated to chromosomes treated the same way. There is consequently some question as to whether or not the HCl treatment totally extracts the histones from condensed chromosomes (van Duijn et al., 1985). Clearly, this point needs to be reexamined in order to establish the activity or passivity of histones in mechanisms of banding.

IMPLICATIONS AND MECHANISMS OF R-BANDING

R-bands are produced by exposing chromosome preparations to a buffered salt solution at high temperature (~86°C) followed by staining with Giemsa (Dutrillaux and Lejeune, 1971; Sehested, 1974) or acridine orange (Verma and Lubs, 1975). Although R-bands are the reverse of G- or Q-bands, they can be thought of as a variation on the theme since they reflect the same, albeit reciprocal organization of the chromosome. The reciprocal nature of the banding patterns produced by these different techniques is of great interest in relation to mechanisms of chromosome banding since it provides evidence that a specific chromosome segment can fluctuate in its staining potential from one extreme to the other (i.e. from a stained to a relatively unstained state), depending on the particular pretreatment to which the chromosome is subjected.

Electron microscopy of R-banded chromosomes shows that the distribution of aggregated chromatin is reversed relative to G-banding; however, the treatments producing positive R-banding induce less extreme differences in the degree of chromatin condensation between R- and G-bands than those giving rise to positive G-banding. R-banding has extremely adverse effects on all chromatin, leaving the impression that the chromatin fibers have largely fused together into an ill-defined mass (Burkholder, 1981). The structural differences between positive R-bands and negative G-bands are generally subtle, principally involving slight differences in electron-density.

At the molecular level, the extreme heat treatment used in the preparation of R-bands may have deleterious effects on both DNA and proteins. As discussed above, G-band DNA appears to have a relatively higher AT sequence content and lower GC content than R-band DNA. In solution, AT-rich DNA denatures at a lower temperature than GC-rich DNA. Therefore, heat treatment of chromosomes could selectively denature the DNA of G-bands but leave the R-band DNA in a native double-stranded configuration. Acridine orange, which stains single-stranded and double-stranded DNA differentially (de la Chapelle et al., 1971), produces staining patterns in heat-treated R-banded chromosomes that are consistent with this view (Bobrow and Madan, 1973; Comings et al., 1973; Comings and Wyandt, 1976). A monoclonal antibody to double-stranded DNA specific for poly dG, poly dC nucleotides has provided further support for a differential denaturation of DNA during R-banding (Magaud et al., 1985).

Radioisotope analyses of the extraction of chromatin during R-banding demonstrated that only about 3% each of the chromosomal DNA and protein was extracted by the pretreatment (Comings et al., 1973). An electrophoretic study of the extraction of proteins from isolated and fixed nuclei, chromatin, or chromosomes during R-banding failed to demonstrate any specific proteins in the treatment solution (Burkholder and Duczek, 1980b, 1982a,b), but did provide indirect evidence that the pretreatment causes an irreversible aggregation of proteins within the treated chromosomes (Burkholder and Duczek, 1982b).

Although the available data do not provide many details concerning the molecular and structural basis for R-banding, it is possible to speculate on the mechanism involved. The heat pretreatment appears to selectively denature the AT-enriched DNA of G-bands, leaving the GC-enriched DNA of R-bands in a native configuration. The heat treatment may also induce protein denaturation. If different proteins are associated with G- and R-band regions, there may be regional differences in their susceptibility to denaturation or in their ability to aggregate with denatured vs. native DNA. Although there is currently no information in this regard, it seems plausible that the denaturation of both DNA and protein in G-bands might cause a complete degeneration of chromatin structure followed by nonspecific aggregation. At the molecular level, this may involve the formation of strong interactions between the denatured protein and DNA. Protein/DNA interactions are probably altered in R-bands as well, but the retention of native DNA, perhaps combined with a lower degree of protein denaturation might help to preserve chromatin relationships relatively better than in G-bands. At the structural level, this could be reflected in the marginally-increased electron-density of the R-band compared to G-band regions (Burkholder, 1981).

Cytologically, R-banded chromosomes are usually observed by Giemsa staining of heat-treated chromosomes, although phase contrast optics are commonly used to enhance the image. Although the thiazin dyes in Giemsa appear to bind equally well to native and denatured DNA (Sumner and Evans, 1973; Comings and Avelino, 1975), they may be inhibited from interacting with the denatured G-band DNA by the DNA/protein coagulation occurring in these regions. The thiazins would consequently stain the R-bands more effectively since DNA/protein relationships are somewhat better preserved in these regions. Alternatively, the denatured DNA might be partially extracted so there are fewer binding sites for the dye components in G-band as compared to R-band regions (Comings, 1978).

IMPLICATIONS AND MECHANISMS OF C-BANDING

C-banding is technically the most complex of all the banding techniques, and is based on a method originally developed for *in situ* hybridization (Pardue and Gall, 1970). The chromosome preparations are pretreated with HCl, exposed to NaOH or $Ba(OH)_2$, and then incubated in warm SSC before staining in Giemsa (Arrighi and Hsu, 1971; Sumner, 1972). The resulting intensely-stained C-bands correspond to the chromosomal sites of constitutive heterochromatin (Hsu, 1974). This type of heterochromatin frequently, but not always, contains highly-repetitious satellite DNA (Comings and Mattoccia, 1972; Arrighi et al., 1974) and is sometimes enriched in methylcytosine (Miller et al., 1974). It is usually located in the centromeric regions of chromosomes but may be found at non-centromeric sites as well (Hsu and Arrighi, 1971). It is thought to be permanently genetically-repressed.

Since the method employs conditions which are known to denature (alkali treatment) and renature (SSC incubation) DNA, the original explanation of C-banding presumed that there was widespread denaturation of DNA followed by selective reannealing of the highly-repetitious sequences in constitutive heterochromatin. It was assumed that the Giemsa produced C-bands by interacting with the renatured double-stranded DNA but not with the denatured DNA in the chromosome arms. Although denaturation-renaturation of DNA may in fact occur during C-banding, it is now apparent that this is not the primary mechanism involved. Studies utilizing acridine orange indicate that there is no correlation between the strandedness of the DNA and C-bands (Comings et al., 1973; Lubs et al., 1973). In addition, some C-band regions do not contain highly-repetitious DNA (Arrighi et al., 1974), and Giemsa is equally effective in binding to double-stranded and single-stranded DNA (Comings and Avelino, 1975; Sumner and Evans, 1973).

Several lines of evidence indicate that a major effect of the C-banding technique is the extraction of non-C-band chromatin and DNA. C-band chromatin is resistant to extraction and remains within the chromosome. Biochemical and radioisotope analyses reveal that the alkali and salt incubations both extract chromosomal constituents (Comings et al., 1973); collectively, the whole procedure removes about 60% of the chromosomal DNA. Autoradiography of ^{3}H-thymidine-labelled chromosomes (Pathak and Arrighi, 1973) and Feulgen staining (Comings et al., 1973) have each demonstrated that the loss of DNA involves the non-C-band regions. Electron microscopy has shown varying degrees of chromatin extraction from non-C-band regions of chromosomes while C-band chromatin is highly resistant to extraction and remains condensed even when non-C-band chromatin is totally extracted (Burkholder, 1975). The resulting quantitative difference in DNA in C-band and non-C-band regions is subsequently visualized by Giemsa staining.

What is the mechanism by which DNA is extracted during C-banding? Holmquist (1979) obtained evidence indicating that each of the three steps in the procedure plays a role in this process. The acid (HCl) pretreatment acts to depurinate the DNA without breaking the phosphate-sugar backbone of the molecule. The subsequent alkali treatment produces DNA denaturation which facilitates DNA solubilization, and the incubation in SSC breaks the DNA backbone and allows the fragments to elute into solution.

What is responsible for the resistance of C-band DNA to extraction? It is unclear whether or not different degrees of DNA denaturation, or subsequent renaturation, can affect DNA loss.

Whether or not both C-band and non-C-band DNA are depurinated to the same extent is also not known. However, it appears that the rate of depurination and extraction is sensitive to several factors including protein-DNA interactions and the degree of chromatin condensation (Holmquist, 1979).

This, and other circumstantial evidence suggests that DNA-protein interactions may be intimately involved in determining the differential resistance of C-band and non-C-band chromatin to extraction. The DNA in unfixed, isolated heterochromatin is more resistant than euchromatin to digestion with DNase I, and this appears to be due to a tighter binding of protein to heterochromatin DNA rather than to differential condensation of chromatin (Burkholder and Weaver, 1977). In chromosomes, the DNA in C-band regions is also more resistant to DNase I digestion than non-C-band regions (Burkholder and Weaver, 1977), most logically as a result of tighter protein interactions with DNA in the C-band heterochromatin. In support of this view, the constitutive heterochromatin of African green monkey (*Cercopithecus aethiops*) cells seems to have an altered DNA-protein configuration, part of which may be due to tightly-bound nonhistones present in increased amounts relative to bulk chromatin (Musich et al., 1977). Protein-specific dansyl chloride staining of histone-depleted mouse chromosomes also suggests that constitutive heterochromatin may be enriched in nonhistone proteins compared to the chromosome arms (Matsukuma and Utakoji, 1977). Finally, constitutive heterochromatin may be transiently resistant to destruction with protease since extended trypsin treatment, beyond that which normally produces G-banding, results in the production of C-banding (Burkholder, 1975).

To determine precisely why the DNA in constitutive heterochromatin is protected during C-banding or DNase digestion, a clear understanding of the protein constituents of this chromatin is required. Numerous attempts have been made to biochemically fractionate heterochromatin from interphase nuclei in order to analyse the constituent proteins (Comings and Harris, 1975; Comings et al., 1977; Gottesfeld, 1977; Mazrimas et al., 1979; Zhang and Horz, 1982). These have not been very informative for it appears that the specific proteins associated with heterochromatin are highly dependent on the method of fractionating the chromatin (cf. Comings and Harris, 1975; Comings et al., 1977). At present, no specific proteins have been identified that might be considered as representative of constitutive heterochromatin, or that might account for its resistance to nuclease digestion or C-banding treatments.

Another approach to investigate the protein involvement in C-banding has been to electrophoretically examine the proteins extracted at each step of the procedure as well as the residual proteins in the C-banded chromosomes (Burkholder and Duczek, 1982b). The HCl and NaOH or $Ba(OH)_2$ treatments extract characteristic subsets of the histone and nonhistone proteins while little additional intact protein is found in the SSC extract. The residual chromosomal proteins consist of a few characteristic nonhistones and some of each of the five major histones. There was also evidence for the presence of degraded proteins in the residual chromosomes particularly after the $Ba(OH)_2$ C-banding method. Although many of the residual proteins may be located in constitutive heterochromatin, and could potentially determine the resistance of this chromatin to extraction, these can not yet be identified from proteins associated with residual non-C-band chromatin. The close similarity of the residual proteins observed after either the NaOH or $Ba(OH)_2$ C-banding techniques indicate that these methods have reproducible, common effects on the protein constituents of the chromosomes.

Although the involvement of DNA loss in C-banding is well-established, and the evidence suggests that DNA-protein interactions in constitutive heterochromatin produce a configuration that is resistant to extraction, the details remain uncertain. It is probable that a clearer picture of the involvement of proteins in C-banding will only be obtained when the constitutive heterochromatin-specific proteins are identified and their role in protecting the DNA is directly investigated. This may eventually be achieved through the development of improved methods for biochemically fractionating the constitutive heterochromatin from other chromatin, or by obtaining informative antibodies to constitutive heterochromatin-specific proteins which will allow further exploration of specific protein function.

ENDONUCLEASE-INDUCED CHROMOSOME BANDING

Several studies have shown that digestion of fixed chromosomes with endonucleases can produce banding. Micrococcal nuclease induces G-bands in fixed mouse chromosomes (Sahasrabuddhe et al., 1978). Certain restriction enzymes, which recognize and cut DNA at sequence-specific restriction sites, induce banding patterns in human and mouse chromosomes ranging from G- to C-bands (Mezzanotte et al., 1983; Miller et al., 1983; Kaelbling et al., 1984; Burkholder and Schmidt, 1986).

G- and/or C-banding has also been produced *in vitro* by digesting isolated, unfixed mouse chromosomes with appropriate endonucleases (Burkholder and Schmidt, 1986; Burkholder, 1987). Time course digestions have been used to investigate the range of banding effects. Micrococcal nuclease, which cleaves the internucleosomal DNA, initially produces C-bands, but eventually destroys the whole chromosome, including the constitutive heterochromatin. This is not unexpected since all nuclear chromatin, including heterochromatin, is organized as nucleosomes. However, the progressive degeneration of the chromosome indicates that the nucleosomal DNA in constitutive heterochromatin is transiently less vulnerable to attack, once again emphasizing the relative inaccessibility of this DNA.

Digestion of isolated mouse chromosomes with Alu I or EcoR I, which have few restriction sites in satellite DNA, produce C-banding. With time, there is a progressive decrease in the Giemsa-stainability of the chromosome arms but no changes occur in the characteristic C-banded appearance of the constitutive heterochromatin, even in extended digests. Hae III initially produces G-bands which give way to C-bands after longer digestion. BstN I, Ava II, and Sau96 I each produce G-banding.

Biochemical analyses have demonstrated that Alu I-induced C-banding extracts up to 70% of the chromosomal DNA, but no detectable satellite DNA (Burkholder, 1987). Extracted fragments ranged up to about 1000 base-pairs in length, although larger fragments were detected within the residual chromosomes. This suggests that there is a limit to the size of the fragments solubilized. Larger fragments are probably retained in the chromosome through interactions with other chromosomal constituents (Miller et al., 1984; Bianchi et al., 1985). The mechanism of the C-banding produced by Alu I appears to involve DNA fragmentation at restriction sites in non-C-band (i.e. non-satellite) DNA, followed by solubilization of the smaller fragments. The net result is a decrease in DNA content and correlated decrease in Giemsa staining relative to the unaffected satellite-containing constitutive heterochromatin. Alu I digestion consequently appears to mimic the overall effects of the regular C-banding procedure, even though the mechanisms of DNA cleavage are totally different.

If restriction enzymes cut the DNA within chromosomes as effectively as they cleave free DNA (Bianchi et al., 1985), then the banding patterns produced should directly reflect the distribution of restriction sites within the chromosome. Chromosome regions which have many restriction sites and incur substantial DNA loss should become palely-stained, while stained bands presumably represent regions with a low frequency of restriction sites (Miller et al., 1984). Although this provides an adequate explanation of Alu I-induced C-banding, there is some question as to whether or not this is universally the explanation for all restriction endonuclease-induced banding (Mezzanotte et al., 1985).

An alternative possibility is that the higher order chromosome organization may control the accessibility of an enzyme to its target sequences and may not necessarily permit all of the potential restriction sites within the chromosome to be cut (Mezzanotte et al., 1983; Lica and Hamkalo, 1983). In this case, the resulting banding patterns may identify regional variations in DNA-protein interactions rather than a regional distribution of restriction sites. Biochemical studies have shown that G-banding induced in isolated mouse chromosomes by BstN I extracts about 50% of the chromosomal DNA (Burkholder, 1987). This enzyme has a restriction site in the mouse satellite DNA repeat (Southern, 1975); however, in digests of isolated chromosomes producing G-banding, these restriction sites are not cleaved and no satellite DNA is extracted from the chromosomes. If the DNA from the G-banded chromosomes is isolated and redigested with BstN I, cleavage at the specific restriction sites in satellite DNA does occur. This demonstrates that in unfixed chromosomes, the configuration of the centromeric heterochromatin protects the satellite DNA from cleavage by this enzyme.

From these experiments, it is clear that endonucleases provide an alternate means to investigate various aspects of chromosome banding and to further probe the regional properties and organization of chromosomes. Unlike the usual banding methods, endonucleases have well-defined and precise effects on DNA, and digestions can be performed on unfixed chromosomes under relatively physiological ionic conditions. Restriction enzymes may be particularly useful in identifying regional differences in the distribution of restriction sites along the chromosome, or perhaps more importantly, in identifying regional variations in restriction site accessibility to an enzyme.

EMERGING CONCEPTS OF MAMMALIAN CHROMOSOME ORGANIZATION

The eukaryotic chromosome has often appeared to be more complex than the simple sum of its component parts. Even though the individual biochemical constituents have been defined and the general morphology is apparent, it has proven difficult to correlate the structure of the chromosome with its molecular organization. The experimental manipulation of chromosomes has been a very successful approach in investigations of the interrelationship between the structural and molecular organization of chromosomes, but has proved to be a double-edged sword. On the one hand, these manipulations have been instrumental in providing most of our current information on chromosome organization, but on the other, the relationship of the organization of manipulated chromosomes to that existing *in vivo* is tenuous, until proven otherwise by alternate methods.

Although the definitive organization of the native chromosome is still uncertain, it is possible to synthesize the accumulated data into a relatively coherent model which is consistent with much of the current evidence. The DNA, existing as a single molecule for each chromatid, is organized in affiliation with the histones into a nucleosomal chain, which in turn is supercoiled to form the

30 nm fiber. The subsequent levels of organization, culminating in a fully-condensed metaphase chromosome, are less clear. It is apparent that the 30 nm fiber is organized into looped domains with the base of each loop lying close to the central axis of the chromosome. These loops appear to be twisted into a superhelical configuration and are consequently foreshortened so that they can be tucked into the condensing chromatid. The location of the base of each loop in the central axis of the chromosome is consistent with the appearance of both the meiotic pachytene chromosomes and lampbrush chromosomes. The loops are most likely organized in clusters, giving rise to the chromomeres.

Since the reality of core elements _in vivo_ is suspect, the bases of the loops may not be attached to, nor held in place by a defined scaffold structure, as originally proposed by Paulson and Laemmli (1977). Rather, the loops may be sealed at their bases with topo II and held together by fiber-fiber interactions. If topo II proves to have a structural function in addition to an enzymatic one, these molecules could play a direct role in the fiber-fiber linkages. Alternately, there may be other proteins involved, possibly in conjunction with topo II. In this model, the chromatin fiber itself is responsible for the linear continuity of the chromosome and has an inherent dynamic organization based on fiber interactions as opposed to an independent structural framework. This form of organization is consistent with the genetic and cytogenetic data.

As a result of the loop organization, the axial region of a chromatid is expected to have an extremely high concentration of converging chromatin fibers and associated nonhistones, including topo II. During the extraction of the histones, these nonhistones are likely to aggregate, giving rise to the observed scaffold network. In many studies, the scaffold appears to contain residual chromatin (Okada and Comings, 1979; Hadlaczky et al., 1981a; Burkholder, 1983). This may represent the interfiber contacts remaining as a result of incomplete dehistonization of the axial region. In support of this view, axial chromatin in scaffold preparations is frequently peripherally-continuous with looping DNA (Hadlaczky et al., 1981a; Burkholder, 1983).

The organization of the 30 nm chromatin into radiating loops and subsequent loop condensation appears to form a fundamental chromatid fiber 200-300 nm in diameter. At a final level of organization, most likely developing during prophase, the chromatid fibers undergo quaternary coiling to form the compact and fully-condensed metaphase chromosome. The mechanism of this coiling is completely unknown but presumably involves chromosomal proteins.

One of the major impacts of the banding techniques has been the demonstration of a regional structural and biochemical order within chromosomes. It is apparent that Q- and G-bands reflect the existence of chromomeres in mitotic chromosomes. This indicates not only that mitotic and meiotic chromosomes have the same fundamental organization but also that Q- and G-bands disclose an inherent partitioning of the chromosome related to chromatin condensation. The mechanism responsible for arranging the chromatin fiber into individual chromomeres is not known, nor what causes the coalescence of small chromomeres during chromosome condensation, forming the larger chromomeres visible as bands at metaphase. In this context, it would be useful to know the distribution of topo II in relation to chromomeres in both meiotic pachytene chromosomes and banded mitotic prophase chromosomes. In swollen or dehistonized chromosomes, topo II sometimes appears to be localized in discrete foci along the chromosome (Earnshaw and Heck, 1985; Gasser et al.,1986), but whether or not these correspond to chromomeres has not been pursued.

From a variety of studies, it can be concluded that there are specific regional fluctuations in both DNA and protein composition along the chromosome. Although the chromosomal DNA extends uninterruptedly from one end of a chromatid to the other, Q-banding demonstrates that the DNA is segmentally organized along the chromosome in terms of its base composition; AT-enriched Q(G)-bands alternate with GC-enriched R-bands. G- and R-banding imply that the DNA segments of differing base composition are associated with different spectra of nonhistone proteins. These so-far unidentified proteins presumably have different biochemical properties, and behave differently during G- and R-banding treatments. The functional role of such proteins within the intact chromosome is not known.

The organization of the chromosome into G- and R-bands has functional significance in terms of the timing of DNA replication. The events triggering DNA synthesis in R-band replicons occur early during the S phase, while those involved in the synthesis of G-band replicons occur later. Replicons are known to be activated for DNA synthesis in clusters (Huberman and Riggs, 1968; Hand, 1975); thus, the G- and R-bands may represent such clusters differing in their time of initiation. Whether or not the banding patterns also represent an organization in terms of the expression of the genome is not convincingly decided. The R-bands may contain proportionally more active genes than G-bands, but it is doubtful that the latter can be considered to be genetically-silent.

C-banding demonstrates the location of constitutive heterochromatin which is usually partitioned within the centromeric region of chromosomes. The effects of the C-banding treatments indicate that the DNA in this chromatin, which is often highly-repetitious, is complexed with protein in a manner that differs from other DNA-protein interactions elsewhere in the chromosome. The proteins involved remain elusive, but their interaction with DNA appears to protect constitutive heterochromatin against the adverse effects of C-banding treatments. This unique configuration of DNA and protein presumably plays a structural role in the centromere, strengthening this region and possibly serving as a base for the interaction responsible for holding sister chromatids together and for the organization of the kinetochore. The constitutive heterochromatin may also form a protective buffer zone between the segregation and spindle-related activities of the centromere and the genetically-important chromatin in the chromosome arms. The mechanism responsible for holding the sister chromatids together has not been identified, although antibodies to proteins possibly involved in this function have recently been described (Kingwell and Rattner, 1986). Specific kinetochore-associated proteins have also been identified (Earnshaw and Rothfield, 1985; Valdivia and Brinkley, 1985) but the role of these proteins and the molecular organization of this structure is still in question.

FUTURE DIRECTIONS AND PROSPECTS

Many outstanding questions concerning chromosome organization remain to be answered. A few of these include the reality of the scaffold *in vivo*, the intrachromosomal role of topo II, and the identification and functions of proteins that are regionally-distributed along the chromosome or involved in other aspects of chromosome organization.

Experimental manipulations are certain to have ongoing roles in future work on chromosomes, but there is a definite need for the development and utilization of methodologies that minimally-distort normal intrachromosomal relationships while still allowing the chromosome to be experimentally dissected and analyzed. In this regard, the use of specific endonucleases as chromosome probes, coupled with investigations of their effects on chromosome structure and biochemistry may facilitate the elucidation of some of the regional properties of chromosomes. Another approach, still in its infancy, which appears to have a bright future in chromosome research is the use of immunological probes acquired from the sera of patients with auto-immune diseases or produced by monoclonal techniques. Coupled with the use of appropriate morphological or biochemical immuno-detection techniques, these antibodies are likely to have a significant impact in determining the distribution and function of specific proteins or other antigens within chromosomes.

REFERENCES

Abuelo, J. G., and Moore, D. E., 1969, The human chromosome. Electron microscopic observations on chromatin fiber organization, *J. Cell Biol.*, 41:73-90.

Adolph, K. W., 1980a, Organization of chromosomes in mitotic HeLa cells, Exp. Cell Res., 125:95-103.

Adolph, K. W., 1980b, Isolation and structural organization of human mitotic chromosomes, Chromosoma, 76:23-33.

Adolph, K. W., 1981, A serial sectioning study of the structure of human mitotic chromosomes, Eur. J. Cell Biol., 24:146-153.

Adolph, K. W., Cheng, S. M., and Laemmli, U. K., 1977a, Role of nonhistone proteins in metaphase chromosome structure, Cell, 12:805-816.

Adolph, K. W., Cheng, S. M., Paulson, J. R., and Laemmli, U. K., 1977b, Isolation of a protein scaffold from mitotic HeLa cell chromosomes, Proc. Natl. Acad. Sci. USA, 74:4937-4941.

Adolph, S., and Hameister, H., 1985, In situ nick translation of metaphase chromosomes with biotin-labeled d-UTP, Human Genet., 69:117121.

Anderson, L. K., Stack, S. M., and Mitchell, J. B., 1982, An investigation of the basis of a current hypothesis for the lack of G-banding in plant chromosomes, Exp. Cell Res., 138:433-436.

Arrighi, F. E., and Hsu, T. C., 1971, Localization of heterochromatin in human chromosomes, Cytogenetics, 10:81-86.

Arrighi, F. E., Hsu, T. C., Pathak, S., and Sawada, H., 1974, The sex chromosomes of the Chinese hamster: constitutive heterochromatin deficient in repetitive DNA sequences, Cytogenet. Cell Genet., 13:268-274.

Bajer, A., and Mole-Bajer, J., 1957, Mitosis in Endosperm, I., 16 mm film.

Bak, A. L., Zeuthen, J., and Crick, F. H. C., 1977, Higher-order structure of human mitotic chromosomes, Proc. Natl. Acad. Sci. USA, 74:1595-1599.

Benyajati, C., and Worcel, A., 1976, Isolation, characterization, and structure of the folded interphase genome of Drosophila melanogaster, Cell, 9:393-407.

Bianchi, M. S., Bianchi, N. O., Pantelias, G. E., and Wolff, S., 1985, The mechanism and pattern of banding induced by restriction endonucleases in human chromosomes, Chromosoma, 91:131-136.

Bobrow, M., and Madan, K., 1973, The effects of various banding procedures on human chromosomes, studied with acridine orange, Cytogenet. Cell Genet., 12:145-156.

Bosman, F. T., and Nakane, P. K., 1978, Immunoelectronmicroscopy of metaphase chromosomes, J. Histochem. Cytochem., 26:217.

Bostock, C. J., and Prescott, D. M., 1971, Buoyant density of DNA synthesized at different stages of the S phase of mouse L cells, Exp. Cell Res., 64:267-274.

Brody, Th., 1974, Histones in cytological preparations, Exp. Cell Res., 85:255-263.

Brown, R. L., Pathak, S., and Hsu, T. C., 1975, The possible role of histones in the mechanism of chromosomal G banding, Science, 189:1090-1091.

Burkholder, G. D., 1974, Electron microscopic visualization of chromosomes banded with trypsin, Nature, 247:292-294.
Burkholder, G. D., 1975, The ultrastructure of G- and C-banded chromosomes, Exp. Cell Res., 90:269-278.
Burkholder, G. D., 1981, The ultrastructure of R-banded chromosomes, Chromosoma, 83:473-480.
Burkholder, G. D., 1983, Silver staining of histone-depleted metaphase chromosomes, Exp. Cell Res., 147:287-296.
Burkholder, G. D., 1987, Endonuclease-induced chromosome banding: Structural and biochemical effects, in preparation.
Burkholder, G. D., and Comings, D. E., 1972, Do the Giemsa-banding patterns of chromosomes change during embryonic development? Exp. Cell Res., 75:268-271.
Burkholder, G. D., and Weaver, M. G., 1977, DNA-protein interactions and chromosome banding, Exp. Cell Res., 110:251-262.
Burkholder, G. D., and Duczek, L. L., 1980a, Proteins in chromosome banding. I. Effect of G-banding treatments on the proteins of isolated nuclei, Chromosoma, 79:29-41.
Burkholder, G. D., and Duczek, L. L., 1980b, Proteins in chromosome banding. II. Effect of R- and C-banding treatments on the proteins of isolated nuclei, Chromosoma, 79:43-51.
Burkholder, G. D., and Duczek, L. L., 1982a, The effect of the chromosome banding techniques on the histone and nonhistone proteins of isolated chromatin, Can. J. Biochem., 60:328-337.
Burkholder, G. D., and Duczek, L. L., 1982b, The effect of chromosome banding techniques on the proteins of isolated chromosomes, Chromosoma, 87:425-435.
Burkholder, G. D., and Kaiserman, M. Z., 1982, Electron microscopy of silver-stained core-like structures in metaphase chromosomes, Can. J. Genet. Cytol., 24:193-199.
Burkholder, G. D., and Schmidt, G. J., 1986, Endonuclease banding of isolated mammalian metaphase chromosomes, Exp. Cell Res., 164:379-387.
Bustin, M., Yamasaki, H., Goldblatt, D., Shani, M., Huberman, E., and Sachs, L., 1976, Histone distribution in chromosomes revealed by antihistone sera, Exp. Cell Res., 97:440-444.
Buys, C. H. C. M., and Osinga, J., 1980, Abundance of protein-bound sulfhydryl and disulfide groups at chromosomal nucleolus organizing regions. A cytochemical study on the selective silver staining of NORs, Chromosoma, 77:1-11.
Caspersson, T., Zech, L., Johansson, C., and Modest, E. J., 1970, Identification of human chromosomes by DNA-binding fluorescent agents, Chromosoma, 30:215-227.
Chandley, A. C., 1986, A model for effective pairing and recombination at meiosis based on early replicating sites (R-bands) along chromosomes, Human Genet., 72:50-57.
Cockerill, P. N., and Garrard, W. T., 1986, Chromosomal loop anchorage of the kappa immunoglobulin gene occurs next to the enhancer in a region containing topoisomerase II sites, Cell, 44:273-282.

Comings, D. E., 1972a, The structure and function of chromatin, in "Adv. in Human Genet.", H. Harris and K. Hirschhorn, ed., 3:237-431.
Comings, D. E., 1972b, Methylation of euchromatic and heterochromatic DNA, Exp. Cell Res., 74:383-390.
Comings, D. E., 1975, Mechanisms of chromosome banding. IV. Optical properties of the Giemsa dyes, Chromosoma, 50:89-110.
Comings, D. E., 1978, Mechanisms of chromosome banding and implications for chromosome structure, Ann. Rev. Genet., 12:25-46.
Comings, D. E., and Okada, T. A., 1970, Whole-mount electron microscopy of the centromere region of metacentric and telocentric mammalian chromosomes, Cytogenetics, 9:436-449.
Comings, D. E., and Okada, T. A., 1971, Fine structure of kinetochore in Indian muntjac, Exp. Cell Res., 67:97-110.
Comings, D. E., and Okada, T.A., 1972, Electron microscopy of chromosomes, in: "Perspectives in Cytogenetics. The Next Decade," S. W. Wright, B. F. Crandall, and L. Boyer, ed., Charles C. Thomas, Springfield, pp. 223-250.
Comings, D. E., and Mattoccia, E., 1972, DNA of mammalian and avian heterochromatin, Exp. Cell Res., 71:113-131.
Comings, D. E., and Avelino, E., 1974, Mechanisms of chromosome banding. II. Evidence that histones are not involved, Exp. Cell Res., 86:202-206.
Comings, D. E., & Avelino, E., 1975, Mechanisms of chromosome banding. VII. Interaction of methylene blue with DNA and chromatin, Chromosoma, 51:365-379.
Comings, D. E., and Harris, D. C., 1975, Nuclear proteins. I. Electrophoretic comparison of mouse nucleoli, heterochromatin, euchromatin and contractile proteins, Exp. Cell Res., 96:161-179.
Comings, D. E., and Okada, T. A., 1975, Mechanisms of chromosome banding VI. Whole mount electron microscopy of banded metaphase chromosomes and comparison to pachytene chromosomes, Exp. Cell Res., 93:267-274.
Comings, D. E., and Drets, M. E., 1976, Mechanisms of chromosome banding. IX. Are variations in DNA base composition adequate to account for Quinacrine, Hoechst 33258 and Daunomycin banding? Chromosoma, 56:199-211.
Comings, D. E., and Wyandt, H. E., 1976, Reverse banding of Japanese quail microchromosomes, Exp. Cell Res., 99:183-185.
Comings, D. E., Avelino, E., Okada, T. A., and Wyandt, H. E., 1973, The mechanism of C- and G-banding of chromosomes, Exp. Cell Res., 77:469-493.
Comings, D. E., Kovacs, B. W., Avelino, E., and Harris, D. C., 1975, Mechanisms of chromosome banding. V. Quinacrine banding, Chromosoma, 50:111-145.
Comings, D. E., Harris, D. C., Okada, T. A., and Holmquist, G., 1977, Nuclear proteins. IV. Deficiency of non-histone proteins in condensed chromatin of Drosophila virilis and mouse, Exp. Cell Res., 105:349-365.

Cook, P. R., and Brazell, I. A., 1978, Spectrofluorometric measurement of the binding of ethidium to superhelical DNA from cell nuclei, Eur. J. Biochem., 84:465-477.

Corneo, G., Ginelli, E., and Polli, E., 1970, Repeated sequences in human DNA, J. Mol. Biol., 48:319-327.

Curtis, D., and Horobin, R. W., 1982, Chromosome banding: specification of structural features of dyes giving rise to G-banding, Histochem. J., 14:911-928.

Daskal, Y., Mace, M. L., Wray, W., and Busch, H., 1976, Use of direct current sputtering for improved visualization of chromosome topology by scanning electron microscopy, Exp. Cell Res., 100:204-212.

de la Chapelle, A., Schroder, J., and Selander, R.-K., 1971, Repetitious DNA in mammalian chromosomes, Hereditas, 69:149-153.

Dev, V. G., Warburton, D., Miller, O. J., Miller, D. A., Erlanger, B. F., and Beiser, S. M., 1972, Consistent pattern of binding of anti-adenosine antibodies to human metaphase chromosomes, Exp. Cell Res., 74:288-293.

DiNardo, S., Voelkel, K., and Sternglanz, R., 1984, DNA topoisomerase II mutant of Saccharomyces cerevisiae: Topoisomerase II is required for segregation of daughter molecules at the termination of DNA replication, Proc. Natl. Acad. Sci. USA, 81:2616-2620.

DuPraw, E. J., 1965, Macromolecular organization of nuclei and chromosomes: A folded fibre model based on whole-mount electron microscopy, Nature, 206:338-343.

DuPraw, E. J., 1966, Evidence for a 'folded-fibre' organization in human chromosomes, Nature, 209:577-581.

DuPraw, E. J., 1970, "DNA and Chromosomes", Holt, Rinehart and Winston, Inc., New York.

Dutrillaux, B., and Lejeune, J., 1971, Sur une nouvelle technique d'analyse du caryotype humain, C.R. Acad. Sci., (Paris), 272(D):2638-2640.

Dutrillaux, B., and Lejeune, J., 1975, New techniques in the study of human chromosomes: Methods and applications, in "Adv. Human Genet.", H. Harris and K. Hirschhorn, ed., 5:119-156.

Dutrillaux, B., Rethore, M.-O., and Lejeune, J., 1975, Comparaison du caryotype de l'orang-outang (Pongo pygmaeus) a celui de l'homme, du chimpanze et du gorille, Ann. Genet., 18:153-161.

Earnshaw, W. C., and Laemmli, U. K., 1983, Architecture of metaphase chromosomes and chromosome scaffolds, J. Cell. Biol., 96:84-93.

Earnshaw, W. C., and Laemmli, U. K., 1984, Silver staining the chromosome scaffold, Chromosoma, 89:186-192.

Earnshaw, W. C., and Heck, M. M. S., 1985, Localization of topoisomerase II in mitotic chromosomes, J. Cell Biol., 100:1716-1725.

Earnshaw, W. C., and Rothfield, N., 1985, Identification of a family of human centromere proteins using autoimmune sera from

patients with scleroderma, Chromosoma, 91:313-321.
Earnshaw, W. C., Halligan, N., Cooke, C., and Rothfield, N., 1984, The kinetochore is part of the metaphase chromosome scaffold, J. Cell Biol., 98:352-357.
Earnshaw, W. C., Halligan, B., Cooke, C. A., Heck, M. M. S., and Liu, L. F., 1985, Topoisomerase II is a structural component of mitotic chromosome scaffolds, J. Cell Biol., 100:1706-1715.
Ellison, J. R., and Barr, H. J., 1972, Quinacrine fluorescence of specific chromosome regions. Late replication and high A:T content in Samoaia leonensis, Chromosoma, 36:375-390.
Epplen, J. T., Siebers, J.-W., and Vogel, W., 1975, DNA replication patterns of human chromosomes from fibroblasts and amniotic fluid cells revealed by a Giemsa staining technique, Cytogenet. Cell Genet., 15, 177-185.
Felsenfeld, G., 1978, Chromatin, Nature, 271:115-122.
Ganner, E., and Evans, H. J., 1971, The relationship between patterns of DNA replication and of Quinacrine fluorescence in the human chromosome complement, Chromosoma, 35:326-341.
Gasser, S. M., Laroche, T., Falquet, J., Boy de la Tour, E., and Laemmli, U. K., 1986, Metaphase chromosome structure. Involvement of topoisomerase II, J. Mol. Biol., 188:613-629.
Gazit, B., Cedar, H., Lerer, I., and Voss, R., 1982, Active genes are sensitive to deoxyribonuclease I during metaphase, Science, 217:648-650.
Goldman, M. A., Holmquist, G. P., Gray, M. C., Caston, L. A., and Nag, A., 1984, Replication timing of genes and middle repetitive sequences, Science, 224:686-692.
Golomb, H. M., and Bahr, G. F., 1974, Correlation of the fluorescent banding pattern and ultrastructure of a human chromosome, Exp. Cell Res., 84:121-126.
Goodpasture, C., and Bloom, S. E., 1975, Visualization of nucleolar organizer regions in mammalian chromosomes using silver staining, Chromosoma, 53:37-50.
Gormley, I. P., and Ross, A., 1972, Surface topography of human chromosomes examined at each stage during ASG banding procedure, Exp. Cell Res., 74:585-587.
Gormley, I. P., and Ross, A., 1976, Studies on the relationship of a collapsed chromosomal morphology to the production of Q- and G-bands, Exp. Cell Res., 98:152-158.
Gottesfeld, J, M., 1977, Methods for fractionation of chromatin into transcriptionally active and inactive segments, in: "Methods in Cell Biol.", D. M. Prescott, ed., XVI. Chromatin and Chromosomal Protein Research I, G. Stein, J. Stein, and L. J. Kleinsmith, ed., pp. 421-436, Academic Press, New York.
Gottesfeld, J. M., Bonner, J., Radda, G. K., and Walker, I. O., 1974, Biophysical studies on the mechanism of Quinacrine staining of chromosomes, Biochem., 13:2937-2945.
Greilhuber, J., 1977, Why plant chromosomes do not show G-bands, Theor. Appl. Genet., 50:121-124.
Hadlaczky, G., Sumner, A. T., and Ross, A., 1981a, Protein-depleted

chromosomes. I. Structure of isolated protein-depleted chromosomes, Chromosoma, 81:537-555.

Hadlaczky, G., Sumner, A. T., and Ross, A., 1981b, Protein-depleted chromosomes. II. Experiments concerning the reality of chromosome scaffolds, Chromosoma, 81:557-567.

Hadlaczky, G., Praznovszky, T., and Bisztray, G., 1982, Structure of isolated protein-depleted chromosomes of plants, Chromosoma, 86:643-659.

Hancock, J. M., and Sumner, A. T., 1982, The role of proteins in the production of different types of chromosome bands, Cytobios, 35:37-46.

Hand, R., 1975, Regulation of DNA replication on subchromosomal units of mammalian cells, J. Cell Biol., 64:89-97.

Hanks, S. K., Gollin, S. M., Rao, P. N., Wray, W., and Hittelman, W. N., 1983, Cell cycle-specific changes in the ultrastructural organization of prematurely condensed chromosomes, Chromosoma, 88:333-342.

Harrison, C. J., Britch, M., Allen, T. D., and Harris, R., 1981, Scanning electron microscopy of the G-banded human karyotype, Exp. Cell Res., 134:141-153.

Harrison, C. J., Allen, T. D., Britch, M., and Harris, R., 1982, High-resolution scanning electron microscopy of human metaphase chromosomes, J. Cell Sci., 56:409-422.

Harrison, C. J., Allen, T. D., and Harris, R., 1983, Scanning electron microscopy of variations in human metaphase chromosome structure revealed by Giemsa banding, Cytogenet. Cell Genet., 35:21-27.

Holliday, R., 1964, A mechanism for gene conversion in fungi, Genet. Res., 5:282-304.

Holmquist, G., 1979, The mechanism of C-banding: Depurination and β-elimination, Chromosoma, 72:203-224.

Holmquist, G. P., and Comings, D. E., 1976, Histones and G banding of chromosomes, Science, 193:599-602.

Holmquist, G., Gray, M., Porter, T., and Jordan, J., 1982, Characterization of Giemsa dark and light band DNA, Cell, 31:121-129.

Horz, W., and Altenburger, W., 1981, Nucleotide sequence of mouse satellite DNA, Nucleic Acids Res., 9:683-696.

Howell, W. M., and Hsu, T. C., 1979, Chromosome core structure revealed by silver staining, Chromosoma, 73:61-66.

Howell, W.M., Denton, T. E., and Diamond, J. R., 1975, Differential staining of the satellite regions of human acrocentric chromosomes, Experientia, 31:260-262.

Hsu, T. C., 1974, Longitudinal differentiation of chromosomes, Ann. Rev. Genet., 7:153-176.

Hsu, T. C., and Arrighi, F. E., 1971, Distribution of constitutive heterochromatin in mammalian chromosomes, Chromosoma, 34:243-253.

Hsu, T. C., Pathak, S., and Shafer, D. A., 1973, Induction of chromosome crossbanding by treating cells with chemical agents

before fixation, Exp. Cell Res., 79:484-487.
Huberman, J. A., and Riggs, A. D., 1968, On the mechanism of DNA replication in mammalian chromosomes, J. Mol. Biol., 32:327-341.
Igo-Kemenes, T., and Zachau, H.. G, 1978, Domains in chromatin structure, Cold Spring Harbor Symp. Quant. Biol., 42:109-118.
Igo-Kemenes, T., Horz., W., and Zachau, H. G., 1982, Chromatin, Ann Rev. Biochem., 51:89-121.
Jhanwar, S. C., Burns, J. P., Alonso, M. L., Hew, W., and Chaganti, R. S. K., 1982, Mid-pachytene chromomere maps of human autosomes, Cytogenet. Cell Genet., 33:240-248.
Jones, K. W., and Corneo, G., 1971, Location of satellite and homogeneous DNA sequences on human chromosomes, Nature New Biol., 233:268-271.
Kaelbling, M., Miller, D. A., and Miller, O. J., 1984, Restriction enzyme banding of mouse metaphase chromosomes, Chromosoma, 90:128-132.
Kaiserman, M. Z., and Burkholder, G. D., 1980, Silver stained core-like structures in Chinese hamster metaphase chromosomes, Can. J. Genet. Cytol., 22:627-632.
Kato, H., and Moriwaki, K., 1972, Factors involved in the production of banded structures in mammalian chromosomes, Chromosoma, 38:105-120.
Kavenoff, R., and Zimm, B. H., 1973, Chromosome-sized DNA molecules from Drosophila, Chromosoma, 41:1-27.
Kerem, B-S., Goitein, R., Diamond, G., Cedar, H., and Marcus, M., 1984, Mapping of DNAase I sensitive regions on mitotic chromosomes, Cell, 38:493-499.
Kingwell, B., and Rattner, J.B., 1986, Sister chromatids are associated together at discrete points in metaphase chromosomes, J. Cell Biol., 103:492a.
Kitchin, R. M., and Loudenslager, E. J., 1976, An in vivo Giemsa chromosome banding technique, Stain Technol., 50:371-374.
Korenberg, J. R., and Engels, W. R., 1978, Base ratio, DNA content, and quinacrine-brightness of human chromosomes, Proc. Natl. Acad. Sci. USA, 75:3382-3386.
Korenberg, J. R., Therman, E., and Denniston, C., 1978, Hot spots and functional organization of human chromosomes, Human Genet., 43:13-22.
Labhart, P., Koller, T., and Wunderli, H, 1982, Involvement of higher order chromatin structures in metaphase chromosome organization, Cell, 30:115-121.
Laemmli, U. K., Cheng, S. M., Adolph, K. W., Paulson, J. R., Brown, J. A., and Baumbach, W. R., 1978, Metaphase chromosome structure: The role of nonhistone proteins, Cold Spring Harbor Symp. Quant. Biol., 42:351-360.
Laird, C. D., 1971, Chromatid structure: Relationship between DNA content and nucleotide sequence diversity, Chromosoma, 32:378-406.
Latt, S. A., 1973, Microfluorometric detection of deoxyribonucleic

acid replication in human metaphase chromosomes, Proc. Natl. Acad. Sci., USA, 70:3395-3399.

Latt, S. A., 1976, Optical studies of metaphase chromosome organization, Ann. Rev. Biophys. & Bioeng., 5:1-37.

Laughlin, T. J., Wilkinson-Singley, E., Olins, D. E., and Olins, A. L., 1982, Stereo electron microscope studies of mitotic chromosomes from Chinese hamster ovary cells, Eur. J. Cell Biol., 27:170-176.

Lebkowski, J., and Laemmli, U. K., 1982, Evidence for two levels of DNA folding in histone-depleted HeLa interphase nuclei, J. Mol. Biol., 156:309-324.

Lepault, J., Bram, S., Escaig, J., and Wray, W., 1980, Chromatin freeze fracture electron microscopy: A comparative study of core particles, chromatin, metaphase chromosomes, and nuclei, Nucleic Acids Res., 8:265-278.

Lewis, C. D., and Laemmli, U. K., 1982, Higher order metaphase chromosome structure: Evidence for metalloprotein interactions, Cell, 29:171-181.

Lica, L., and Hamkalo, B., 1983, Preparation of centromeric heterochromatin by restriction endonuclease digestion of mouse L929 cells, Chromosoma, 88:42-49.

Lubs, H. A., McKenzie, W. H., and Merrick, S., 1973, Comparative methodology and mechanisms of banding, in: "Chromosome Identification--Techniques and Applications in Biology and Medicine," T. Caspersson and L. Zech, ed., pp. 315-322, Academic Press, New York.

Magaud, J.-P., Rimokh, R., Brochier, J., Lafage, M., and Germain, D., 1985, Chromosomal R-banding with a monoclonal antidouble-stranded DNA antibody, Human Genet., 69:238-242.

Marsden, M. P. F., and Laemmli, U. K., 1979, Metaphase chromosome structure: Evidence for a radial loop model, Cell, 17:849-858.

Matsui, S., Chai, L., Tsui, S., and Sandberg, A. A., 1985, Chromosome structure in the absence of non-histone chromosomal proteins (NHCP), J. Cell Biol., 101:75a.

Matsukuma, S., and Utakoji, T., 1976, Uneven extraction of protein in Chinese hamster chromosomes during G-staining procedures, Exp. Cell Res., 97:297-303.

Matsukuma, S., and Utakoji, T., 1977, Non-histone protein associated with centromeric heterochromatin in the mouse chromosome, Exp. Cell Res., 105:217-222.

Mattern, M. R., Paone, R. F., and Day III, R. S., 1982, Eukaryotic DNA repair is blocked at different steps by inhibition of DNA topoisomerases and of DNA polymerase α and β, Biochim. Biophys. Acta, 697:6-13.

Mayfield, J. E., and McKenna, J. F., 1978, A-T rich sequences in vertebrate DNA. A possible explanation of Q-banding in metaphase chromosomes, Chromosoma, 67:157-163.

Mazrimas, J. A., Balhorn, R., and Hatch, F. T., 1979, Separation of satellite DNA chromatin and main band DNA chromatin from mouse brain, Nucleic Acids Res., 7:935-946.

McGhee, J. D., and Felsenfeld, G., 1980, Nucleosome structure, Ann. Rev. Biochem., 49:1115-1156.

McKay, R. D. G., 1973, The mechanism of G and C banding in mammalian metaphase chromosomes, Chromosoma, 44:1-14.

Mezzanotte, R., Bianchi, U., Vanni, R., and Ferrucci, L., 1983, Chromatin organization and restriction endonuclease activity on human metaphase chromosomes, Cytogenet. Cell Genet., 36:562-566.

Mezzanotte, R., Ferrucci, L., Vanni, R., and Sumner, A. T., 1985, Some factors affecting the action of restriction endonucleases on human metaphase chromosomes, Exp. Cell Res., 161:247-253.

Miller, D. A., Choi, Y.-C., and Miller, O. J., 1983, Chromosome localization of highly repetitive human DNA's and amplified ribosomal DNA with restriction enzymes, Science, 219:395-397.

Miller, D. A., Gosden, J. R., Hastie, N. D., and Evans, H. J., 1984, Mechanism of endonuclease banding of chromosomes, Exp. Cell Res., 155:294-298.

Miller, O. J., Schnedl W., Allen, J., and Erlanger, B. F., 1974, 5-methylcytosine localised in mammalian constitutive heterochromatin, Nature, 251:636-637.

Miller, D. A., Dev, V. G., Tantravahi, R., and Miller, O. J., 1976, Suppression of human nucleolus organizer activity in mouse-human somatic hybrid cells, Exp. Cell Res., 101:235-243.

Mirkovitch, J., Mirault, M.-E., and Laemmli, U. K., 1984, Organization of the higher-order chromatin loop: Specific DNA attachment sites on nuclear scaffold, Cell, 39:223-232.

Mirkovitch, J., Spierer, P., and Laemmli, U. K., 1986, Genes and loops in 320,000 base-pairs of the Drosophila melanogaster chromosome, J. Mol. Biol., 190:255-258.

Moliter, H., Drahovsky, D., and Wacker, A., 1974, Structural integrity of chromatid DNA in mouse L cells, J. Mol. Biol., 86:161-163.

Musich, P. R., Brown, F. L., and Maio, J. J., 1977, Subunit structure of chromatin and the organization of eukaryotic highly repetitive DNA: Nucleosomal proteins associated with a highly repetitive mammalian DNA, Proc. Natl. Acad. Sci., USA, 74:3297-3301.

Nasedkina, T. V., and Slesinger, S. I., 1982, The structure of partly decondensed metaphase chromosomes, Chromosoma, 86:239-249.

Nash, W. G., and O'Brien, S. J., 1982, Conserved regions of homologous G-banded chromosomes between orders in mammalian evolution: Carnivores and primates, Proc. Natl. Acad. Sci. USA, 79:6631-6635.

Nelson, W. G., Liu, L. F., and Coffey, D. S., 1986, Newly replicated DNA is associated with DNA topoisomerase II in cultured rat prostatic adenocarcinoma cells, Nature, 322:187-189.

Ohnuki, Y., 1968, Structure of chromosomes. I. Morphological studies of the spiral structure of human somatic chromosomes, Chromosoma, 25:402-428.

Okada, T. A., & Comings, D. E., 1974, Mechanisms of chromosome banding. III. Similarity between G-bands of mitotic chromosomes and chromomeres of meiotic chromosomes, Chromosoma, 48:65-71.

Okada, T.A., and Comings, D. E., 1979, Higher order structure of chromosomes, Chromosoma, 72:1-14.

Okada, T. A., and Comings, D. E., 1980, A search for protein cores in chromosomes: Is the scaffold an artifact? Am. J. Hum. Genet., 32:814-832.

Pachmann, U., and Rigler, R., 1972, Quantum yield of acridines interacting with DNA of defined base sequence. A basis for the explanation of acridine bands in chromosomes, Exp. Cell Res., 72:602-608.

Pardue, M. L., and Gall, J. G., 1970, Chromosomal localization of mouse satellite DNA, Science, 168:1356-1358.

Pathak, S., and Arrighi, F. E., 1973, Loss of DNA following C-banding procedures, Cytogenet. Cell Genet., 12:414-422.

Paulson, J. R., and Laemmli, U. K., 1977, The structure of histone-depleted metaphase chromosomes, Cell, 12:817-828.

Pothier, L,, Gallagher, J. F., Wright, C. E., and Libby, P. R., 1975, Histones in fixed cytological preparations of Chinese hamster chromosomes demonstrated by immunofluorescence, Nature, 255:350-352 .

Rattner, J. B., and Hamkalo, B. A., 1978a, Higher order structure in metaphase chromosomes. I. The 250 A^{o} fiber, Chromosoma, 69:363-372.

Rattner, J. B., and Hamkalo, B. A., 1978b, Higher order structure in metaphase chromosomes. II. The relationship between the 250 A^{o} fiber, superbeads,and beads-on-a-string, Chromosoma, 69:373-379.

Rattner, J. B., and Lin, C. C., 1985, Radial loops and helical coils coexist in metaphase chromosomes, Cell, 42:291-296.

Rattner, J. B., Krystal, G., and Hamkalo, B. A., 1978, Selective digestion of mouse metaphase chromosomes, Chromosoma, 66:259-268.

Retief, A. E., and Ruchel, R., 1977, Histones removed by fixation. Their role in the mechanism of chromosomal banding, Exp. Cell Res., 106:233-237.

Roos, U.-P., 1973, Light and electron microscopy of rat kangaroo cells in mitosis. II. Kinetochore structure and function, Chromosoma, 41:195-220.

Ross, A., and Gormley, I. P., 1973, Examination of surface topography of Giemsa-banded human chromosomes by light and electron microscopic techniques, Exp. Cell Res., 81:79-86.

Rowley, J. D., and Bodmer, W. F., 1971, Relationship of centromeric heterochromatin to fluorescent banding patterns of metaphase chromosomes in the mouse, Nature, 231:503-506.

Sahasrabuddhe, C. G., Pathak, S., and Hsu, T. C., 1978, Responses of mammalian metaphase chromosomes to endonuclease digestion, Chromosoma, 69:331-338.

Sawyer, J. R., and Hozier, J. C., 1986, High resolution of mouse chromosomes: Banding conservation between man and mouse, Science, 232:1632-1635.
Schweizer, D., 1976, Reverse fluorescent chromosome banding with chromomycin and DAPI, Chromosoma, 58:307-324.
Sedat, J., and Manuelidis, L., 1978, A direct approach to the structure of eukaryotic chromosomes, Cold Spring Harbor Symp. Quant. Biol., 42:331-350.
Sehested, J., 1974, A simple method for R-banding of human chromosomes, showing a pH-dependent connection between R and G bands, Humangenetik, 21:55-58.
Sivak, A., and Wolman, S. R., 1974, Chromosomal proteins in fixed metaphase cells, Histochem., 42:345-349.
Southern, E. M., 1975, Long range periodicity in mouse satellite DNA, J. Mol. Biol., 94:51-69.
Sperling, K., and Rao, P. N., 1974, The phenomenon of premature chromosome condensation: Its relevance to basic and applied research, Humangenetik, 23:235-258.
Steck, T. R., and Drlica, K., 1984, Bacterial chromosome segregation: Evidence for DNA gyrase involvement in decatenation, Cell, 36:1081-1088.
Stubblefield, E., and Wray, W., 1971, Architecture of the Chinese hamster metaphase chromosome, Chromosoma, 32:262-294.
Sumner, A. T., 1972, A simple technique for demonstrating centromeric heterochromatin, Exp. Cell Res., 75:304-306.
Sumner, A. T., 1974, Involvement of protein disulphides and sulphydryls in chromosome banding, Exp. Cell Res., 83:438-442.
Sumner, A. T., 1980, Dye binding mechanisms in G-banding of chromosomes, J. Micros., 119:397-406.
Sumner, A. T., 1981, The distribution of quinacrine on chromosomes as determined by X-ray microanalysis. I. Q-bands on CHO chromosomes, Chromosoma, 82:717-734.
Sumner, A. T., 1982, The nature and mechanisms of chromosome banding, Cancer Genet. & Cytogenet., 6:59-87.
Sumner, A. T., 1985, The distribution of quinacrine on chromosomes as determined by X-ray microanalysis. II. Comparison of heterochromatic and euchromatic regions of mouse chromosomes, Chromosoma, 91:145-150.
Sumner, A. T., and Evans, H. J., 1973, Mechanisms involved in the banding of chromosomes with Quinacrine and Giemsa. II. The interaction of the dyes with the chromosomal components, Exp. Cell Res., 81:223-236.
Sumner, A. T., Evans, H. J., and Buckland, R. A., 1971, New technique for distinguishing between human chromosomes, Nature New Biol., 232:31-32.
Sumner, A. T., Evans, H. J., and Buckland, R. A., 1973, Mechanisms involved in the banding of chromosomes with Quinacrine and Giemsa. I. The effects of fixation in methanol-acetic acid, Exp. Cell Res., 81:214-222.
Takayama, S., 1976, Configurational changes in chromatids from

helical to banded structures, Chromosoma, 56:47-54.
Taniguchi, T., and Takayama, S., 1986, High-order structure of metaphase chromosomes: Evidence for a multiple coiling model, Chromosoma, 93:511-514.
Taylor, J. H., 1958, The duplication of chromosomes, Sci. Amer., 198:36-42.
Tobia, A., Schildkraut, C. L., and Maio, J. J., 1970, Deoxyribonucleic acid replication in synchronized cultured mammalian cells. I. Time of synthesis of molecules of different average guanine + cytosine content, J. Mol. Biol., 54:499-515.
Uemura, T., and Yanagida, M., 1984, Isolation of type I and II DNA topoisomerase mutants from fission yeast: single and double mutants show different phenotypes in cell growth and chromatin organization, EMBO J., 3:1737-1744.
Utsumi, K. R., 1982, Scanning electron microscopy of Giemsa-stained chromosomes and surface-spread chromosomes, Chromosoma, 86:683-702.
Valdivia, M. M., and Brinkley, B. R., 1985, Fractionation and initial characterization of the kinetochore from mammalian metaphase chromosomes, J. Cell Biol., 101:1124-1134.
van de Sande, J. H., Lin, C. C., and Jorgenson, K. F., 1977, Reverse banding on chromosomes produced by a guanosine-cytosine specific DNA binding antibiotic: olivomycin, Science, 195:400-402.
van Duijn, P., van Prooijen-Knegt, A. C., and van der Ploeg, M., 1985, The involvement of nucleosomes in Giemsa staining of chromosomes. A new hypothesis on the banding mechanism, Histochem., 82:363-376.
Verma, R. S., and Lubs, H. A., 1975, A simple R banding technic, Am. J. Hum. Genet., 27:110-117.
Vogel, W., Faust, J., Schmid, M., and Siebers, J.-W., 1974, On the relevance of non-histone proteins to the production of Giemsa banding patterns on chromosomes, Humangenetik, 21:227-236.
Vogelstein, B., Pardoll, D. M., and Coffey, D. S., 1980, Supercoiled loops and eukaryotic DNA replication, Cell, 22:79-85.
Wang, H. C., 1987, personal communication.
Wang, H. C., and Fedoroff, S., 1972, Banding in human chromosomes treated with trypsin, Nature New Biol., 235:52-53.
Wang, J. C., 1985, DNA topoisomerases, Ann. Rev. Biochem., 54:665-697.
Weintraub, H., 1985, Assembly and propagation of repressed and derepressed chromosomal states, Cell, 42:705-711.
Weisblum, B., and deHaseth, P. L., 1972, Quinacrine, a chromosome stain specific for deoxyadenylate-deoxythymidylate-rich regions in DNA, Proc. Natl. Acad. Sci. USA, 69:629-632.
Weisblum, B., and de Haseth, P. L., 1973, Nucleotide specificity of the quinacrine staining reaction for chromosomes, Chromosomes Today, 4:35-51.
Wunderli, H., Westphal, M., Armbruster, B.., and Labhart, P, 1983, Comparative studies on the structural organization of mem-

brane-depleted nuclei and metaphase chromosomes, Chromosoma, 88:241-248.

Yunis, J. J., 1981, Mid-prophase human chromosomes. The attainment of 2000 bands, Human Genet., 56:293-298.

Yunis, J. J., and Sanchez, O., 1973, G-banding and chromosome structure, Chromosoma, 44:15-23.

Yunis, J. J., and Prakash, O., 1982, The origin of man: A chromosomal pictorial legacy, Science, 215:1525-1530.

Yunis, J. J., Kuo, M. T., and Saunders, G. F., 1977, Localization of sequences specifying messenger RNA to light-staining G-bands of human chromosomes, Chromosoma, 61:335-344.

Yunis, J. J., Sawyer, J. R., and Ball, D. W., 1978, The characterization of high-resolution G-banded chromosomes of man, Chromosoma, 67:293-307.

Zhang, X. Y., and Horz, W., 1982, Analysis of highly purified satellite DNA containing chromatin from the mouse, Nucleic Acids Res., 10:1481-1494.

Zheng, H.-Z., and Burkholder, G. D., 1982, Differential silver staining of chromatin in metaphase chromosomes, Exp. Cell Res., 141:117-125.

CHROMOSOMES AND KINETOCHORES DO MORE IN MITOSIS THAN PREVIOUSLY THOUGHT

R. Bruce Nicklas

Department of Zoology
Duke University
Durham, North Carolina 27706, USA

INTRODUCTION

Chromosomes do more to promote their own distribution to the daughter cells in mitosis than we thought. A role of chromosomes in their own movement was made clear in the 1930's when it was recognized that a site on each chromosome -- the kinetochore or centromere -- is essential for normal chromosome movement in mitosis and meiosis (review: Schrader, 1953). Since then, one certain role of the kinetochore has been identified: the mechanical attachment of each chromosome to the spindle. Now, new aspects of attachment have been revealed as well as less expected chromosomal activities. My aim is an informal account of recent surprises followed by some remarks on the nature of the kinetochore as a gene and its evolutionary origin. More extensive reviews of all but the most recent work include Rieder (1982), McIntosh (1985), and Brinkley et al. (1985).

KINETOCHORES AND CHROMOSOME ATTACHMENT TO THE SPINDLE

Chromosomes are individually attached to the spindle by protein fibers called microtubules which extend between the kinetochore and a spindle pole (Fig. 1). A chromosomal site for attachment is an essential condition for directed chromosome movement, so this role alone would amply justify the existence of the kinetochore. Chromosomes lacking a kinetochore are not attached to the spindle and are lost. Proper attachment results in distribution of one copy of each chromosome to each daughter cell, and thus attachment makes possible the perpetuation of the chromosome complement (review: Nicklas, 1985).

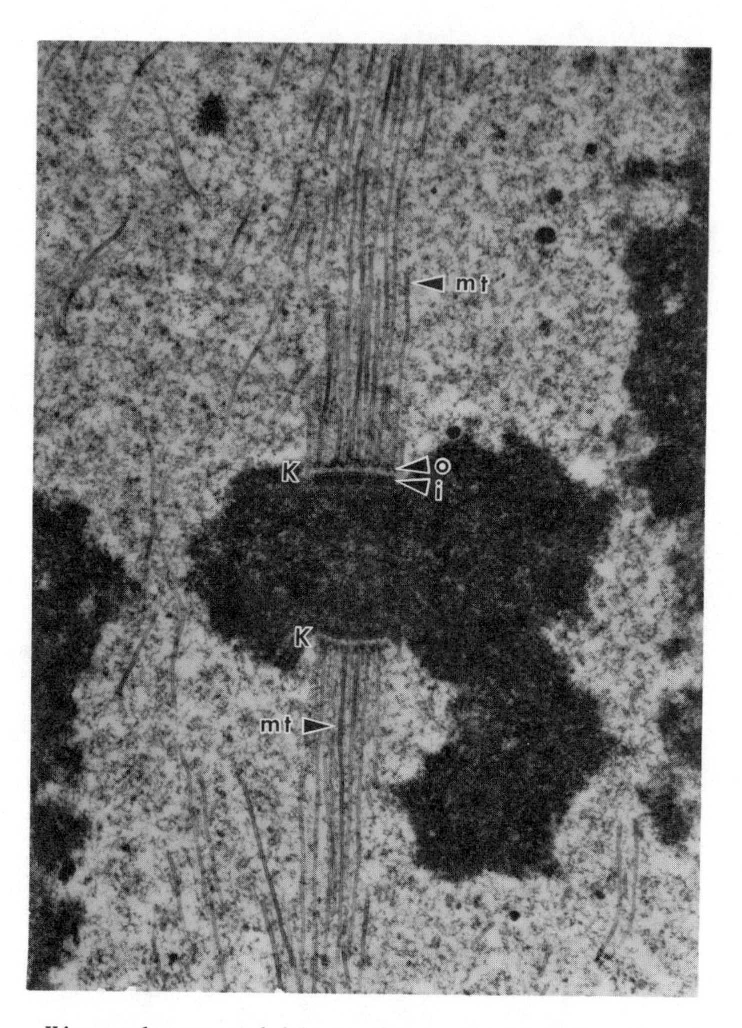

Fig. 1. Kinetochores and kinetochore microtubules in a metaphase chromosome of the alga Oedogonium. Two sister kinetochores ("K") are seen, each with several kinetochore microtubules ("mt") extending toward a pole (the poles are not shown). The kinetochore consists of dense inner ("i") and outer ("o") plates separated by a nearly clear zone; kinetochore microtubules insert in the outer plate. From Schibler, M. J. and J. D. Pickett-Heaps, 1987, Protoplasma, in press; used with permission.

The mechanical attachment of chromosomes to the spindle is made visible in living cells by micromanipulation. When chromosomes are pulled with a microneedle, the kinetochore remains tethered to the pole as the chromosome is stretched (for review see Ellis and Begg, 1981). Circumstantial but rich evidence identifies kinetochore microtubules as the tether, linked to the kinetochore at one end and to other spindle microtubules near the pole (Ellis and Begg, 1981; Nicklas et al., 1982). The mechanical role of the kinetochore, then, is to link the chromosome to microtubules. More generally, the kinetochore is built for specific interactions with microtubules, as the origin of kinetochore microtubules reveals.

Kinetochores Initiate and Capture Microtubules

Two alternatives for the origin of kinetochore microtubules are apparent: kinetochores might grow microtubules by initiating the assembly of new ones or they might capture preexisting spindle microtubules (Nicklas, 1971). In fact, kinetochores can do both. The origin of kinetochore microtubules has been studied in vitro, using isolated chromosomes and microtubule subunits or microtubules, as well as in living cells, during normal spindle formation or during recovery after drug treatment. In vitro, the kinetochores of isolated chromosomes can grow new microtubules in the presence of microtubule subunits (review: McIntosh, 1985; new observations: Mitchison and Kirschner 1985a), or they can capture microtubules when whole microtubules are present (Mitchison and Kirschner 1985b). The recent evidence for capture is illustrated in Fig. 2. When living cells are studied, there is unambiguous evidence for initiation after drug treatment, when subunits abound, and there is copious but somewhat equivocal evidence for capture during normal spindle formation, when whole microtubules are plentiful (reviews: Rieder, 1982; De Brabander et al., 1985; McIntosh, 1985).

The normal origin of kinetochore microtubules in living cells is uncertain. I would bet that capture usually predominates because (1) the kinetochore initiates assembly of new microtubules relatively poorly (at least in vitro), but captures microtubules efficiently (Mitchison and Kirschner, 1985a,b), and because (2) during normal spindle formation, by the time kinetochore microtubules appear, whole microtubules are plentiful while subunits are presumably in relatively short supply. However, remembering the cell's flair for exploiting every molecular capability, it is also a safe bet that the kinetochores sometimes grow their own kinetochore microtubules (e.g., when chromosomes interact precociously with a poorly developed spindle, before much microtubule assembly has occurred; then, the chromosomes are presented with a rich supply of subunits).

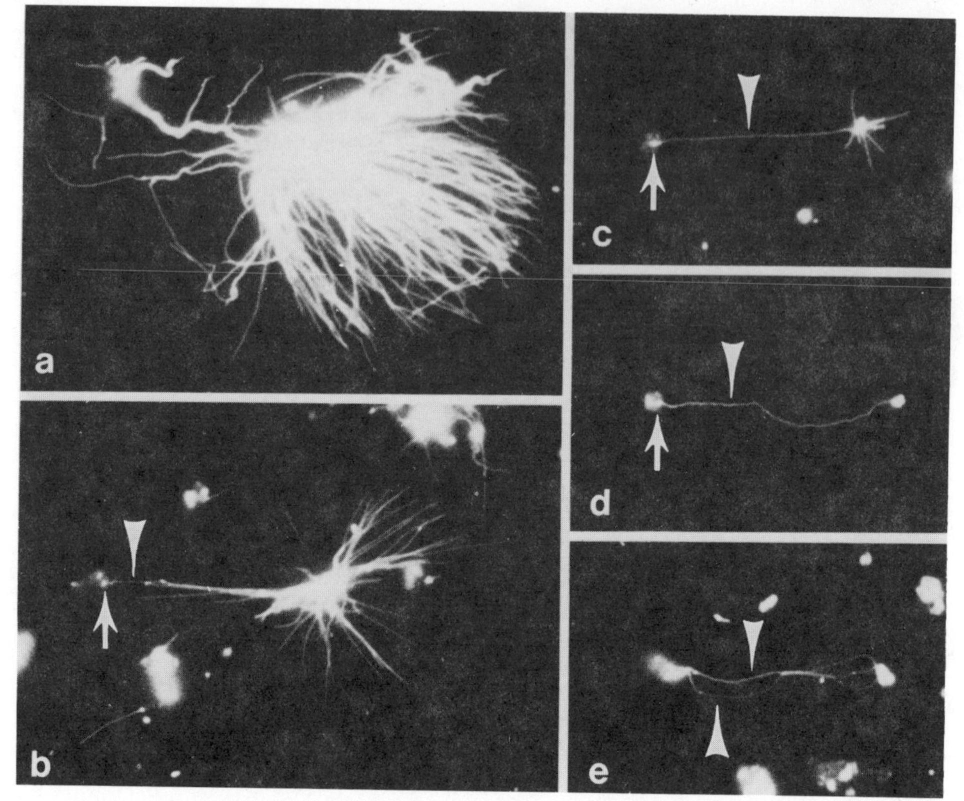

Fig. 2. Capture and stabilization of microtubules by kinetochores in vitro. Microtubules were made visible by fluorescent antibody staining. In each print, one chromosome is present (on the left), but is not visible in these pictures (the chromosomes can be seen when different optics are used); kinetochores are sometimes visible as bright dots (arrows).
a. A large array of microtubules was grown on a polar organizer, and then chromosomes were added to the preparation; a chromosome to the left is obscured by the microtubules.
b–e. After buffer was added; the resulting dilution of subunits caused unstable microtubules to disassemble.
b. Forty seconds after dilution. A single captured microtubule (arrowhead) runs to one kinetochore (arrow). The other, uncaptured microtubules have already shrunk to about half their original length.
c–e. Ninety seconds (c), and 10 minutes (d,e) after dilution. Only the microtubules stabilized by connection to a kinetochore remain (arrowheads). From Mitchison and Kirschner, 1985b; used with permission.

The major point, however, is not whether capture or initiation of microtubules predominates. It is simply that kinetochores can do both.

Kinetochores Are Efficient: the Speed of Microtubule Growth or Capture

Kinetochores grow or capture microtubules with impressive speed. We have begun to characterize the process by combining micromanipulation experiments with electron microscopy (Nicklas and Kubai, 1985). Just as attachment to the spindle is obvious when a chromosome is pulled with a microneedle, so is detachment: when pulled harder, the chromosome suddenly pops free of the spindle and can be placed wherever desired in the cell. Detached chromosomes invariably reattach to the spindle and move back to it. The immediate consequences of detachment and the earliest stages in reattachment are diagrammed in Fig. 3. A chromosome originally on a metaphase spindle (as shown to the right in Fig. 3a) is detached and placed in the cytoplasm, far from the spindle (Fig. 3a, left). Detached chromosomes lack kinetochore microtubules: at most only a single one is seen when cells are fixed immediately after detachment (20 to 40 kinetochore microtubules are present at each kinetochore before detachment). After only a minute or two, the chromosome begins to move back toward the spindle. Cells fixed at this time invariably show that a few, long kinetochore microtubules have already reappeared (Fig. 3b).

The capture of microtubules by the kinetochore in vitro (Fig. 2) is also fast. It is too fast for molecular models postulating only a few capture sites per kinetochore or requiring a precise angle of microtubule-kinetochore interaction (Mitchison and Kirschner, 1985b; see also Hill, 1985).

The dynamic properties of microtubules also contribute to fast capture. Spindle microtubules are even more dynamic than previously supposed (e.g., Salmon et al., 1984; Mitchison et al., 1986). In vitro, microtubules growing from a polar center exhibit an entirely unexpected behavior: some continue to grow while simultaneously others shorten with catastrophic rapidity (review: Kirschner and Mitchison, 1986; Horio and Hotani, 1986). New microtubules continually grow out from the center, replacing others that have vanished; Kirschner and Mitchison (1986) calculate that 1000 microtubules per pole per minute may be generated! Microtubules are envisaged as continuously probing the environment by extending in all directions; some encounter a kinetochore and are captured and stabilized (Kirschner and Mitchison, 1986). This surely sounds like an evolutionary design for providing kinetochores with a generous supply of microtubules to capture, no matter where the kinetochores may be (Mitchison and Kirschner, 1985b; Kirschner and Mitchison, 1986; Nicklas, 1985).

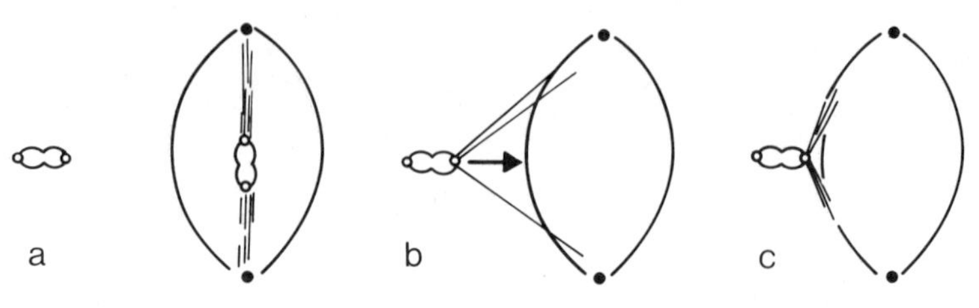

Fig. 3. Diagrams of chromosome movement and kinetochore microtubules following chromosome detachment from the spindle. A detached chromosome (3a, left) lacks microtubules, but kinetochore microtubules soon reappear and the chromosome moves back to the spindle (b and c) Kinetochores are depicted as open circles and the poles as filled circles; kinetochore microtubules are shown as thin lines and the spindle outline is indicated by a heavier line. From Nicklas and Kubai, 1985; used with permission.

Capture is so swift that still another idea is worth considering: active movement of microtubule ends. It is a natural starting assumption that the free ends of the microtubules that radiate from the poles move only by diffusion, but more active movement certainly is possible. Polar microtubules may continuously wave around so that their ends may have a greatly enhanced chance of encountering a kinetochore. Advances in methods for seeing microtubules in living cells may soon reveal the movement of microtubules and perhaps even their capture by kinetochores. The reality thus revealed may make present speculations seem unimaginative.

Kinetochores Are Promiscuous: the Quality of Initial Attachment

Kinetochore microtubules appear very quickly, but the *quality* of the initial array is not so impressive. For instance, if the arrangement of kinetochore microtubules shown in Fig. 3c persisted, meiotic segregation would be irregular, and the gametes would receive abnormal chromosome complements. Just about every conceivable inappropriate microtubule arrangement has been seen early in normal mitosis and meiosis (review: Rieder, 1982; additional observations: Church and Lin, 1982, 1985; Jensen, 1982). Inappropriate arrays of kinetochore microtubules almost never persist, however, and the one arrangement consistent with genetically appropriate chromosome distribution eventually arises (review: Nicklas, 1985). The point worth emphasizing here is the importance the cell puts on chromosome attachment to the spindle -- even a flawed initial attachment is better than none because flaws can be rectified, but failure to attach can only lead to random chromosome distribution.

In summary, attachment to the spindle is vital, and kinetochores are the means for that attachment. Whatever else they may do, kinetochores interact vigorously with the microtubules and microtubule subunits that link them to the spindle.

CHROMOSOMES AND SPINDLE ORGANIZATION

Kinetochores Stabilize Microtubules

Unstable microtubules can be stabilized by binding to a kinetochore, as Salmon (1975) first proposed and as the experiments of Mitchison and Kirschner (1985b) vividly demonstrate (Fig. 2). Kirschner and Mitchison (1986) have proposed that important features of spindle organization can be explained simply by combining dynamic instability of polar microtubules with selective stabilization of those which encounter kinetochores -- at least the origin of the general classes and distribution of microtubules would be accounted for. Here again, though, real surprises can be expected: the apparent stability of kinetochore microtubules may be more complicated, and hence more interesting, than we thought, according to a preliminary account (McIntosh et al, 1986).

Chromosomes Enhance the Extent of Microtubule Assembly

A previously unsuspected ability of chromatin to enhance microtubule assembly has been discovered: microtubules form around nuclei or DNA injected into *Xenopus* eggs (Karsenti et al., 1984). Most remarkably, chromatin activates otherwise inactive polar organizers, so that new arrays of microtubules appear. As Karsenti and his colleagues say, the results strongly suggest a novel and important role of chromosomes in spindle formation. An enhancement of microtubule assembly near chromosomes has also been reported for mouse oocytes (Maro et al., 1986).

Equally direct evidence for a potent effect of chromosomes on microtubule assembly or stability has been obtained by using micromanipulation to alter the number of chromosomes present (Nicklas and Gordon, 1985). Chromosomes were detached from the spindle and then simply pulled out of grasshopper (*Melanoplus differentialis*) spermatocytes. Later, the cells were fixed for electron microscopy and the total length of microtubules in the spindle was determined. The expectation from in vitro studies (Bryan, 1976) is clear: if no microtubules or subunits are removed along with the chromosomes, then the total length of spindle microtubules should be constant, regardless of the number of chromosomes present. In fact, however, spindle microtubule length decreased by a constant amount for each chromosome removed; it was estimated that in the total absence of chromosomes, only 40% of the original microtubule length would remain (Nicklas and Gordon, 1985).

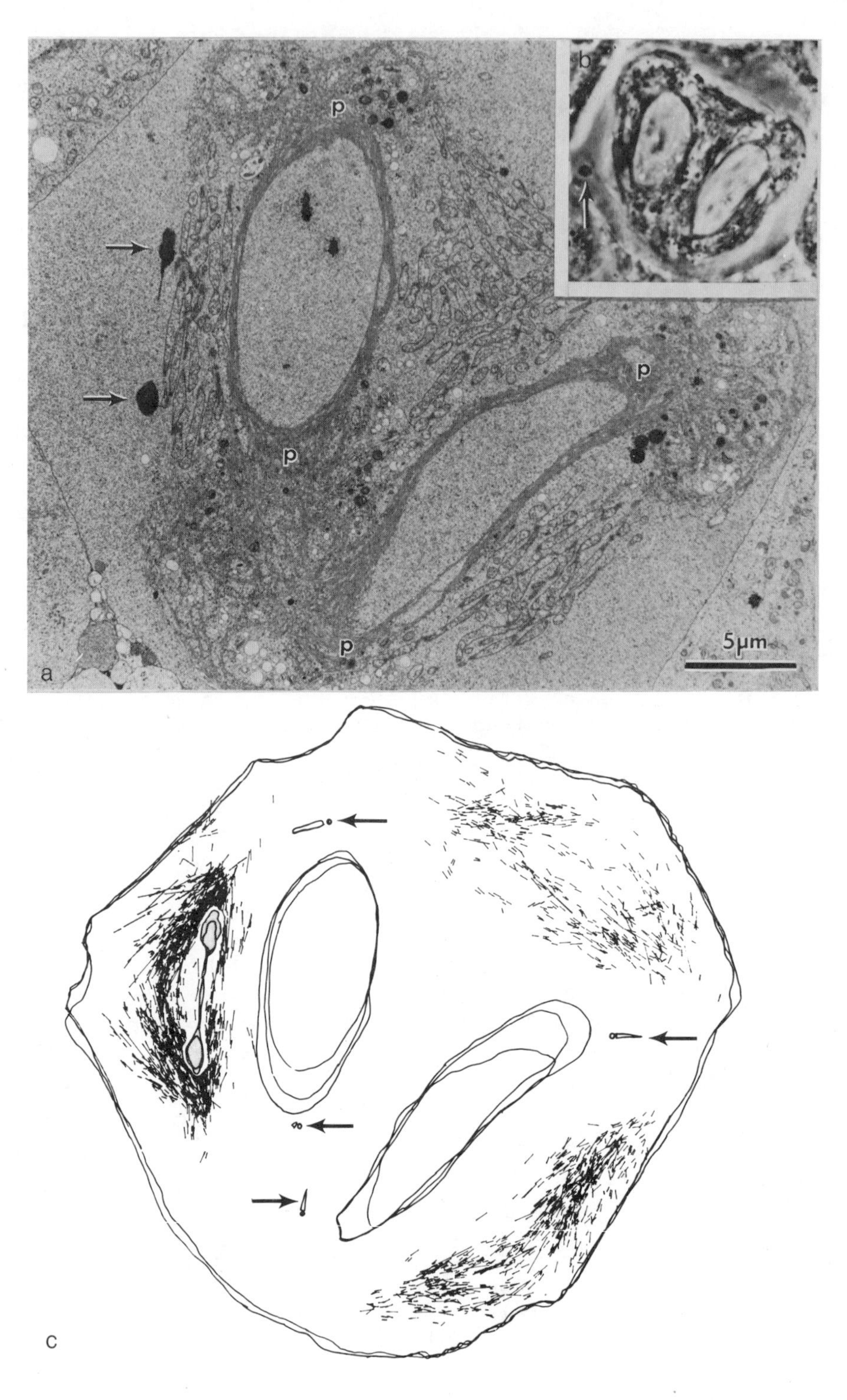
b
p
p
p
p
5μm
a
c

We concluded that the assembly of a large fraction of spindle microtubule length is strictly dependent on the presence of chromosomes (for a discussion of alternative interpretations that probably can be ruled out, see Nicklas and Gordon, 1985).

A Single Chromosomes Can Trigger Formation of a Complete Spindle

A combined micromanipulation and electron microscope study of *Drosophila* spermatocytes is illustrated in Fig. 4 (Nicklas et al., 1985; Church et al., 1986). In these cells, the spindle is largely isolated from the cytoplasm by several layers of membranes. Chromosomes detached from the spindle by micromanipulation and placed beyond the membranes (Fig. 4a) cannot rejoin the spindle. Remarkably, a miniature spindle very rapidly forms around the chromosome in the cytoplasm, and the displaced chromosome divides normally in anaphase on this mini-spindle (Fig. 4b,c). The presence of the chromosome in the cytoplasm greatly enhances microtubule density in its vicinity (Fig. 4c). More than four times greater microtubule length has been measured in some mini-spindles than in comparable, but chromosome-free, regions of the cytoplasm (Church et al., 1986).

Unexpectedly, mini-spindles often have a pole unrelated to any pole of a normal spindle -- a pole has been induced to form where none was evident before the chromosome was moved to the cytoplasm. Thus, the presence of a chromosome in the cytoplasm not only enhances microtubule assembly but also triggers the formation of a functionally normal spindle, complete with two poles, at a site where a spindle never occurs normally (Nicklas et al., 1985; Church et al., 1986).

Fig. 4 (facing page). Mini-spindle formation in *Drosophila*.

4a. A living spermatocyte with two ordinary spindles (more than one spindle is not rare in cultured spermatocytes). A chromosome was detached from one of the spindles, pulled through the surrounding membranes, and placed in the cytoplasm (arrow).

4b. Survey electron micrograph. The cell was fixed after the detached chromosome had divided in anaphase as if on a spindle of its own. Note the separated halves of the chromosome (arrows) and the thick layer of membranes and mitochondria that separate the chromosome from the ordinary spindles. The poles of the ordinary spindles are labeled "p".

4c. Reconstruction of microtubules in three regions; membrane outlines and centrioles (arrows) are also shown. A mass of microtubules defines a mini-spindle around the manipulated chromosome (left). Microtubule density is much lower in the other areas reconstructed. From Nicklas et al., 1985; used with permission.

Chromosomes by Themselves, without Polar Organizers, May Sometimes Form a Spindle

Chromosomes make a contribution to spindle formation, but in most materials, polar organizers are also essential (Mazia et al., 1981; Mazia, 1984). In *Drosophila* mini-spindles, for instance, the presence of the chromosome may activate polar organizers (Church et al., 1986). Recent, elegant experiments conclusively demonstrate that in sea urchin (*Lytechinus* spp.) eggs, chromosomes by themselves are impotent (Sluder and Rieder, 1985). In one organism, however, a case is being made that chromosomes can act alone.

Dietz (1966) discovered, in crane fly (*Pales ferrugenia*) spermatocytes, functionally normal spindles which lacked centrioles and asters at one or both poles. Careful observations on these poles have disclosed no detectable microtubule organizing material ("pericentriolar material") by either electron microscopy or immumochemical tests (Bastmeyer et al., 1986; Steffen et al., 1986). This is not proof, of course, that *no* polar organizer was present. Also, in the recent work, it was not directly demonstrated that polar organizers were absent when the spindle formed (as opposed to a later separation of poles from an already-formed spindle), although absence beforehand is likely on the basis of Deitz' (1966) results. At the least, however, the crane fly spermatocyte is an excellent candidate for a cell in which chromosomes by themselves can organize a spindle.

Implications

The major biological implications of chromosomal contributions to spindle organization are the following: chromosomes can help regulate the interphase/mitosis transition in microtubule assembly and arrangement (Karsenti et al., 1984), and chromosomes can make the spindle self-adjusting -- microtubule length may be perfectly and immediately adjusted to the number of chromosomes present (Nicklas et al., 1985). For example, evolutionary increases in the number of chromosomes could automatically result in the assembly of exactly the additional microtubule length necessary to accommodate the additional chromosomes on the spindle (longer-term adaptation through genetic change is also likely, of course). This presumes that microtubule subunits in excess of those necessary for mitosis are always present because, for instance, they are needed in interphase. That may or may not be true, but it provides an excuse to notice another biological potential of chromosomal participation in spindle organization. A difference between interphase and mitosis is precisely the sort of control that a chromosomal impact on the extent of microtubule assembly in mitosis permits.

A fresh respect for chromosomes in spindle formation is just a beginning, of course. We know nothing about the molecular mechanisms involved, but we are highly motivated to learn.

KINETOCHORES AS SITES FOR MICROTUBULE ASSEMBLY AND DISASSEMBLY

Kinetochore microtubules change in length during mitosis, and exactly where the associated assembly and disassembly occurs is an important question. Microtubule assembly at the kinetochore in metaphase has been demonstrated directly both in vitro and in living cells (Mitchison and Kirschner, 1985b; Mitchison et al., 1986). This is not entirely unexpected because the microtubule end most active in assembly terminates at the kinetochore (McIntosh and Euteneuer, 1984). However, it might be supposed that the kinetochore would cover the end of an attached microtubule, thus preventing subunit addition. A model permitting assembly and disassembly at the kinetochore is considered below.

The study showing assembly at the kinetochore in metaphase also provides evidence that the kinetochore is a site of microtubule disassembly in anaphase (Mitchison et al., 1986). Cells in metaphase were injected with labeled microtubule subunits and later were fixed for electron microscopy. Labeled microtubule segments extending poleward from the kinetochores were seen. If the cells were allowed to proceed into anaphase before fixation, the labeled microtubule segments were shorter, identifying the kinetochore as a site for microtubule disassembly as well as assembly. Forer and Schaap (Forer, 1976; Schaap and Forer, 1984) similarly concluded that the kinetochore is the site of birefringent fiber disassembly in anaphase, based on observations in living cells that as the kinetochore moves, certain fiber markers remain fixed in position. However, their observations did not decisively pinpoint the kinetochore as the site of disassembly because reliable markers close to the kinetochore were lacking.

Disassembly at the kinetochore in anaphase suggests that kinetochore microtubules remain in place while shortening occurs at the kinetochore: the kinetochore might move poleward independent of the microtubules rather than being pulled poleward along with the microtubules. The evidence on movement is not conclusive, however, and a more direct approach is described below.

In conclusion, there is good evidence that the kinetochore is a site of microtubule assembly in metaphase and of disassembly in anaphase, but it should be kept in mind that additional sites nearer the pole have not been ruled out.

THE KINETOCHORE MAY BE THE MOTOR FOR CHROMOSOME MOVEMENT

It is a commonplace observation that we have a surfeit of speculations about chromosome movement in mitosis, but no generally accepted theory. That is not quite true. Most workers have accepted the "traction fiber theory", that chromosomes are pulled to the poles in anaphase by forces acting on their kinetochore microtubules. The checkered past of the traction theory is interesting in itself and provides a good introduction to recent developments (the more extended accounts of Schrader, 1953, and Inoué, 1981, are highly recommended). The idea that spindle fibers pull chromosomes to the pole is over a century old, originating with van Beneden in 1883, only four years after the first definitive descriptions of mitosis. The traction fiber theory was popular until the early years of this century when doubts about the reality of fibers lead to its partial eclipse. A plethora of theories followed, most of which now seem strange, to say the least (Schrader, 1953). But by the 1930's and 40's, doubts about fiber reality were being resolved, and the traction fiber theory was revived (Cornman, 1944); it became dominant when proof of fiber reality was in hand from the work of Inoué and of Mazia and Dan (review: Inoué, 1981). A trustworthy sign of wide acceptance is that it now seems pretentious to call the traction fiber idea a "theory" -- it is so elementary, so well-established. Well, those of us in this camp, including Nicklas, may well have been wrong. The traction fiber may be doomed, to be replaced by a kinetochore that participates actively in its own movement.

The first in vitro movements of chromosomes relative to microtubules have just been reported by Mitchison and Kirschner (1985b). Their innovative experiments strongly suggest that the kinetochore has (or activates) an ATP-driven motor that, on the spindle, would move microtubules poleward and the chromosome farther from the pole -- exactly in the wrong direction to explain poleward movement in anaphase! Mitchison (1986) cautions that further work is needed, especially on the polarity of microtubule movement, but even if, as now seems likely, the microtubules are pushed poleward, there is ample scope for such an activity in chromosome movements before metaphase (Mitchison and Kirschner, 1985b). In any case, an important kinetochore capability has probably been seen for the first time. But what about anaphase?

Gorbsky and co-workers (1987) have just reported direct evidence for kinetochore movement on stationary kinetochore microtubules. They injected cells with fluorescent microtubule subunits; these were incorporated into spindle microtubules, making them fluorescent. Then, as diagrammed in Fig. 5, the spindle was irradiated with a band of light to bleach the fluorescence locally without harming the microtubules (Gorbsky et al. and others have shown that photobleaching destroys the fluorescent tag but not the

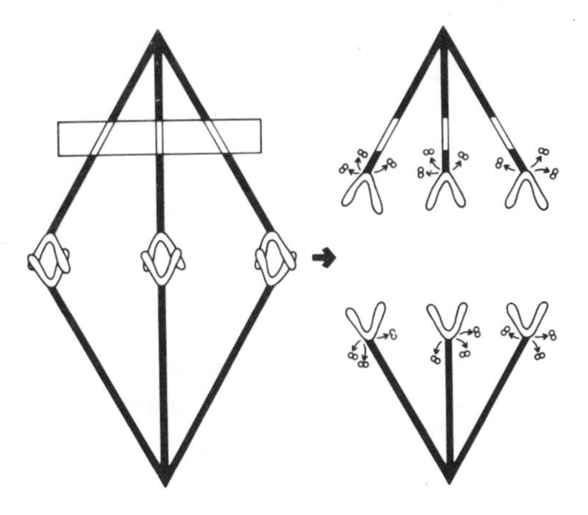

Fig. 5. Diagrammatic representation of a photobleaching experiment performed during anaphase chromosome movement. Fluorescent kinetochore microtubule bundles are represented by black bars. The rectangular box in the left spindle (early anaphase) represents the photobleaching beam; bleached microtubule segments are represented by open bars. As anaphase movement proceeds (right), kinetochore microtubules disassemble at the kinetochores (the small double circles represent microtubule subunits) and the chromosomes approach the bleached region, which does not move. From Gorbsky et al., 1987; used by permission.

microtubules). Thus, many microtubules are marked by a band of low fluorescence (Fig. 5), and movement of the band will betray the movement of those microtubules. So now, Gorbsky and co-workers can ask: do the marked microtubules move poleward in anaphase along with the chromosome, as a traction fiber would, or is the band immobile as the chromosome moves? The observations are clear: the bleached region stays in place as the chromosomes move to it (Fig. 5). In different cells, the bleached band can be placed near or far from the chromosome, and anaphase movement can be allowed to continue until the chromosome has moved part way to the band, to it, or beyond it. Uniformly, no movement of the band was seen.

The obvious interpretation is that kinetochore microtubules are not pulled poleward and do not pull the chromosomes poleward. Instead, the kinetochore is the motor for chromosome movement in anaphase or activates a motor on the microtubules at the kinetochore (Gorbsky et al., 1987). (These possibilities are considered in more detail below.)

The major complication in this experiment is that many of the microtubules are not marked by a photobleached band (see Fig. 6 in

Gorbsky et al., 1987). Gorbsky and co-workers make a plausible argument that kinetochore microtubules are preferentially marked, and that the movement of unmarked, i.e., fully fluorescent, microtubules would be detected because they would move into the band, increasing the fluorescence there. In the end, this is a question of signal detection, of sensitivity, and certainly the movement of some kinetochore microtubules is not yet ruled out. Some caution is suggested, particularly because the spindle is a very capable force producer (Nicklas, 1983), and putative force producers associated with just one kinetochore microtubule, moving that microtubule and pulling on the chromosome, could provide many times the force required for normal anaphase movement. I conclude that these experiments make a good, though not a conclusive case, for the active participation of the kinetochore in its own movement. This is a very exciting area for further work.

KINETOCHORE MODELS

How might a kinetochore function simultaneously as an attachment site, a microtubule assembly/disassembly site, and a mover of microtubules? More specifically, how can the attachment of microtubule to kinetochore be maintained despite assembly/disassembly reactions at the end of the microtubule and despite microtubule movement relative to the kinetochore? Structural models of the class first proposed by Margolis and Wilson (1981) provide a simple and appealing general solution. The kinetochore is viewed as providing a number of sleeves in which microtubules insert. A single sleeve/microtubule unit is diagrammed in Fig. 6. Attachment of a microtubule to the kinetochore is provided by lateral interactions of some sort between them, in the sleeve, leaving the microtubule end free for assembly or disassembly. Two general models of how such a kinetochore might move along the microtubules in its sleeves have been proposed.

One model invokes attractive interactions (e.g., electrostatic attraction) between the microtubule and the wall of the sleeve to allow the kinetochore to remain attached to a depolymerizing microtubule and thus to follow it poleward (Hill, 1985). Depolymerization from the chromosome side toward the pole reduces the length of microtubule in the sleeve and the kinetochore then moves poleward so that more microtubule length is included in the sleeve. The movement is driven by the attraction between the microtubule and the wall; this maintains the microtubule at a certain depth in the sleeve. On this model, chromosome movement and microtubule depolymerization are strictly coupled (cf. Inoué, 1981).

In a second general model, the kinetochore is part of a classical biological motor -- an ATP-driven motor that is either part of the kinetochore, lining the sleeves and pulling on the

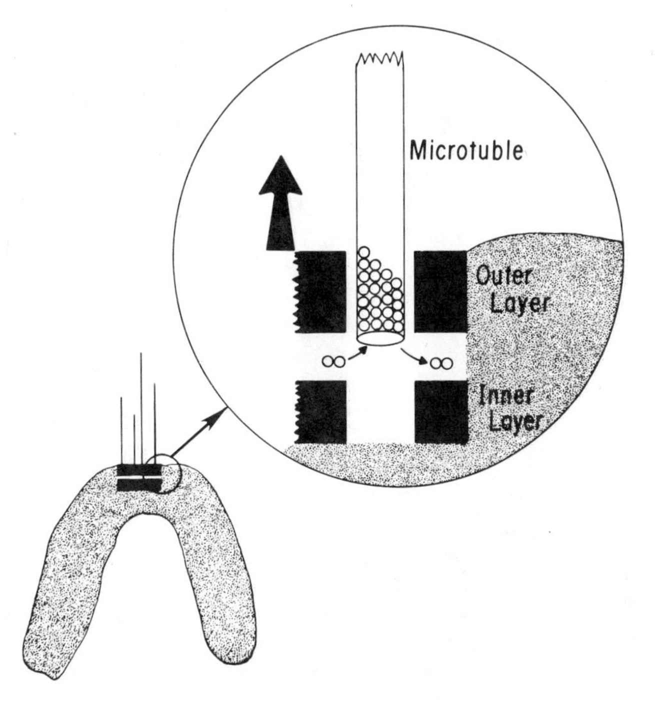

Fig. 6. A generalized model of the kinetochore. Left: an entire anaphase chromosome with kinetochore microtubules is shown for orientation. Right: an enlarged view of one unit of the kinetochore in cross-section: a microtubule protrudes from a sleeve in the outer layer of the kinetochore. The arrow shows the direction of chromosome movement toward the pole (not shown). Subunits (double circles) can be added to or lost from the end of the microtubule.

microtubules, or is linked to the microtubules and activated by the kinetochore (Hill, 1986; Mitchison, 1986).

The microtubule-in-a-sleeve model for the mature kinetochore postulates a precise kinetochore/microtubule geometry that appears to be incompatible with the speed of microtubule capture, as mentioned above. Electron microscopic observations of the initial association of microtubules with kinetochores certainly fit with varied rather than tightly constrained angles of association (Church and Lin, 1982, 1985; Nicklas and Kubai, 1985). Similarly, in the initial stages of microtubule initiation, the nascent kinetochore microtubules do not have the same, precise structural association with the kinetochore that is seen later -- even which part of the kinetochore is involved is different (De Brabander et al., 1981; Witt et al., 1980; this leads to contrasting models of initiation: cf. McIntosh, 1985 and De Brabander et al., 1985). Thus, initial irregularity and precise arrangement later on is not a problem for a model of the final arrangement, it is a fact of life requiring explanation on any model. Understanding the remarkable transformation from the apparent irregularity seen initially to the structural precision seen later (e.g., Fig. 1) is an important challenge for the future.

A different model has been proposed by Pickett-Heaps (review: 1986). He has for some years been suggesting that the kinetochore plays an active part in chromosome movement, based mainly on what may be a special spindle component in some diatoms. He prefers a model in which an elastic, non-microtubular, component attaches at the kinetochore and pulls the chromosome poleward (Pickett-Heaps, 1986; cf. McIntosh, 1981).

MOLECULAR BIOLOGY AND STRUCTURE OF THE KINETOCHORE

Inventive work with yeast has made kinetochore genetics and molecular biology a reality. Mini-chromosomes capable of participating in mitosis and meiosis have been constructed that are small enough that critical portions can be sequenced (reviews: Carbon, 1984; Blackburn and Szostak, 1984; Bloom et al., 1986). Short DNA sequences essential for proper chromosome distribution have been identified, and a distinctive DNA-protein organization in the presumptive kinetochore is being characterized. The goal of identifying and isolating all the proteins and then of total reconstruction of the kinetochore seems within grasp: to start with the DNA, add proteins, and end with bound microtubules.

At the same time, structural work on the kinetochores of higher eukaryotes is progressing. Kinetochore organization has been partly disrupted to reveal that this structure does not consist of protein plates appliquéd onto ordinary chromatin. Instead, the kinetochore is composed of chromatin fiber loops, containing DNA as well as protein, all the way out to where microtubules are attached (review: Rieder, 1982; Ris and Witt, 1981; Rattner, 1986). Several promising approaches take advantage of human auto-antibodies that react with kinetochores. These have been used to identify putative kinetochore proteins (e.g., Palmer et al. 1986) and the genes that code for them (Valdiva et al., 1986), and they have also been used in attempts to isolate the kinetochore (Brinkley et al., 1985).

A union of studies on yeast and higher eukaryotes is widely anticipated. The kinetochore of higher organisms is more complicated structurally, but fundamental features of DNA sequence and kinetochore proteins may well be conserved. If so, the rapidly advancing work on yeast will provide relevant information and molecular probes to be used for studies of other eukaryotes -- for example, protein probes that will identify related proteins and conserved DNA sequences. A simple quantitative consideration makes comparison of the yeast kinetochore with those of other eukaryotes especially interesting. Yeast chromosomes probably have only one microtubule per kinetochore (Peterson and Ris, 1976), while the number of microtubules per kinetochore is more than 10 in most higher eukaryotes and up to 100 in some plants (review: Rieder, 1982). Therefore, yeast will provide the baseline for an important correlation of the number of copies of a putative kinetochore DNA sequence with the number of microtubules per kinetochore.

THE KINETOCHORE AS A GENE AND ITS EVOLUTIONARY ORIGIN

An uncomplicated view of the kinetochore is useful to clarify thought and to guide speculation about origins. In essence, the kinetochore is a specific DNA sequence that binds specific proteins which in turn link the chromosome to the division apparatus (Nicklas, 1971). This view makes the kinetochore sequence merely a special example of DNA sequences that function by binding specific proteins, i.e., like promoter sequences, for instance, rather than by coding for RNA or protein. In higher organisms, the fundamental sequence is likely to be repeated several times, with the number of copies related to the number of kinetochore microtubules per chromosome. This helps make sense of two features of kinetochore evolution in eukaryotes. First, having all the copies in one site would characterize the familiar localized kinetochore, while their dispersion to multiple sites all along the chromosome would produce "holokinetic" chromosomes, which have evolved independently in several plant and animal groups (review: Rieder, 1982). Second, the number of copies of the sequence would be expected to vary, as is true of other repetitive sequences. Therefore, the number of copies should easily respond to selection following changes in chromosome number, increasing after a fission of one chromosome into two and decreasing after the fusion of two chromosomes into one.

The evolutionary origin of the kinetochore is straightforward, on the above view. The kinetochore of eukaryotic cells differs from its prokaryotic antecedent only by what it binds to -- the spindle components in eukaryotes, the cell cortex in prokaryotes (Nicklas, 1971). The kinetochore originated in prokaryotes as a sequence devoted to protein binding. It may even have represented little evolutionary novelty: selection for reliable and controlled replication and transcription may well have led to protein binding as a common DNA property very early in evolution, when chromosome distribution was still a haphazard but tolerable matter of many DNA copies randomly included in the daughter cells at division (chloroplasts and mitochondria may have managed no better to this day). Thereafter, co-evolution of the kinetochoric DNA and the protein products of other loci provides a conceptually simple route from prokaryotic division to eukaryotic mitosis and meiosis. From this perspective, the true complexities of this evolutionary transformation of one and a half billion years ago will at least in part be appreciated when the DNA/division apparatus link is fully understood in contemporary organisms.

This simple speculation has been restated in part because of the persistent promotion of a far more complicated origin for mitosis and the kinetochore (e.g., Margulis, 1970; Margulis and Sagan, 1986). Margulis proposes fundamentally different modes of chromosome distribution in pro- and eukaryotes, bridged by symbiotic spirochetes. A symbiotic origin for some organelles clearly has merit but a symbiotic origin of mitosis is contrived and without significant supporting evidence. Interested biologists may wish to weigh the Margulis account against the less exotic evolutionary pathway envisioned above.

SUMMARY

Recent work discloses a remarkable number of unexpected activities of chromosomes and kinetochores in mitosis. The list now looks like this:

Attachment of Chromosome to Spindle
- Kinetochores:
  1. Initiate new microtubules
  2. Capture preexisting microtubules

Spindle Organization
- Kinetochores:
  3. Stabilize microtubules
- Chromosomes:
  4. Enhance microtubule assembly
  5. Trigger formation of a complete spindle
  6. May sometimes induce a spindle all by themselves

Chromosome Movement
- Kinetochores:
  7. Are a site of microtubule assembly and disassembly
  8. May be the motor or activate the motor

One thing is not surprising: devotees of the cell in mitosis are in a buoyant mood, delighting in the present and confidently expecting further refreshing surprises in the future.

ACKNOWLEDGEMENTS

I am grateful to Donna Kubai for a sharp editorial pencil and to Gary Gorbsky and co-workers, Marc Kirschner, Tim Mitchison, and Matthew Schibler for generous permission to use illustrations. Our work reviewed here was supported in part by grants from the National Institutes of Health (GM 13745, to RBN) and the National Science Foundation (DCB 8207538, to Kathleen Church).

REFERENCES

Bastmeyer, M., Steffen, W., and Fuge, H., 1986, Immunostaining of spindle components in tipulid spermatocytes using a serum against pericentriolar material, Eur. J. Cell Biol., 42:305-310.

Blackburn, E. H., and Szostak, J. W., 1984, The molecular structure of centromeres and telomeres, Ann. Rev. Biochem., 53:163-194.

Bloom, K., Hill, A., and Yeh, E., 1986, Structural analysis of a yeast centromere, BioEssays, 4:100-104.

Brinkley, B. R., Tousson, A., and Valdivia, M. M., 1985, The kinetochore of mammalian chromosomes: structure and function in normal mitosis and aneuploidy, in: "Aneuploidy," V. L. Dellarco, P. E. Voytek, and A. Hollaender, eds., Plenum Press, New York. pp. 243-267.

Bryan, J., 1976, A quantitative analysis of microtubule elongation, J. Cell Biol., 71:749-767.

Carbon, J., 1984, Yeast centromeres: structure and function, Cell, 37:351-353.

Church, K., and Lin, H.-P. P., 1982, Meiosis in Drosophila melanogaster. II. The prometaphase-I kinetochore microtubule bundle and kinetochore orientation in males, J. Cell Biol., 93:365-373.

Church, K., and Lin, H.-P. P., 1985, Kinetochore microtubules and chromosome movement during prometaphase in Drosophila melanogaster spermatocytes studied in life and with the electron microscope, Chromosoma, 92:273-282.

Church, K., Nicklas, R. B., and Lin, H.-P. P., 1986, Micromanipulated bivalents can trigger mini-spindle formation in Drosophila melanogaster spermatocyte cytoplasm, J. Cell Biol., 103: 2765-2773.

Cornman, I., 1944, A summary of evidence in favor of the traction fiber in mitosis, Amer. Naturalist, 78:410-422.

De Brabander, M., Aerts, F., De Mey, J., Geuens, G., Moeremans, M., Nuydens, R., and Willebrords, R., 1985, Microtubule dynamics and the mitotic cycle: a model, in: "Aneuploidy," V. L. Dellarco, P. E. Voytek, and A. Hollaender, eds., Plenum Press, New York. pp. 269-278.

De Brabander, M., Geuens, G., De Mey, J., and Joniau, M., 1981, Nucleated assembly of mitotic microtubules in living PTK_2 cells after release from nocodazole treatment, Cell Motility, 1:469-483.

Dietz, R., 1966, The dispensability of the centrioles in the spermatocyte divisions of Pales ferruginea (Nematocera), in:

"Chromosomes Today," vol. 1, C. D. Darlington and K. R. Lewis, eds., Oliver and Boyd, Edinburgh. pp. 161-166.

Ellis, G. W., and Begg, D. A., 1981, Chromosome micromanipulation studies, in: "Mitosis/Cytokinesis," A. M. Zimmerman and A. Forer, eds., Academic Press, New York. pp. 155-179.

Forer, A., 1976, Actin filaments and birefringent spindle fibers during chromosome movements, in: "Cell Motility, Book C," R. Goldman, T. Pollard, and J. Rosenbaum, eds., Cold Spring Harbor Laboratory. pp. 1273-1293.

Gorbsky, G. J., Sammak, P. J., and Borisy, G. G., 1987, Chromosomes move poleward in anaphase along stationary microtubules that coordinately disassemble from their kinetochore ends, J. Cell Biol., 104:9-18.

Hill, T. L., 1985, Theoretical problems related to the attachment of microtubules to kinetochores, Proc. Natl. Acad. Sci. USA, 82:4404-4408.

Hill, T. L., 1986, Kinetic diagram and free energy diagram for kinesin in microtubule-related motility, Proc. Natl. Acad. Sci. USA, 83:3326-3330.

Horio, T., and Hotani, H., 1986, Visualization of the dynamic instability of individual microtubules by dark-field microscopy, Nature, 321:605-607.

Inoue, S., 1981, Cell division and the mitotic spindle, J. Cell Biol., 91:131s-147s.

Jensen, C. G., 1982, Dynamics of spindle microtubule organization: kinetochore fiber microtubules of plant endosperm, J. Cell Biol., 92:540-558.

Karsenti, E., Newport, J., and Kirschner, M., 1984, Respective roles of centrosomes and chromatin in the conversion of microtubule arrays from interphase to metaphase, J. Cell Biol., 99:47s-54s.

Kirschner, M., and Mitchison, T., 1986, Beyond self-assembly: from microtubules to morphogenesis, Cell, 45:329-342.

Margolis, R. L., and Wilson, L., 1981, Microtubule treadmills - possible molecular machinery, Nature, 293:705-711.

Margulis, L., 1970, "Origin of Eukaryotic Cells," Yale University Press, New Haven.

Margulis, L., and Sagan, D., 1986, "Origins of Sex," Yale University Press, New Haven.

Maro, B., Johnson, M. H., Webb, M., and Flach, G., 1986, Mechanism of polar body formation in the mouse oocyte: an interaction between the chromosomes, the cytoskeleton and the plasma membrane, J. Embryol. Exp. Morph., 92:11-32.

Mazia, D., 1984, Centrosomes and mitotic poles, Exp. Cell Res., 153:1-15.

Mazia, D., Paweletz, N., Sluder, G., and Finze, E.-M., 1981, Cooperation of kinetochores and pole in the establishment of monopolar mitotic apparatus, Proc. Natl. Acad. Sci. USA, 78:377-381.

McIntosh, J. R., 1981, Microtubule polarity and interaction in mitotic spindle function, in: "International Cell Biology 1980-81," H. G. Schweiger, ed., Springer Verlag, Berlin. pp. 359-368.

McIntosh, J. R., 1985, Spindle structure and the mechanisms of chromosome movement, in: "Aneuploidy," V. L. Dellarco, P. E. Voytek, and A. Hollaender, eds., Plenum Press, New York, pp. 197-229.

McIntosh, J. R., and Euteneuer, U., 1984, Tubulin hooks as probes for microtubule polarity: An analysis of the method and an evaluation of data on microtubule polarity in the mitotic spindle, J. Cell Biol., 98:525-533.

McIntosh, J. R., Saxton, W. M., and Stemple, D. L., 1986, Differential stability of mitotic microtubules as measured by fluorescence redistribution after photobleaching (FRAP), J. Cell Biol., 103:270a (abstr.).

Mitchison, T. J., 1986, The role of microtubule polarity in the movement of kinesin and kinetochores, J. Cell Sci., Suppl. 5:121-128.

Mitchison, T., Evans, L., Schulze, E., and Kirschner, M., 1986, Sites of microtubule assembly and disassembly in the mitotic spindle, Cell, 45:515-527.

Mitchison, T. J., and Kirschner, M. W., 1985a, Properties of the kinetochore in vitro. I. Microtubule nucleation and tubulin binding. J. Cell Biol., 101:755-765.

Mitchison, T. J., and Kirschner, M. W., 1985b, Properties of the kinetochore in vitro. II. Microtubule capture and ATP-dependent translocation, J. Cell Biol. 101:766-777.

Nicklas, R. B., 1971, Mitosis, in: "Advances in Cell Biology," vol. 2, D. M. Prescott, L. Goldstein, and E. McConkey, eds., Appleton-Century-Crofts, New York. pp. 225-297.

Nicklas, R. B., 1983, Measurements of the force produced by the mitotic spindle in anaphase, J. Cell Biol., 97:542-548.

Nicklas, R. B., 1985, Mitosis in eukaryotic cells: an overview of chromosome distribution, in: "Aneuploidy," V. L. Dellarco, P. E. Voytek, and A. Hollaender, eds., Plenum Press, New York. pp. 183-195.

Nicklas, R. B., Church, K., Lin, H.-P. P., and Gordon, G. W., 1985, Chromosomes enhance microtubule assembly or stability, in: "Microtubules and Microtubule Inhibitors 1985," M. De Brabander and J. De Mey, eds., Elsevier Science Publishers B. V., Amsterdam. pp. 261-268.

Nicklas, R. B., and Gordon, G. W., 1985, The total length of spindle microtubules depends on the number of chromosomes present, J. Cell Biol., 100:1-7.

Nicklas, R. B., and Kubai, D. F., 1985, Microtubules, chromosome movement, and reorientation after chromosomes are detached from the spindle by micromanipulation, Chromosoma, 92:313-324.

Nicklas, R. B., Kubai, D. F., and Hays, T. S., 1982, Spindle microtubules and their mechanical associations after micromanipulation in anaphase, J. Cell Biol., 95:91-104.

Palmer, D. K., O'Day, K., and Margolis, R. L., 1986, The 17 kilodalton centromere specific human autoantigen is a histone-like protein, J. Cell Biol., 103:412a (abstr.)

Peterson, J. B., and Ris, H., 1976, Electron-microscopic study of the spindle and chromosome movement in the yeast Saccharomyces cervisiae, J. Cell Sci., 22:219-242.

Pickett-Heaps, J., 1986, Mitotic mechanisms: an alternative view, Trends in Biochem. Sci., 11:504-507.

Rattner, J. B., 1986, Organization within the mammalian kinetochore, Chromosoma, 93:515-520.

Rieder, C. L., 1982, The formation, structure, and composition of the mammalian kinetochore and kinetochore fiber, Int. Rev. Cytol. 79:1-58.

Ris, H., and Witt, P. L., 1981, Structure of the mammalian kinetochore, Chromosoma, 82:153-170.

Salmon, E. D., 1975, Spindle microtubules: thermodynamics of in vivo assembly and role in chromosome movement, Ann. N.Y. Acad. Sci., 253:383-406.

Salmon, E. D., Leslie, R. J., Saxton, W. M., Karow, M. L., and McIntosh, J. R., 1984, Spindle microtubule dynamics in sea urchin embryos: analysis using a fluorescein-labeled tubulin and measurements of fluorescence redistribution after laser photobleaching, J. Cell Biol., 99:2165-2174.

Schaap, C. J., and Forer, A., 1984, Video digitizer analysis of birefringence along the lengths of single chromosomal spindle fibres, J. Cell Sci., 65:21-40.

Schrader, F., 1953, "Mitosis," 2nd ed., Columbia University Press, New York.

Sluder, G., and Rieder, C. L., 1985, Centriole number and the reproductive capacity of spindle poles, J. Cell Biol., 100:887-896.

Steffen, W., Fuge, H., Dietz, R., Bastmeyer, M., and Müller, G., 1986, Aster-free spindle poles in insect spermatocytes: evidence for chromosome-induced spindle formation?, J. Cell Biol., 102:1679-1687.

Valdivia, M. M., Maul, G. G., Jimenez, S. A., Kidd, V., and Brinkley, B. R., 1986., A putative kinetochore cDNA clone isolated by using human autoantibodies from scleroderma crest patients, J. Cell Biol., 103:491a (abstr.).

Witt, P. L., Ris, H., and Borisy, G. G., 1980, Origin of kinetochore microtubules in Chinese hamster ovary cells, Chromosoma, 81:483-505.

CHROMOSOMAL REPLICONS OF HIGHER PLANTS

Jack Van't Hof

Biology Department
Brookhaven National Laboratory
Upton, New York 11973

INTRODUCTION

Twenty-four years have passed since it was noted that the cell cycle duration of diploid root meristematic cells of unrelated plant species is positively correlated with the amount of nuclear DNA (Van't Hof and Sparrow, 1963) and it has been 22 years since it was shown that the duration of S phase is likewise dependent on the genome size of higher plants (Van't Hof, 1965). These correlations were confirmed independently by others (Bennett, 1972; Evans et al., 1971, 1972) and extended to the duration of meiosis by Bennett and colleagues (Bennett, 1971; Bennett and Smith, 1972; Bennett et al., 1972). The earlier work, summarized by Van't Hof (1974), yielded statistical correlations that become more precise as more recently gathered measurements from other species are added (Grif and Ivanov, 1975; Price and Bachmann, 1976).

Why S phase and the cell cycle lengthen in diploid cells with more nuclear DNA remained obscure for many years. Consideration of plausible mechanisms responsible for the correlation had to wait until methods capable of viewing chromosomal DNA replication at the molecular level were developed. The successful application of these methods to plants revealed that their chromosomes are replicated by a structurally and temporally ordered process. The discussion that follows concerns mostly the temporal aspects of the process. It focuses first on the replicon, the replication unit of chromosomal DNA, its organization and its temporal activity during the S phase of the cell cycle. Genetic effects on replicon size, replication fork rate and the S phase duration are mentioned and examples are provided showing how the pattern of chromosomal DNA replication changes before and during cell differentiation. The replication of plant genes and their relation to replicons is discussed and this is followed, lastly, by a few general comments.

REPLICON SIZE AND REPLICATION FORK RATE IN MOST HIGHER PLANTS ARE SIMILAR AND INDEPENDENT OF GENOME SIZE AND S PHASE DURATION.

Most dicotyledonous plants have similar sized replicons and replication forks rates. Measurements of chromosomal DNA of nine plant species representing an 82-fold range in genome size show that replicon size, i.e., the origin to origin distance, and the average fork rate are independent of genome size (Van't Hof and Bjerknes, 1981). The pooled data from these species give an average replicon size of 66 $\pm$ 10.2 kb and an average replication fork rate of 24 $\pm$ 4.2 kb per hour. Work by others indicates that these values hold in general for dicotyledonous plants (Cress et al., 1978; Francis et al., 1985; Ormrod and Francis, 1986).

Replicon sizes among monocots average 44.2 $\pm$ 15.6 kb, 22 kb less than those of dicots (Francis et al., 1985). While smaller replicons may be more frequent in monocots, there are exceptions, as both *Allium cepa*, *Secale cereale* and others have large replicons that are similar in size to those of dicots (Francis and Bennett, 1982; Van't Hof and Bjerknes, 1981). There is evidence that monocots also have slower replication forks than dicots (Francis et al., 1985). However, like dicots, neither the size of replicons nor the replication fork rates of monocots correlates positively with their genome size and S phase duration (Evans et al., 1971; Van't Hof and Bjerknes, 1981; Francis et al., 1985).

THE DURATION OF S PHASE IS DETERMINED BY TEMPORALLY ORDERED REPLICONS AND THE TEMPO-PAUSE: THE RATIONALE.

If neither replication fork rate nor replicon size are positively correlated with genome size and S phase duration, what does determine the time required by a cell to replicate its chromosomal complement? To answer this question one must consider how long it takes an individual replicon to replicate its allotment of DNA. This measurement is obtained by dividing replicon size by twice the single fork rate (Blumenthau et al., 1974). For example, a replicon of a dicotyledonous plant that has an average size of 66 kb, an average fork rate of 24 kb per hour, and bidirectional fork movement, requires about 1.4 hours to replicate its DNA. Given this information one can consider how the S phase duration is determined by temporally ordered replicons by examining two hypothetical cells each with the same genome size and each with two sets of replicons. The first cell is one in which both sets begin and end replication simultaneously. The S phase of this cell is 1.4 hours, the same length of time required for its replicons to replicate their DNA. The second cell is one in which the two sets replicate their DNA sequentially, i.e., the second set begins replication immediately following the first. The S phase of this cell is the sum of the time needed for each replicon set to replicate its DNA, or 2.8 hours. These two examples suggest that the minimum number of sequentially active replicon sets can be estimated if the duration of S phase, the replicon size and the fork rate are known. The estimation is obtained by dividing the duration of S phase by the length of time needed by a single replicon to replicate its DNA (Van't Hof and Bjerknes, 1981). The estimation assumes that no time passes between the ending of replication of one replicon set and the beginning of another. The only pause that occurs in the hypothetical cells is that corresponding to the initiation of replication of the second replicon set. The pause is zero minutes in the first instance and in the second, it is equal to the time interval separating the beginning of replication of the first set and the beginning of the second replicon set. This pause is termed the tempo-pause and it is defined as the length of time between the initiation of replication of one replicon family and that of its successor.

PLANT REPLICONS HAVE A HIERARCHICAL ORGANIZATION: MOLECULAR AND CYTOLOGICAL EVIDENCE.

The minimum number of serially activated replicons during S phase must be viewed in the context of the organization of chromosomal DNA replication in higher plants. In higher plants this organization has characteristics of a three-unit hierarchy. The elementary replication unit is the single replicon which is a member of a group of replicons arranged end-to-end along the DNA duplex. Such a group of tandem replicons, as defined by Blumenthau et al. (1974), reviewed by Edenberg and Huberman (1975) and discussed by Hand (1978), is called a cluster, because its members replicate DNA nearly simultaneously. The third unit, called a family or bank, consists of many clusters. A family is operationally defined as a group of clusters that replicates DNA at a given time during S phase. The temporal order of DNA replication is, therefore, a reflection of the temporal order of replicon families. Since, by definition, individual replicons of a family are nearly simultaneously active, measurement of one representative replicon constitutes a measurement of the family of which it is a member.

The idea that plant chromosomes are replicated by several temporally ordered replicon families had its beginning in the early cytological work of Taylor (1958), Lima-De-Faria (1959), Wimber (1961) and Tanaka (1968). Using four different species these workers agreed that the diploid chromosomal complement is replicated simultaneously at multiple sites on several chromosomes. At the resolution of light microscopy, the level at which these observations were made, the multiple sites reflect activity of replicon clusters. When pulse labeled with [^{3}H]-thymidine, these sites identify clusters of a given replicon family. For example, a 30 minute pulse of [^{3}H]-thymidine labels only diffuse non-heteropycnotic chromatin in *Haplopappus* (2n = 4) indicating that in this plant highly condensed chromatin is not replicating DNA (Tanaka, 1968). (For exceptions to this finding in other species see Nagl, 1977). Viewing the chromosomes of cells previously labeled in late S phase Tanaka (1968) observed that each homologue of chromosome 1 had three heavily labeled patches localized in its three large heterochromatic regions. Besides these heterochromatic regions, two other euchromatic regions on the short arm of chromosome 1 were lightly labeled. Similar observations on chromosome 2 show conclusively that in *Haplopappus* heterochromatin replicates simultaneously in late S phase at different sites on different chromosomes and that two sites of euchromatin also replicate late. In the context of our discussion here, the separate but simultaneously replicated heterochromatic and euchromatic sites in *Haplopappus* chromosomes are member clusters of the same replicon family. The presence of radioactive sites on different chromosomes demonstrates that the clusters of a replicon family are scattered about the genome and that they are not confined to homologues. This observation further implies that the factors responsible for the temporal order of replication of replicon families are independent of chromosomal location and that a component of this regulatory process resides within the clusters themselves, i.e., at the nucleotide sequence level. It is plausible that these sequences interact with factors that are family-specific proteins such as those postulated to be operative in *Physarum* (Muldoon et al., 1971, Wille and Kauffman, 1975).

THE DURATION OF S PHASE IN PLANTS IS DETERMINED BY THREE VARIABLES

Since replicon size is relatively constant in most plants within a group it contributes the least to the duration of S phase. Elimination of replicon size reduces the number of factors that determine the length of S phase to three. Those remaining are (i) the number of serially active replicon families, (ii) the tempo-pause, and (iii) the replication fork rate. The first factor listed, the number of serially active replicon families increases with genome size. Thus, *Arabidopsis thaliana* is estimated to have 2 replicon families while *Vicia faba* has 9 (Van't Hof and Bjerknes, 1981). Further, since neither *A. thaliana* nor *V. faba* has an S phase corresponding in duration to that expected if all their replicons begin and end replication simultaneously, their replicon families replicate DNA sequentially. While this statement may be true conceptually, it requires experimental data for validation. In fact, two sets of data are needed, one from an experiment designed to measure the tempo-pause holding both the fork rate and genome size constant and another in which genome size is held constant and the fork rate and tempo-pause are varied. The first experiment, the measurement of tempo-pause, was done in *A. thaliana* DNA (Van't Hof et al., 1978a) and the second was done with *Helianthus annuus* (Van't Hof et al., 1978). The results obtained with *A. thaliana* show that this species has two replicon families, one of about 687 and another of 1888 replicons per genome and that they initiate replication 36 minutes apart. The sum of the 36 minute tempo-pause and the time needed for each of the two families to replicate their DNA accounts for 95% of the S phase in this species.

The results with *H. annuus* on the other hand, show that its S phase duration is a function of the replication fork rate at moderate temperatures (20 to 35° C) but at extreme temperatures (10 or 38° C) an expanded tempo-pause is responsible for a longer S phase. These findings indicate that under certain circumstances the tempo-pause is uncoupled from the replication fork rate and expands independently resulting in a longer S phase. They also demonstrate that factors responsible for the tempo-pause differ from those concerned with fork rate on the grounds of temperature sensitivity.

In both experiments the size of replicons remained constant. Consequently, this property did not contribute to the results. Instead, these experiments provide evidence that the duration of the S phase of higher plants is determined by replication fork rate and serially active replicon families whose initiation of replication are separated by a tempo-pause.

REPLICON SIZE, FORK RATE AND S PHASE DURATION ARE GENETICALLY CONTROLLED.

Aspects of premeiotic DNA replication are modified by sex related factors.

There is no information about the premeiotic replicon properties in plants but there are data from mouse (Jagiello et al., 1983; Sung et al., 1986). These workers looked at the premeiotic DNA replication of oocytes and spermatocytes in embryos of the same strain of mice and detected sex linked differences in replicon properties. Besides sex linked characteristics the premeiotic S phase is of interest because it is longer than that of somatic cells and because of its possible contribution to genetic consequences at meiosis. In mice, the premeiotic S phase of both oocytes and spermatocytes is 14 hours (Crone et al., 1965; Monesi, 1962). Similar S phases, however, does not mean that the two cell types have similar sized replicons or replication fork rates. For example, oocytes have replicons and a replication fork rate that are three times larger and three times faster than those of spermatocytes and both cells replicate an average replicon in about 30 minutes (Jagiello et al., 1983; Sung et al., 1986). The combination of larger replicons and faster forks or smaller replicons and slower forks plus a constant tempo-pause results in equivalent S phase durations.

Work with mouse somatic cells supports this conclusion. Mouse somatic cells have a 7 hour S phase (Quastler and Sherman, 1959), an average replicon size of 60 kb (Cohen et al., 1979; see their figure 6) and a replication fork rate of about 126 kb per hour (Hand and Tamm, 1972). These numbers indicate that the replicons of somatic cells replicate their DNA in a little more than 14 minutes, about one-half the time needed for oocytes and spermatocytes. The importance of this difference is apparent when the replication characteristics of somatic cells are compared with those of spermatocytes. The two cell types have similar sized replicons but their fork rates differ by a factor of 2.6 and there is a two-fold difference in the duration of S phase. This suggests that the shorter S phase of somatic cells is the result of a higher fork rate. This view is not at odds with the results from oocytes if replicons are organized in clusters and families. Both oocytes and spermatocytes will replicate a given cluster in the same length of time but somatic cells will do it twice as fast. Consequently, the duration of S phase in somatic cells is half that of the germ line cells.

The findings on mouse chromosomal DNA replication are significant to the theme of this paper for three reasons. First, they suggest the existence of sex related factors that affect the replicon size and replication fork rate but do not change the temporal pattern of premeiotic chromosomal DNA replication. Second, they indicate that the length of S phase is attributable to factors other than replicon size, just as in the case of higher plants, and third, they confirm the conclusions derived from work with plants that fork rate influences the S phase duration when the tempo-pause is constant.

Replicon size and replication fork rate in higher plants are controlled genetically.

One conclusion derived from the experiments with mouse oocytes and spermatocytes is that certain genetic factors can change replicon properties and this conclusion, as shown by Francis et al. (1985), also applies to higher plants. These authors examined the replicon size, fork rate and S phase duration of triticale (*Triticosecale* Wittmark), an interspecific hybrid with 28 wheat (*Triticum aestivum*) chromosomes and 14 rye (*S. cereale*) chromosomes. The parental plants are sufficiently dissimilar in all three aspects of chromosomal DNA replication so that recognition of the predominant wheat phenotype in the hybrid was possible. Francis et al. (1985) found that triticale resembles wheat in replicon size and S phase duration and neither parent in replication fork rate. This latter aspect is consequential, since it shows that triticale replicons need 2.5 hours to replicate their DNA, an hour longer than either parent. The dominant wheat phenotype in triticale is good evidence that the replication properties of higher plants are genetically controlled. A change in these properties, however, may be deleterious particularly at certain developmental stages. In triticale, for example, deleterious effects are seen as shrivelled grains and aberrant endosperm nuclei (Bennett, 1977; 1980). Bennett (1977; 1980) and Francis et al. (1985) postulate that these effects result from the compliance of the rye chromosomes (S phase of 6.6 hours) to a S phase of 4.7 hours in triticale leaving insufficient time for the replication of late replicating heterochromatin in the rye chromosomes.

THE INTRODUCTION OF ADDITIONAL HETEROCHROMATIN IN PLANT GENOMES CAUSES GENETIC AND CYTOLOGICAL EFFECTS.

If the mixing of two genomes in a interspecific hybrid alters certain replication characteristics, then what is the effect of additional less diverse DNA? The answer to this question comes from work with maize and rye each of which have variable amounts of heterochromatin in the form of knobs or B chromosomes. In both species, additional heterochromatin, either as knobs or as supernumerary B chromosomes, produces genetic effects and alters the temporal pattern of DNA replication (Abraham and Smith, 1966; Jones and Rees, 1967, 1969; Rhoades and Dempsey, 1972, 1973; Pryor et al., 1980).

The replication of the B chromosome shows that even heterochromatin is replicated temporally. This demonstration is possible because the B chromosomes are not uniformly heterochromatic. Besides large segments of heterochromatin, they also have segments of euchromatin and heterochromatic knobs that can be followed autoradiographically. Using a pulse-label protocol, Pryor et al. (1980) showed that the order of B chromosomal DNA is euchromatin first, during early S phase, large heterochromatin segments next, in late S phase, and finally knobbed heterochromatin.

Further, the number of B chromosomes in rye and maize can be varied by selection and with each added B chromosome the duration of S phase lengthens (Ayonoadu and Rees, 1968; Pryor et al., 1980). In maize, most heterochromatin is late replicating but different classes of heterochromatin have their own time of replication. The proportion of asynchronous late replication in the S phase is directly dependent on the knob and B-heterochromatin content of the nucleus. Also, in maize, the genetic effects of the additional late replicating DNA is expressed as an enhancement of recombinational frequencies and an induced loss of chromosomal segments from knobbed A chromosomes during the second microspore division (Rhoades and Dempsey, 1972; 1973). In rye, additional B chromosomes increase chiasma frequency and possibly their distribution (Ayonoadu and Rees, 1968a). The rye phenotype differs depending on whether the B chromosomes are in odd or even numbers (Jones and Rees, 1969), an effect also recorded in maize (Rhoades and Dempsey, 1972). These cytological and genetical findings from maize, rye and other species (Rees, 1974), and those on replicon properties of triticale provide strong evidence that the addition of certain DNAs into the nucleoplasm of cells can change the pattern of chromosomal DNA replication. Such a change in replication complies with the idea that the added DNA introduces more replicon families and that these additional families may disrupt the normal temporal order of chromosomal DNA replication.

A CHANGE IN TEMPO-PAUSE PRECEDES DIFFERENTIATION OF SPECIFIC CELLS IN PEA ROOTS.

The fact that the mixing of genomes in triticale and the addition of B chromosomes in maize and rye produced genetic effects at specific stages of cell development indicates that the timing and order of chromosomal DNA replication is crucial at certain steps during a cell's lifetime. An example of a normal change in tempo-pause preceding differentiation is seen in pea-root meristem cells (Van't Hof et al., 1986). The diploid precursors of vascular parenchyma differ from other cells in the meristem because they stop temporarily in late S phase after replicating about 80% of their DNA. While replicating the remaining 20%, these cells produce replicon sized molecules of extrachromosomal DNA (Van't Hof and Bjerknes, 1982; Krimer and Van't Hof, 1983; Van't Hof et al., 1983). The extrachromosomal DNA, which contains late replicating rDNA and other repeated sequences (Kraszewska et al., 1985), is currently viewed by the authors as a by-product of genomic rearrangements that precede the differentiation of vascular parenchyma cells. If this view is correct, then a change in tempo-pause is one of the first steps taken by meristematic cells as they go from a proliferative to a differentiated state.

PLANTS HAVE PRIMARY AND SECONDARY REPLICATION INITIATION SITES.

The classical work of Blumenthau et al. (1974) demonstrated that *Drosophila* DNA has primary and secondary replication initiation sites. Which and how many of the sites are used by the cells depends on their developmental stage. Though analogous experiments in higher plants are lacking, there is evidence that they too have primary and secondary preferred sites for replication initiation. Francis et al. (1985a) showed that pea DNA crosslinked by psoralen initiates replication at additional sites producing replicons that are smaller than those of untreated cells. This result raises the interesting possibility that higher plants may, like *Drosophila*, have smaller replicons during the earlier stages of embryogenesis.

DURING DEVELOPMENT THE CONTROL OF SPECIFIC REPLICONS IS RELAXED.

It is apparent from the foregoing discussion that the plasticity of the replication process in eukaryotes is used by cells at specific steps in development. Another example of this plasticity is naturally occurring and induced amplification of genes in certain replicons. When either the tempo-pause or the temporal order of replication is relaxed, amplification can occur. The replicon in which the gene resides is free to initiate and complete more than one round of nascent DNA. Studies by Spradling and colleagues (Spradling and Mahowald, 1980; Spradling et al., 1980; Spradling, 1981) demonstrate that the stage specific amplification of the chorion genes involves replicon sized molecules, not small molecules the size of genes. Amplification occurs in non-dividing follicle cells undergoing polyploidization by endoreduplication. In these cells the normal temporal pattern of replication of replicon families may be inoperative. Nevertheless, the concomitant amplification of a contiguous chromosomal region of about 90 kb in the X chromosome and an even larger segment on chromosome 3 demonstrates that replicons containing the same gene sequences but located on different chromosomes respond similarly to the same signals. This finding agrees with the idea that chorion genes, though positioned on different chromosomes, are located in replicons that are members of the same family. Amplification of the chorion replicon is achieved by repeated initiation and chain elongation producing multi-fork configurations (Nelson-Olsheim and Miller, 1983). Recently, de Cicco and Spradling (1984), using P-element transformation, traced the sequence responsible for regulation of amplification to a 3.8 kb genomic segment that contains the origin for disproportionate gene replication.

GENE REPLICATION IN PLANTS IS TEMPORALLY ORDERED.

If genes, the units of heredity, reside within replicons, the units of chromosomal DNA replication, the two units are inseparable. This linkage predicts that the replication of plant genes, like plant replicons, is temporally ordered. The replication patterns of four genes in synchronized pea-root meristems support this prediction (Van't Hof et al., 1987 and unpublished results), since the patterns of the four genes are different. The rRNA genes replicate throughout S phase but more, 65%, replicate in late S phase. The legumin genes replicate in a wave-like pattern peaking in early S phase at the third hour and again in late S phase at the eighth hour. The small subunit of ribulose-1,5-bisphosphate carboxylase genes replicates in the early half of S phase except during the first hour, while the chlorophyll a/b binding protein genes also replicate in early S phase, peaking at the second and third hours. After the fifth hour more a/b binding protein genes replicate again, peaking at the end of S phase. The temporal bimodality of replication of the legumin and chlorophyll a/b binding proteins genes suggests that homologous sequences, possibly pseudogenes, replicate at the end of S phase along with other DNA presumed to be genetically inactive. More importantly, in terms of the theme of this discussion, the temporal bimodality demonstrates that early and late replicating sequences are located in different replicon families.

The rRNA genes, likewise, reside in different replicon families. In fact, there may be rDNA in every replicon family because some rRNA genes are replicated during each hour of the S phase. This feature of the rRNA genes is not unexpected considering there are 3,900 copies (Ingle and Sinclair, 1972; Cullis and Davies, 1975) representing 27,000 kb of chromosomal DNA. This rDNA exists as tandem repeats of two size classes, one of 9 and another of 8.6 kb (Ellis et al., 1984). Given a replicon size of 54 to 72 kb (Van't Hof and Bjerknes, 1977; Schvartzman et al., 1984), the rDNA of pea resides in 481 to 659 replicons and about 26 clusters (Van't Hof, 1980). With even 17 clusters replicating in late S phase, the probability is high that a few of the remaining clusters would be incorporated in each replicon family. Their abundance and their membership in each replicon family make the rRNA genes useful subjects for the analysis of plant replicons, replicon clusters and the regulatory factors governing their temporal pattern of replication.

SUMMARY AND GENERAL COMMENTS

This brief discussion of replicons of higher plants offers a glimpse into the properties of chromosomal DNA replication. It gives evidence that the S phase of unrelated plant species is comprised of temporally ordered replicon families that increase in number with genome size. This orderly process, which assures a normal inheritance of genetic material to recipient daughter cells, is maintained at the level of replicon clusters by two mutually exclusive mechanisms, one involving the rate at which single replicons replicate their allotment of DNA, and another by means of the tempo-pause. The same two mechanisms are used by cells to alter the pattern of chromosomal DNA replication just prior to and during normal development. Both mechanisms are genetically determined and produce genetic effects when disturbed or disrupted by additional non-conforming DNAs. Further insight into how these two mechanisms operate requires more molecular information about the nature of replicons and the factors that govern when a replicon family replicates. Plant material is a rich and ideal source for this information just awaiting exploitation.

ACKNOWLEDGMENT

Research supported by the Office of Health and Environment Research, U. S. Department of Energy.

REFERENCE LIST

Abraham, S., and Smith, H. H., 1966, DNA synthesis in the B chromosomes of maize, J. Heredity, 57: 78-80.

Ayonoadu, U. W., and Rees, H., 1968, The regulation of mitosis by B chromosomes in rye, Exptl. Cell Res., 52: 284-290.

Ayonoadu, U. W., and Rees, H., 1968a, The influence of B chromosomes on chiasma frequencies in black Mexican sweet corn, Genetica, 39: 75-81.

Bennett, M. D., 1971, The duration of meiosis, Proc. R. Soc. Lond. Ser. B. Biol. Sci., 178: 277-299.

Bennett, M. D., 1972, Nuclear DNA content and minimum generation time in herbaceous plants, Proc. R. Soc. Lond. Ser. B. Biol. Sci., 181: 109-135.

Bennett, M. D., 1977, Heterochromatin, aberrant endosperm nuclei and grain shrivelling in wheat-rye genotypes, Heredity, 39: 411-419.

Bennett, M. D., 1980, Theoretical and applied DNA studies and triticale breeding, Hodowla Roslin Aklimatyzacja I Nasiennictwo, 24: 289-298.

Bennett, M. D., and Smith, J. B., 1972, The effects of polyploidy on meiotic duration and pollen development in cereal anthers, Proc. R. Soc. Lond. Ser. B. Biol. Sci., 181: 81-107.

Bennett, M. D., Chapman, V., and Riley, R., 1971, The duration of meiosis in pollen mother cells of wheat, rye, and triticale, Proc. R. Soc. Lond. Ser. B. Biol. Sci., 178: 259-275.

Blumenthau, A. B., Kriegstein, H. J., and Hogness, D. S., 1974, The units of DNA replication in Drosophila melanogaster chromosomes. Cold Spring Harbor Symp. Quant. Biol., 38: 205 -223.

Cohen, J. E., Jasney, B. R., and Tamm, I., 1979, Spatial distribution of initiation sites for mammalian DNA replication: a statistical analysis, J. Mol. Biol., 128: 219-245.

Cress, D. E., Jackson, P. J., Kadouri, A., Chu, Y. E., and Lark, K. G., 1978, DNA replication in soybean protoplasts and suspension-cultured cells: comparison of exponential and fluorodeoxyuridine synchronized cultures, Planta, 143: 241-253.

Crone, M., Levy, E., and Peters, H., 1965, The duration of the premeiotic DNA synthesis in mouse oocytes, Exptl. Cell Res., 39: 678-688.

Cullis, C. A., and Davies, R. D., 1975, Ribosomal DNA amounts in Pisum sativum, Genetics, 81: 485-492.

de Cicco, D. V., and Spradling, A. C., 1984, Localization of a cis-acting element responsible for the developmentally regulated amplification of Drosophila chorion genes, Cell, 38: 45-54.

Edenberg, J. H., and Huberman, J. A., 1975, Eukaryotic chromosomal replication, Ann. Rev. Genet., 9: 245-284.

Ellis, T. H. N., Davies, D. R., Castleton, J. A., and Bedford, I. D., 1984, The organization and genetics of rDNA length variants in peas, Chromosoma, 91: 74-81.

Evans, G. M., and Rees, H., 1971, Mitotic cycles in dicotyledons and monoctyledons, Nature, 233: 350-351.

Evans, G. M., Rees, H., Snell, C. L., and Sun, S., 1972, The relationship between nuclear DNA amount and the duration of the mitotic cycle, in: "Chromosomes Today," C. D. Darlington and K. R. Lewis, eds., Hafner Publishing Co.,New York pp. 24-31.

Francis, D., and Bennett, M. D., 1982, Replicon size and mean rate of DNA synthesis in rye (Secale cereale L. cv. Petkus Spring), Chromosoma, 86: 115-122.

Francis, D., Kidd, A. D., and Bennett, M. D., 1985, DNA replication in relation to DNA c values, in: "The Cell Division Cycle in Plants," J. A. Bryant and D. Francis, eds., Cambridge University Press, Cambridge, pp 61-82.

Francis, D., Davies, N. D., Bryant, J. A., Hughes, S. G., Sibson, D. R., and Fitchett, P. N., 1985a, Effects of psoralen on replicon size and mean rate of DNA synthesis in partially synchronized cell of Pisum sativum L., Exptl. Cell Res., 158: 500-508.

Grif, V. G., and Ivanov, V. B., 1975, Temporary pattern of the mitotic cycle in flowering plants, Tsitilogia, 17(6): 694-717 (In Russian).

Hand, R., 1978, Eukaryotic DNA: organization of the genome for replication, Cell, 15: 317-325.

Hand, R., and Tamm, I., 1972, Rate of DNA chain growth in mammalian cells infected with cytocidal RNA viruses, Virology, 47: 331-337.

Ingle, J., and Sinclair, J., 1972, Ribosomal RNA genes and plant development, Nature, 235: 30-32.

Jagiello, G., Sung, W. K., and Van't Hof, J., 1983, Fiber DNA studies of premeiotic mouse spermatogenesis, Exptl. Cell Res., 146: 281-287.

Jones, N. R., and Rees, H., 1967, Genotypic control of chromosomal behaviour in rye. XI. The influence of B chromosomes on meiosis, Heredity, 22: 333-347.

Jones, N. R., and Rees, H., 1969, An anomalous variation due to B chromosomes in rye, Heredity, 24: 265-271.

Kraszewska, E. K., Bjerknes, C. A., Lamm, S. S., and Van't Hof, J., 1985, Extrachromosomal DNA of pea-root (Pisum sativum) has repeated sequences and ribosomal genes, Plant Mol. Biol., 5: 353-361.

Krimer, D. B., and Van't Hof, J., 1983, Extrachromosomal DNA of pea (Pisum sativum) root-tip cells replicates by strand displacement, Proc. Natl. Acad. Sci. U.S.A., 80: 1933-1937.

Lima-De-Faria, A. J., 1959, Differential uptake of tritiated thymidine into hetero- and euchromatin in Melanoplus and Secale, Biophys. Biochem. Cytol. (J. Cell Biol.), 6: 457-475.

Monesi, V., 1962, Autoradiographic study of DNA synthesis and the cell cycle in spermatogonia and spermatocytes of mouse testis using tritiated thymidine, J. Cell Biol., 14: 1-18.

Muldoon, J., Evens, T. E., Nygaard, O. F., and Evans, H., 1971, Control of DNA replication by protein synthesis at defined times during the S period in Physarum polycephalum, Biochim. Biophys. Acta, 247: 310-321.

Nagl, W., 1977, Nuclear structures during the cell cycle, in: "Mechanisms and Control of Cell Division," T. L. Rost and E. M. Gifford, Jr., eds., Dowden,Hutchinson and Ross, Inc., Stroudsburg, pp. 147-193.

Nelson-Osheim, Y., and Miller, O. L., 1983, Novel amplification and transcription activity of chorion genes in Drosophila melanogaster follicle cells, Cell, 33: 543-553.

Ormrod, J. C., and Francis, D., 1986, Mean rates of DNA replication and replicon size in the shoot apex of Silene coelirosa L. during the initial 120 minutes of the first day of floral induction, Protoplasma, 130: 206-210.

Price, J. H., and Bachmann, K., 1976, Mitotic cycle time and DNA content in annual and perennial Microseridinae (Compositae, Cichoriaceae), Plant Syst. Evol., 126: 323-330.

Pryor, A., Faulkner, K., Rhoades, M. M., and Peacock, W. J., 1980, Asynchronous replication of heterochromatin in maize, Proc. Acad. Sci. U.S.A., 77: 6705-6709.

Quastler, H., and Sherman, F. G., 1959, Cell population kinetics in the intestinal epithelium of the mouse, Exptl. Cell Res., 17: 420-438.

Rees, H., 1974, B chromosomes, Sci. Prog. Oxf., 61: 535-554.

Rhoades, M. M., and Dempsey, E., 1972, On the mechanism of chromatin loss induced by the B chromosome of maize, Genetics, 71: 73-96.

Rhoades, M. M., and Dempsey, E., 1973, Chromatin elimination induced by the B chromosome of maize, J. Heredity, 64: 12-18.

Schvartzman, J. B., Krimer, D. B., and Van't Hof, J., 1984, The effects of different thymidine concentrations on DNA

replication in pea-root cells synchronized by a protracted 5-fluorodeoxyuridine treatment, Exptl. Cell Res., 150: 379-389.
Spradling, A. C., 1981, The organization and amplification of two chromosomal domains containing Drosophila chorion genes, Cell, 27: 193-201.
Spradling, A. C., and Mahowald, A. P., 1980, Amplification of genes for chorion proteins during oogenesis in Drosophila melanogaster, Proc. Natl. Acad. Sci. U.S.A., 77: 1096-1100.
Spradling, A. C., Digan, M. E., Mahowald, A. P., Scott, M., and Craig, E. A., 1980, Two clusters of genes for major chorion proteins of Drosophila melanogaster, Cell, 19: 904-914.
Sung, W. K., Van't Hof, J., and Jagiello, G., 1986, DNA synthesis studies in pre-meiotic mouse oogenesis, Exptl. Cell Res., 163: 370-380.
Tanaka, R., 1968, Asynchronous DNA replication in the chromosomes of Haplopappus gracilis (2n = 4), Cytologia, 33: 520-525.
Taylor, H. J., 1958, The mode of chromosome duplication in Crepis capillaris, Exptl. Cell Res., 15: 350-357.
Van't Hof, J., 1965, Relationships between mitotic cycle duration, S period duration and the average rate of DNA synthesis in the root meristem cells of several plants, Exptl. Cell Res., 39: 48-58.
Van't Hof, J., 1974, The duration of chromosomal DNA synthesis, of the mitotic cycle, and of meiosis of higher plants, in: "Handbook of Genetics," vol. 2, R. C. King, ed., Plenum Press, New York, pp. 363-377.
Van't Hof, J., 1980, Pea (Pisum sativum) cells arrested in G_2 have nascent DNA with breaks between replicons and replication clusters, Exptl. Cell Res., 129: 231-237.
Van't Hof, J., and Bjerknes, C. A., 1977, 18 um replication units of chromosomal DNA fibers of differentiated cells of pea (Pisum sativum), Chromosoma, 64: 287-294.
Van't Hof, J., and Bjerknes, C. A., 1981, Similar replicon properties of higher plant cells with different S periods and genome sizes, Exptl. Cell Res., 136: 461-465.
Van't Hof, J., and Bjerknes, C. A., 1982, Cells of pea (Pisum sativum) that differentiate from G_2 phase have extrachromosomal DNA, Mol. Cell Biol., 2: 339-345.
Van't Hof, J., and Sparrow, A. H., 1963, A relationship between DNA content, nuclear volume, and minimum mitotic cycle time, Proc. Natl. Acad. Sci. U.S.A., 49: 897-902.
Van't Hof, J., Bjerknes, C. A., and Clinton, J. H., 1978, Replicon properties of chromosomal DNA fibers and the duration of DNA synthesis of sunflower root-tip meristem cells at different temperatures, Chromosoma, 66: 161-171.
Van't Hof, J., Kuniyuki, A., and Bjerknes, C. A., 1978a, The size and number of replicon families of chromosomal DNA of Arabidopsis thaliana, Chromosoma, 68: 269-285.
Van't Hof, J., Bjerknes, C. A., and Delihas, N. C., 1983, Excision

and replication of extrachromosomal DNA of pea (*Pisum sativum*), *Mol. Cell. Biol*., 3: 172-181.

Van´t Hof, J., Bjerknes, C. A., and Lamm, S. S., 1986, Meristematic precursors of vascular parenchyma differentiate from G_2 phase after replicating DNA discontinuously, *Amer. J. Bot*., 73: 87-95.

Van´t Hof, J., Hernandez, P., Bjerknes, C. A., Kraszewska, E. K., and Lamm, S. S., 1987, Replication of the rRNA and legumin genes in synchronized root cells of pea (*Pisum sativum*): evidence for transient EcoR I sites in replicating rRNA genes, *Plant Mol. Biol*., 8: 133-143

Wille, J. J., and Kauffman, S. A., 1975, Premature replication of late S period DNA regions in early nuclei transferred to late cytoplasm by fusion in *Physarum polycephalum*, *Biochim. Biophys. Acta*, 407: 158-173.

Wimber, D. E., 1961, Asynchronous replication of deoxyribonucleic acid in root tip chromosomes of *Tradescantia paludosa*, *Exptl. Cell Res*., 23: 402-407.

NEW PERSPECTIVES ON THE GENETICS AND MOLECULAR BIOLOGY OF CONSTITUTIVE HETEROCHROMATIN

Arthur J. Hilliker and Cecil B. Sharp

Department of Molecular Biology and Genetics
The University of Guelph
Guelph, Ontario
Canada

INTRODUCTION

For many years cytogeneticists have been intrigued by a fundamental characteristic of the chromosomes of higher eukaryotes, namely the presence of heterochromatic chromosome regions. Classical, or constitutive (Brown,1966), heterochromatin was first clearly defined by Heitz (1928), and its biological roles have been debated ever since. Constitutive heterochromatin is defined cytologically - whole chromosomes or chromosome segments which maintain a condensed appearance relative to euchromatin throughout most, if not all, of the cell cycle. Given the apparent paucity of Mendelian factors within heterochromatic regions, many geneticists came to hold the view that heterochromatin was genetically inert. This point of view was further supported by the elucidation of the cytogenetic basis of position effect variegation - the repression of expression of euchromatic genes by their translocation to a position adjacent to heterochromatin.

However, there is now a great deal of evidence that heterochromatin is not genetically inert. Phenotypic and meiotic effects associated with specific heterochromatic regions of particular organisms have been noted, for example in *Triticum*, *Secale* and their hybrids (see Gustafson and Lukeszewski, 1985) and in *Zea mays* (see Rhoades, 1978). Further, ordinary Mendelian genes have been found within the heterochromatin of *Drosophila melanogaster*. Coincident with the genetic analysis of *Drosophila* heterochromatin, there has been major progress in elucidating the DNA sequence organization of the constitutive heterochromatin of a variety of eukaryotic organisms (reviewed in Brutlag, 1980).

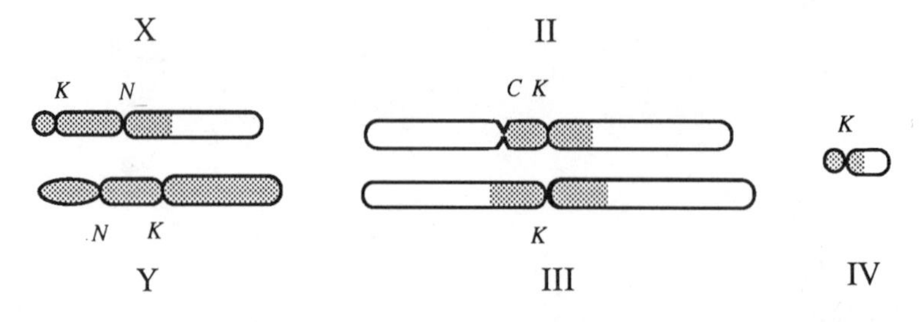

Fig. 1. Diagrammatic representation of the D. melanogaster chromosomal complement. Heterochromatin is stippled. K represents the location of the kinetochore, N is the location of the nucleolar constrictions and C is the location of the prominent secondary constriction at the 2L heterochromatin-euchromatin junction. (Modified from Cooper, 1950; Hilliker et al., 1980.)

In D. melanogaster, the entire Y chromosome, the proximal one-half of the X chromosome, and the proximal one-quarter of each arm in chromosomes 2 and 3 are heterochromatic (Fig. 1) (Heitz, 1933; Kaufmann, 1934). The X and Y chromosome heterochromatin have secondary constrictions within them (reviewed by Cooper, 1959). During the formation of polytene chromosomes the bulk of heterochromatin does not replicate (Heitz, 1934; Rudkin, 1969; Gall, 1973). At the heterochromatic-euchromatic junction there are diffusely banded segments termed β-heterochromatin which do replicate during polytenization (Gall, 1973). It has been proposed that the β-heterochromatin is euchromatin and that its diffuse polytene banding is the result of its juxtaposition to heterochromatin (Hilliker et al., 1980). Additional properties of D. melanogaster heterochromatin include: 1) late replication relative to euchromatin (Barigozzi et al., 1966), characteristic of the constitutive heterochromatin of many species (Lima de Faria and Jaworska, 1968); and 2) little or no meiotic recombination (Muller and Painter, 1932; Roberts, 1965; Schalet and Lefevre, 1976; see also Baker, 1958 for his study in D. virilis).

The highly repeated, or satellite, DNA sequences of D. virilis and D. melanogaster are found in the heterochromatic regions (Rae, 1970; Gall et al., 1971, Peacock et al., 1973). In D. melanogaster, each highly repeated DNA sequence is associated with long tandem repeats (Peacock et al., 1977). There are four major satellite sequences in D. melanogaster (1.672, 1.686, 1.688, and 1.705 gm/cc) and the DNA sequence of the major repeating unit of each satellite has been determined (Peacock et al., 1973, 1977; Endow et al., 1975; Sederoff et al., 1975; Brutlag et al., 1977; Carlson and Brutlag, 1977; Brutlag and Peacock, 1979; Hsieh and Brutlag, 1979). Recently, numerous minor sequence variants, present in tandem arrays, have been found for the three simple sequence satellites (1.672, 1.686,

1.705 gm/cc) (Lohe and Brutlag, 1986). Qualitatively similar variation is observed in the more complex, 254-359 base pair 1.688 gm/cc satellite sequence (Lohe and Brutlag, 1986). Thus, *D. melanogaster* heterochromatin is highly enriched in these four satellite DNA sequences, of which three have repeat lengths of 5-10 base pairs and one has a repeat length of several hundred nucleotides. However, within each of these major satellite groups there are sequence variations which tend to occur in tandem arrays and, therefore, in different locations. Each sequence variant of a particular satellite sequence could be considered as a unique satellite DNA sequence. With the exception of the 1.688 satellite sequence, which has a single site on the X chromosome (Hilliker and Appels, 1982), the major satellite sequences are found in multiple sites within the heterochromatic regions of the *D. melanogaster* genome (Peacock et al., 1977). (For further details on the molecular organization of *D. melanogaster* heterochromatin see Appels and Peacock, 1978 and Brutlag, 1980).

Fluorescent staining studies have demonstrated that *Drosophila* heterochromatin can be resolved into characteristic banded regions (Holmquist, 1975; Gatti et al., 1976; Pimpinelli et al., 1976; Gatti and Pimpinelli, 1983). Thus, at both the cytological level and the DNA sequence level, *Drosophila* heterochromatin is linearly differentiated (i.e. has a substructure).

Although the concept of the genetic inertness of constitutive heterochromatin has been a major tenet of cytogenetic theory, genetic studies in *D. melanogaster* have documented the existence of major gene loci in heterochromatin as well as minor phenotypic effects associated with particular heterochromatic regions. An earlier review (Hilliker et al., 1980) outlines in detail the genetic analysis of *D. melanogaster* heterochromatin up to 1980.

In this paper the genetic analysis of constitutive heterochromatin and related aspects of DNA sequence organization will be discussed. Although this will not be a comprehensive review, we shall discuss many of the major issues raised in these analyses. We shall focus our discussion on the model eukaryote *D. melanogaster*, with some emphasis on our own work.

GENETIC ELEMENTS IN DROSOPHILA AUTOSOMAL HETEROCHROMATIN

Analysis of the second chromosome heterochromatin of *D. melanogaster* has revealed a number of genetic elements within both major heterochromatic blocks. Approximately 20% of 2L and 25% of 2R consist of centromeric heterochromatin (Fig. 1). Analysis of deletions of chromosome 2 heterochromatin (Hilliker and Holm, 1975) and ethyl methane sulfonate (EMS) induced mutations falling within these heterochromatic regions (Hilliker, 1976) identified several vital and visible gene loci and allowed the derivation of a detailed genetic map. Fig. 2 illustrates the genetic map of the chromosome 2 heterochromatin. Thirteen vital gene loci are found within the second chromosome heterochromatin. The genetic loci mapped to the heterochromatin of chromosome 2 occur at a very low density relative to vital gene density in euchromatin - approximately 1% (Hilliker, 1976).

The vital loci identified appear to be unique sequence genes, from the analysis of the EMS-induced alleles. None of the 113 EMS-induced lethals behaved as a deficiency and several complementation groups exhibited complex interallelic complementation maps. It is unlikely that an allele exhibiting complementation with other alleles of the same locus is anything more drastic than a base pair substitution (see Fincham, 1966). Further, from detailed genetic analyses of Drosophila euchromatic regions, EMS appears to act largely as a point mutagen (e.g. Lim and Snyder, 1974; Hochman, 1971; Hilliker et al., 1980).

The vital loci contained within the second chromosome heterochromatin resemble "typical" euchromatic loci. Two of the genes, light and rolled, have visible alleles as well as lethal alleles. The lethal phase associated with all lethal complementation groups is post-embryonic. This is characteristic of the majority of euchromatic lethal mutations. For example, exhaustive analysis of X-chromosome recessive lethal mutations demonstrates that only 15-20% are embryonic lethal (Wieschaus et al., 1984; Eberl, 1986).

However, one noteworthy feature is that three of the second chromosome heterochromatic complementation groups exhibit complex interallelic complementation. One gene exhibited an extremely complex complementation map (l(2R)EMS34-14; Fig. 2). Within this group, complementation between alleles was almost completely unambiguous, i.e. heterozygous combinations of different alleles were either completely viable or inviable. No single allele failed to complement with all other alleles. This locus also exhibits a very high EMS mutability. However, there are euchromatic loci such as rudimentary, dumpy and maroon-like which also exhibit complex interallelic complementation.

More recent analyses have demonstrated that two of the genetic elements important in the segregation distortion (SD) phenomenon are within the second chromosome heterochromatin. Segregation distortion refers to a form of meiotic drive that is caused by certain second chromosomes obtained from natural populations (Sandler et al., 1959) (for reviews see Zimmering et al., 1970; Hartl and Hiraizumi, 1976; Crow, 1979; Sandler and Golic, 1985). Males heterozygous for an SD chromosome and an SD^+ chromosome produce a large excess of SD-bearing sperm, due to the dysfunction of sperm bearing the SD^+ chromosome (Hartl et al., 1967; Nicoletti, 1967). Ultrastructural

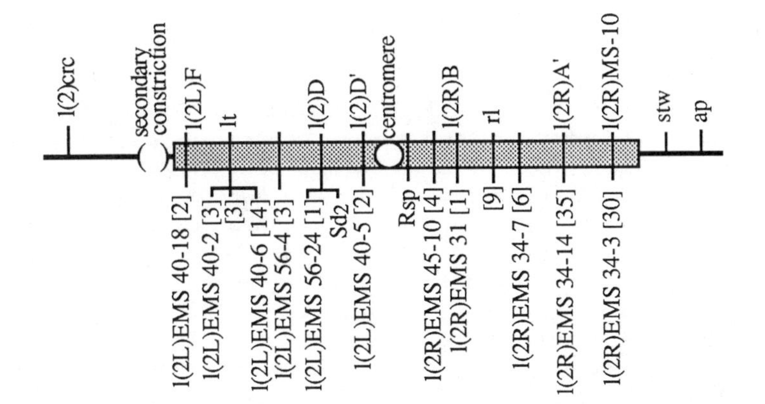

Fig. 2. Summary genetic map of the second chromosome heterochromatin. The vital loci defined by EMS-induced lethals are on the bottom and the correlated genetic sites defined by deletion mapping are on the top. The number of EMS-induced alleles associated with each complementation group is in square brackets. (Modified from Hilliker et al., 1980).

studies demonstrate that the SD^+ sperm in SD/SD^+ males undergo an aberrant spermiogenesis (Tokuyasu et al., 1976). A defect in the normal transition of lysine-rich to arginine-rich sperm histones has been noted in SD^+ -bearing sperm of SD/SD^+ males (Hauschteck-Jungen and Hartl, 1982), suggesting that this histone transition may be what SD directly disrupts in SD^+ -bearing sperm.

SD chromosomes have a number of genes that are involved in bringing about segregation distortion. When the normal complement of genes on an SD chromosome is altered, then the degree of segregation distortion is altered as well. Generally, the degree of segregation distortion is measured by k, the proportion of SD-bearing progeny recovered from an SD/SD^+ male. This measure can be used to detect changes in mean k-values which result from various genetic manipulations. However, because k is simply a proportion, it cannot be used to compare changes at different mean k-values, i.e. it is not a proper interval measure. The ordinal nature of k is also revealed by the observation that the variance of k is highly dependent upon the mean of k, even after the effects of bionomial sampling are taken into consideration (Sandler and Hiraizumi, 1960). A transformation of k (the k-probit transformation) has been proposed as a valid interval measure of SD (Miklos and Smith-White, 1971), but we feel that the transformation is inappropriate as some of the assumptions of the transformation are not generally applicable (Sharp and Hilliker, 1986, 1987). However, k can still be used to detect changes between different SD chromosomes. It simply can't be used to compare changes at different mean k-values as it is an ordinal measure. This confounds still further the analysis of this somewhat cryptic genetic syndrome.

The genetic dissection of SD chromosomes has proved quite difficult, due largely to the genetic complexity of the system. Hartl (1974) showed that SD chromosomes contain at least two major loci, Sd and Rsp (Responder). There are two main alleles of Rsp, Rsp^s (Responder-sensitive) and Rsp^i (Responder-insensitive). SD chromosomes are basically Sd Rsp^i/Sd^+ Rsp^s. In order to obtain segregation distortion, there must be an Sd allele and an allelic difference at Rsp. It has since been shown that a multiple allelic series defines Rsp, with alleles varying from insensitive to moderately sensitive to extremely sensitive (Martin and Hiraizumi, 1979; Hiraizumi et al., 1979; Lyttle et al., 1986).

The Rsp locus is almost certainly located very proximally in the 2R heterochromatin. Ganetzky (1977) showed that gamma ray-induced insensitive cn bw chromosomes (which are normally Rsp^s) had deficiencies in the 2R heterochromatin, but the exact location of Rsp could not be determined as the data were ambiguous. Recombination studies between lt and straw (stw, is located very proximally in 2R euchromatin) located Rsp between those markers (Hiraizumi et al., 1980; Hiraizumi, 1981). Sharp et al. (1985) showed that Rsp is located between lt and rl, by using gamma ray-induced recombination and 2R heterochromatic deletions, thus, definitely locating Rsp in the heterochromatin of chromosome 2. However, Df(2R)M-S10, a deficiency for virtually all of the 2R heterochromatin (Hilliker and Holm, 1975), is not deficient for Rsp (Ganetzky, 1977; Sharp et al., 1985). Thus, Rsp must be located in the 2R heterochromatin, proximal to all known vital loci, very near to the centromere.

There is another component of SD located in heterochromatin. This is an element which has been called both Sd_2 (Sharp, 1977) and E(SD) (Ganetzky, 1977). Ganetzky (1977) located an enhancer of SD, which he designated E(SD), in or near the 2L heterochromatin. Deficiencies in the 2L heterochromatin of SD-5 chromosomes located an element between the lethal complementation groups EMS 56-24 and EMS 56-4 (see Fig. 2) which, when deleted, reduced the k-value from about 0.98 to 0.6 (Sharp, 1977; Sharp et al., 1985). Brittnacher and Ganetzky (1984) showed that for SD-Roma, this site is located between lt and EMS 56-4. We presume that this difference is due to the different origins of SD-5 and SD-Roma or that there is a heterochromatic inversion. This heterochromatic site, however, is not the only Sd site. There is another site in the 2L euchromatin which has been identified by deletions at 37D2-6 (Ganetzky, 1977; Brittnacher and Ganetzky, 1983) and by recombination to a site just distal to purple (Sharp et al., 1985). The removal of this site (Sd_1) has a greater effect than the removal of Sd_2 - it reduces k from 0.98 to about 0.55, when the cn bw Rsp allele is tested and from about 0.999 to 0.7 when a super-sensitive responder allele is tested (Sharp et al.,

1985). However, we still see a significant amount of segregation distortion when Sd_1 has been removed by recombination, due to the Sd_2 site located in the 2L heterochromatin.

Hartl (1980) obtained evidence that SD-36 has an additional Sd site located to the right of pr, but not in heterochromatin. This could be called Sd_3. Recently, it has been suggested that some of the components of SD might be transposable elements (Sandler and Golic, 1985; Hickey et al., 1986). The presence of at least three Sd elements (and different locations of Sd_2 in SD-Roma and SD-5) would certainly appear to be in agreement with such an hypothesis. If this hypothesis is correct, then it is interesting to note that the Sd element can operate just about as well in constitutive heterochromatin as it can in euchromatin. Furthermore, if this hypothesis is true, then there will be no wild-type alleles of Sd_1 and Sd_2. We are currently screening the Sd_2 region for male sterile loci affecting spermiogenesis to see if we can identify a wild-type allele of Sd_2.

Recently, progress has been made toward the cloning of the second chromosome heterochromatic gene the lt locus (see Figure 2) by R. Devlin (personal communication) who has cloned DNA in the vicinity of lt, utilizing a P element induced mutant allele.

A recent genetic analysis of the centromeric heterochromatin of the third chromosome has revealed the presence of 11 vital genes (Marchant, 1986). Thus, the second and third chromosome heterochromatic regions are similar in that both contain about a dozen genes necessary for viability. Further study of the third chromosome heterochromatin may reveal more cryptic genetic elements, such as the components of SD located in the second chromosome heterochromatin (see above).

GENETIC AND MOLECULAR STRUCTURE OF THE X CHROMOSOME HETEROCHROMATIN

The DNA sequence organization of the X chromosome heterochromatin has been analyzed much more extensively than that of other heterochromatic regions of the *Drosophila* genome. Fig. 3 presents a summary map of the DNA sequence organization of the X chromosome heterochromatin (Xh) correlated with the genetic effects associated with the various regions of Xh. The localization of repeated DNA sequences within Xh was accomplished by mapping specific sequences relative to inversion breakpoints that sub-divided Xh. *In situ* hybridization of tritiated cRNA probes to mitotic chromosomes permitted this mapping (Steffensen et al., 1981; Appels and Hilliker, 1982; Lindsley et al., 1982; Hilliker and Appels, 1982).

A number of genetic elements, map to Xh, but only one (the bobbed locus) is a vital locus. All genetic elements identified in Xh are also found on the Y chromosome. This is intriguing as it is

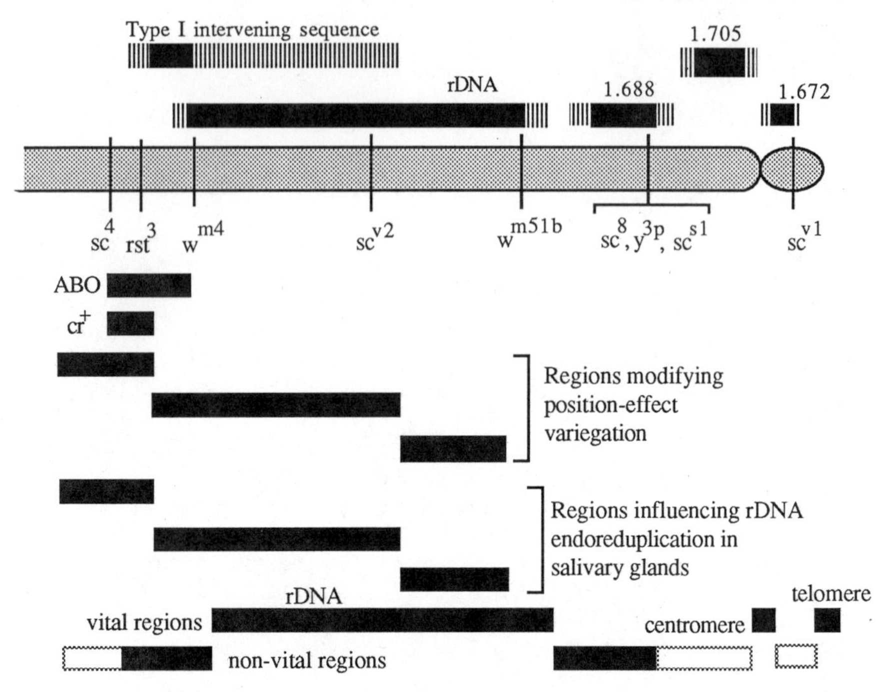

Fig. 3. Repeated DNA sequences and genetic effects associated with particular regions of Xh as defined by the proximal breakpoints of specific inversions. The non-vital regions delineated by the open bar probably lack vital loci. The vertical lines subdividing Xh indicate the proximal breakpoints of the respective inversion. The broken bar indicates uncertainty about the nature of the boundary when one sequence ends and another sequence begins. It should be noted that although the Xh breakpoints of In(1)w^{m4} and In(1)w^{m51b} mark the boundaries of the rDNA region, there may be additional, undefined, DNA sequences in this region. (Modified from Hilliker and Appels, 1982.)

consistent with the notion that the Y chromosome may have evolved from the X (see Muller, 1932). We speculate that the only genetic functions of the X that the Y retained are those within Xh. The Y has, however, acquired some specific loci (not on the X) important for spermatogenesis.

The best characterized genetic locus in Xh is the bobbed locus (bb). This locus corresponds to the nucleolar organizer and is the repetitive structural locus for the 18s and 28s rRNA molecules (Ritossa et al., 1966). The genetic and molecular analyses of bb are reviewed in Hilliker et al., (1980), Long and Dawid (1980) and

Wellauer et al., (1978). The proximal breakpoints of In(1)w^{m4} and In(1)w^{m51b} define the distal and proximal boundaries of the rDNA region of Xh (Fig. 3) (Appels and Hilliker, 1982). Thus, the rDNA and associated sequences occupy about 50% of Xh.

Although the rDNA region of Xh constitutes a vital genetic locus, the region has cytological and genetic characteristics diagnostic of heterochromatin. By late prophase most of the rDNA region is clearly heterochromatic, although at least a portion of it must be relatively decondensed during interphase in order for the rDNA to be transcribed. It may be, however, that only a segment of the rDNA region constitutes the functional nucleolus organizer (see Pimpinelli et al., 1985). There is evidence in *Zea mays* (Givens and Phillips, 1976) that not all rDNA genes are functional in interphase and that a portion remain heterochromatic throughout interphase.

Additionally, the rDNA is able to induce position effect variegation of euchromatic genes placed next to it. Specifically, In(1)w^{51b}, In(1)w^{m4} and, most significantly, In(1)sc^{V2} (see Fig. 3), are associated with position effect variegation. All three rearrangements break within rDNA (Appels and Hilliker, 1982; Lindsley et al., 1982; for further discussion and citation of earlier references see Hilliker and Appels, 1982).

By inducing deletions in In(1)w^{m51b} (Fig. 4) Hilliker and Appels (1982) were able to obtain chromosomes in which the rDNA cluster was free of flanking heterochromatin. Analysis of these deletions demonstrated that the rDNA could function as an active *bb* locus free of its flanking heterochromatin and also maintain its heterochromatic appearance at late prophase of mitosis. Thus, the rDNA locus, a vital repeated gene, has intrinsic heterochromatic properties.

An additional property associated with both the rDNA region and the segment of Xh distal to the rDNA is the modification of position-effect variegation. Deletions of heterochromatin usually enhance position-effect variegation (reviewed in Spofford, 1976) and deletions of the rDNA region of Xh enhance position-effect variegation (Hilliker and Appels, 1982). This further implicates the rDNA region as being in many ways "typically" heterochromatic despite the fact that it is an actively transcribed region of the genome.

Heterochromatin immediately distal to the rDNA region of Xh (Fig. 3) has a number of interesting genetic effects. Two genetic elements have been localized to this region, namely Xh^{abo} (Parry and Sandler, 1974), now termed ABO (Pimpinelli et al., 1985), and the compensation response (*cr*) locus (Procunier and Tartof, 1978). ABO is a region of Xh that interacts with the maternal effect mutation abnormal oocyte (*abo*), which is located in the second chromosome euchromatin (Mange and Sandler, 1973). ABO maps to the region defined by the proximal breakpoints of In(1)sc^{4} and In(1)w^{m4} (Fig. 3)

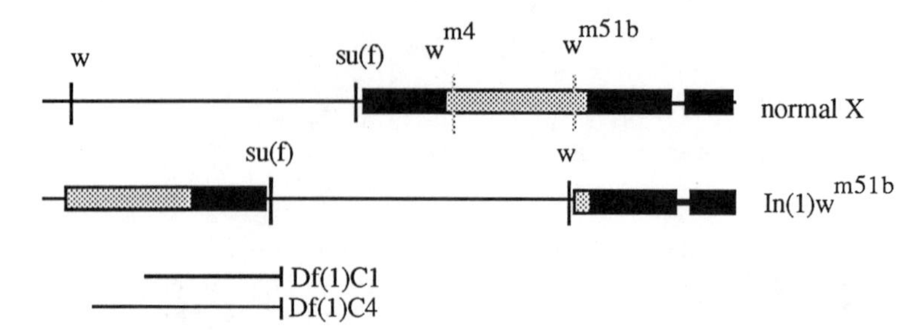

Fig. 4. Xh deletions selected in In(1)w^{m51b}. The top chromosome is a diagrammatic representation of a normal X chromosome. Xh is indicated by the thick line with the position of the rDNA indicated by the lightly stippled portion. Also indicated are the Xh breakpoints of In(1)w^{m4} and In(1)w^{m51b} as well as the location of the w and of the su(f) loci, for reference. Below the normal X chromosome is a diagrammatic representation of In(1)w^{m51b}. The Xh deletions indicated below the In(1)w^{51b} chromosome were recovered by selecting su(f)/In(1)w^{m51b} daughters of irradiated In(1)w^{m51b}/Y males. In addition to the two deletions illustrated, smaller deletions were also recovered. (Modified from Hilliker and Appels, 1982.)

(Sandler, 1970; Parry and Sandler, 1974; Yedvobnick et al., 1980; Pimpinelli et al., 1985). Utilizing the Xh deletions synthesized by Hilliker and Appels (1982), Pimpinelli et al. (1985) have identified a second minor ABO element proximal to In(1)w^{m4} (this minor element has not been included in Fig. 3).

The region corresponding to the major ABO site in Xh is dispensible in abo^+ flies, as females heterozygous for Df(1)C1 (Fig. 4) and In(1)$sc^{4L}sc^{S1R}$ are viable and fertile (Hilliker and Appels, 1982). However, there is a second ABO site in the heterochromatin of chromosome 2 (B. Ganetzky and J. Haemer, in Sandler, 1977).

Although the region between In(1)w^{m4} and In(1)sc^4 (Fig. 3) does not constitute an essential region of Xh, an additional genetic element maps within this region, namely the compensation response (cr) locus (Procunier and Tartof, 1978). Normally, a deficiency of the rDNA gene cluster is not associated with a reduction in rDNA content (Tartof, 1971; Spear and Gall, 1973). However, when the cr locus is relocated away from the rDNA, heterozygosity for a rDNA deletion results in a reduced rDNA content.

Analysis by in situ hybridization of nucleolar DNA replication during polytenization in In(1)w^{m51b}/Y males indicated that in larval salivary gland nuclei the rDNA of In(1)w^{m51b} was replicated preferentially to the Y chromosome rDNA. Deletion derivatives of In(1)w^{m51b} which remove at least part of the region corresponding to cr^{+} (and ABO) do not exhibit preferential replication of In(1)w^{m51b} DNA unless the entire block of Xh adjacent to the rDNA is removed (Hilliker and Appels, 1982).

Studies by Endow and Glover (1979) and Endow (1980) have indicated that at least in some genotypes only one of the two sex chromosome nucleolar organizers is replicated during polytene chromosome formation. Macgregor (1973) proposed this as a mechanism to account for the compensation phenomenon and further proposed that in *Drosophila* polytene nuclei the dominance of one nucleolus organizer over another may be due to a site adjacent to the dominant nucleolus organizer. The observations of Procunier and Tartof (1978), Hilliker and Appels (1982) and Endow (1980) support the Macgregor hypothesis and implicate the region between the proximal breakpoints of In(1)sc^{4} and In(1)w^{m4} (Fig. 3) as being important in the regulation of the replication of rDNA in the polytene nuclei. Yet, this region is dispensible for viability and fertility in a laboratory environment.

The region between the proximal breakpoints of In(1)sc^{4} and In(1)w^{m4} contains a DNA sequence also found within the rDNA region immediately proximal to it. This DNA sequence is known as the Type I rDNA intervening sequence. This is specific to the X chromosome rDNA and was initially found as an insertion sequence within 28s rDNA (reviewed in Wellauer et al., 1978). Dawid and Botchan (1977) and Peacock et al., (1981) biochemically detected Type I intervening sequences which were not inserted into 28s rDNA genes. The non-28s rDNA Type I intervening sequences are very likely present in tandem arrays (Kidd and Glover, 1980; Dawid et al., 1981). *In situ* hybridization studies of chromosomal rearrangements having a breakpoint within Xh have documented that the major site of the non-28s rDNA Type I intervening sequence is a chromosomal region defined by the proximal breakpoints of In(1)rst^{3} and In(1)w^{m4} (Fig. 3) (Hilliker and Appels, 1982; Appels and Hilliker, 1982). Thus, the major site of Type I insertions which are not within rDNA is immediately adjacent to the nucleolus organizer, and within the region which has been implicated in the regulation of rDNA replication. However, as illustrated in Fig. 3, this region is distinct from, although immediately adjacent to, the location of the cr^{+} region as delimited by Procunier and Tartof (1978). If the ABO effect correlates with this block of Type I intervening sequence, then the ABO of chromosome 2 heterochromatin must be correlated with a different repeated sequence, as there does not appear to be Type I intervening sequence in the chromosome 2 heterochromatin (Hilliker and Appels, 1982).

In situ hybridization experiments (Lindsley et al., 1982; Hilliker and Appels, 1982) further demonstrated that the Type I insertion sequence is not uniformly distributed within the rDNA region of Xh. The distribution within the rDNA locus is biased towards the distal half (Fig. 3).

One highly repeated sequence, namely the 1.688 gm/cc satellite (Brutlag et al., 1977) is unique to the X chromosome heterochromatin (Fig. 3) (Hilliker and Appels, 1982). This sequence does not appear to be essential for viability, as rearrangements which remove most, but not all of the 1.688 gm/cc satellite, such as $In(1)sc^{4L}sc^{8R}$ survive in combination with a Y chromosome. However, a small amount of 1.688 gm/cc satellite does remain and it cannot be categorically stated that it is a completely non-essential DNA sequence.

The other two satellite sequences found within Xh, the 1.705 gm/cc and 1.672 gm/cc satellites, are present immediately adjacent to the centromere (Steffensen et al., 1980) and are found in other chromosomal locations (Peacock et al., 1978). Recently, however, Lohe and Brutlag (1986) have found variants within the 1.705 and 1.672 satellites. Although each of the major satellite sequences of *Drosophila* (1.672, 1.688, 1.686 and 1.705) is defined primarily by a single repeating sequence, there are minor variant classes of each of these. Indeed, there are several minor satellites which have not been well characterized and the possibility exists that there may be others to be discovered (Lohe and Brutlag, 1986). Thus, the DNA sequence organization of Xh, as illustrated in Fig. 3, will undoubtedly be elaborated by future studies.

GENETIC STRUCTURE OF THE Y CHROMOSOME

The genetic organization of the heterochromatic Y chromosome has been reviewed by Williamson (1976), Hilliker et al., (1980) and most recently by Pimpinelli et al., (1986). As discussed earlier, a number of genetic elements are common to Xh and Y, including *bb* and ABO. However, there are at least 7 loci unique to the Y chromosome. Six of these genes are male fertility factors (Brosseau, 1960; Williamson, 1972; Kennison, 1981). These fertility loci function in primary spermatocytes and are essential for spermiogenesis. The seventh locus is the *crystal* locus (Hardy et al., 1984; see also Pimpinelli et al., 1986). There is also a genetic element associated with the $Y^{bbSuVar-5}$ chromosome which produces an overall increase in RNA synthesis including rRNA and polyadenylated RNA (Clark and Kiefer, 1977; Clark et al., 1977).

Using a variety of staining techniques, *Drosophila* heterochromatin can be resolved into characteristic banded regions (Holmquist, 1975; Gatti et al., 1976; Pimpinelli et al., 1976). Cytological analysis of the Y chromosome has revealed 25 differentially stained blocks (Gatti and Pimpinelli, 1983). Rearrangements affecting single fertility factors involve very large segments of the Y chromosome (Gatti and Pimpinelli, 1983; S. Bonacorsi and M. Gatti, in

Pimpinelli et al., 1986). Indeed, the kl-5 fertility gene appears to occupy 10% (4000kb) of the Y chromosome. The large size of the fertility genes on the Y chromosome is also indicated by the high rate (approximately 50%) of male sterility associated with Y-autosome translocations (Lindsley et al., 1972). An alternative possibility (discussed by Gatti and Pimpinelli, 1983) is that these fertility factor loci could be very susceptible to long range position-effects associated with autosomal translocations. We speculate that the specific adjacent highly repeated sequences may be very important in the expression of the individual fertility factor loci.

MALE MEIOTIC PAIRING SITES

Although the heterochromatic Y and X chromosomes are morphologically and genetically quite distinct, they successfully pair with and segregate from each other during male meiosis. Genetic and cytological studies have demonstrated that there are localized pairing sites responsible for this meiotic segregation and that they are found within Xh and the Y chromosome (Muller and Painter, 1932; Gershenson, 1940; Sandler and Braver, 1954; Cooper, 1964; Peacock, 1965; Peacock and Miklos, 1973; Yamamota and Miklos, 1977).

Appels and Hilliker (1982) demonstrated that the rRNA coding region contributes to X/Y pairing, however, they also found that no single region of Xh is required for fidelity of male meiotic pairing of the sex chromosomes. They examined a series of deletions which encompassed the entire portion of Xh contributing to pairing with the Y chromosome. Essentially, only very large deletions of this region, such as Df(1)C4 (Fig. 4) and In(1)sc^{4L}sc^{8R} show elevated levels of nondisjunction. Removing a subset of the male meiotic pairing sites within Xh does not result in elevated X/Y nondisjunction. Indeed, certain free X duplications carrying only a portion of the Xh pairing region segregate from attached XY chromosomes in XY/Dup(1), free males (e.g. Lindsley and Sandler, 1958). Analysis of Df(1)C4 (Fig. 4) (Appels and Hilliker, 1982) demonstrated that the rDNA region does contribute to male meiotic pairing, although it and any other single region is not essential.

Cooper (1964) described obvious cytological regions of meiotic pairing between X and Y which he termed collochores. Whether collochores are the visible consequence of meiotic pairing or the only sites responsible awaits further cytogenetic analysis.

The molecular nature of pairing sites is unknown, however, DNA sequence homology per se is not sufficient to account for X Y male meiotic pairing (see Yamamoto and Miklos, 1977). However, the characterization of DNA sequence organization in Xh and the assignment of specific DNA repeated sequences to particular regions will undoubtedly provide a useful starting point for the analysis of pairing sites at the molecular level. Before this can be accomplished, the pairing sites must be further localized by genetic and cytogenetic analyses.

Males that possess an X chromosome that is deficient for most of Xh (e.g. $In(1)sc^{4L}sc^{8R}$) show meiotic drive of the sex chromosomes, in addition to sex chromosome nondisjunction (reviewed in Peacock and Miklos, 1973). The most distorted recovery ratios are observed among nondisjunctional sperm; far fewer XY than nullo sperm are recovered. Among disjunctional progeny, X sperm are recovered somewhat less frequently than the Y sperm. The basis of these distorted segregation ratios is sperm dysfunction (Peacock et al., 1975), similar to that observed with SD. However, McKee (1984) showed that in $In(1)sc^{4L}sc^{8R}$ males there is no specific locus analogous to the Rsp element of SD, which determines sensitivity to distortion. Instead, sensitivity to distortion appears to depend largely upon the amount of chromatin in a sperm.

Unlike the heteromorphic sex chromosomes, the autosomes do not appear to have heterochromatic male meiotic pairing sites. Genetic and cytological analysis of the effects of Df(2R)MS-10, a chromosome deficient for a large heterochromatic block in 2R (Hilliker and Holm, 1975), found no evidence for this region being important in meiotic pairing (Yamamoto, 1979; Hilliker, 1975, 1981). However, if male meiotic pairing sites are distributed throughout autosomal heterochromatin and euchromatin, deleting some or all heterochromatic pairing sites may be insufficient to increase autosomal nondisjunction, by analogy with the X chromosome pairing sites (Appels and Hilliker, 1982).

The analysis of the meiotic segregation of compound 2 autosomes allowed Hilliker et al., (1982) to probe for autosomal male meiotic pairing sites in the heterochromatin. Compound autosomes generally possess no euchromatic homologies, but since they arise by a translocation event involving heterochromatic breakpoints (Hilliker and Holm, 1975; Holm, 1976; Hilliker, 1978; Gibson, 1977), a pair of complementary compound autosomes share heterochromatic homology. Thus, C(2L) and C(2R) chromosomes share varying extents of heterochromatic homology. This should result in non-random segregation in at least some instances if there are male meiotic pairing sites in autosomal heterochromatin. The analysis of male meiotic segregation of compound second chromosomes sharing varying extents of heterochromatic homology, provided no evidence for heterochromatic pairing sites (Hilliker et al., 1982). However, this study did provide evidence for a male meiotic pairing site in the proximal euchromatin of 2L, consistent with the earlier identification of this site by Sandler et al. (1968), Evans, (1971) and Gethmann, (1976).

It should be noted that Falk et al., (1985) have indirect evidence for male meiotic pairing sites in autosomal heterochromatin from the analysis of Y-autosomal translocations. As no detailed cytological analyses were undertaken to determine the breakpoints of the translocations, and as the segregation of chromosome 2 is complicated by interactions between the X and Y chromosomes, we do not consider his conclusions to be definitive, although we do consider the issue merits further investigation on the basis of his findings.

POSITION EFFECT VARIEGATION

Position effect variegation is a well-documented phenomenon (for reviews see Lewis, 1950; Hannah, 1951; Baker, 1968; and especially, Spofford, 1976). In *Drosophila* chromosomal rearrangements which result in the translocation of a euchromatic locus next to constitutive heterochromatin are often associated with the partial suppression of the euchromatic gene. It has recently been shown (Rushlow and Chovnick, 1984; Rushlow et al., 1984; Henikoff, 1981) that the suppression is the result of inhibition of transcription of the gene in at least some of the cells where it is normally expressed. Euchromatic loci do not function well in immediate juxtaposition to heterochromatin.

At least one heterochromatic gene exhibits position effect variegation, the *light* locus (*lt*) (Fig. 2). However, the *lt* variegation effect might not be entirely reciprocal to position effect variegation of euchromatic genes. Hessler (1958) found that *lt* variegating rearrangements always had breaks in the vicinity of *lt* and in distal euchromatin. However, whether the breakpoints of these rearrangements were proximal or distal to *lt* was not known. A chromosomal rearrangement which definitely translocates *lt* and its associated heterochromatic block to the distal region of 3R exhibits *lt* variegation, but vital loci on either side of *lt* in the translocated heterochromatic segment do not variegate (Hilliker, 1975). Indeed, the breakpoint in this rearrangement lies in the proximal region of 2L heterochromatin and *lt* is in the distal 2L heterochromatin, some distance from the juxtaposed euchromatin. Thus, position effect variegation of *lt* would appear to be the result of moving *lt* away from the centromere, rather than removing it from heterochromatin. It appears to be important for the function of the *lt* locus that it be in proximity to a centromere.

Recently, we have synthesized a large number of autosomal 2;3 translocations, some of which have breakpoints in chromosome 2 heterochromatin (Hilliker and Trusis-Coulter, 1987; Sharp and Hilliker, unpublished). We have examined those rearrangements involving chromosome 2 heterochromatin for position effect variegation of second chromosome heterochromatin vital loci. We have observed that *l(2L)EMS56-4* does not complement completely with five of these translocations which are also variegating for *lt*. Thus, the possibility exists that *l(2L)EMS56-4* is variegating in these rearrangements. A similar result has also been observed by P. Hearn and B. Wakimoto (personal communication). We also observed that these five *lt* variegating translocations also failed to complement *Df(2L)25D*, a small heterochromatic deficiency distal to the Group VIII loci containing *lt* (Fig. 2) (Hilliker and Holm, 1975; Hilliker, 1976). These results suggest that heterochromatic genes other than *lt* can variegate due to position effects.

INTERCALARY HETEROCHROMATIN

Various authors have proposed the existence of so-called intercalary heterochromatin. Such regions have been inferred from biological properties such as ectopic pairing and delayed replication (reviewed by Spofford, 1976) as well as apparent sensitivity to radiation induced breakage (reviewed by Hannah, 1951). We do not consider these regions heterochromatic, as they are not cytologically discernable as heterochromatic in mitotic chromosomes, they do not possess the ability to induce position effect variegation and as they are not enriched in highly repeated (satellite) DNA sequences. Intercalary heterochromatin may represent a real phenomenon in the sense that it represents chromosomal regions with unusual properties: however, none of these properties is diagnostic of heterochromatin. We would like to suggest that their classification as heterochromatin is inappropriate.

CONCLUSIONS

Numerous genetic elements are located within the constitutive heterochromatin of *D. melanogaster*. We estimate that there are 44-45 discrete genetic elements in *Drosophila* heterochromatin, 3 genes in Xh, 12 in the heterochromatic Y chromosome, 18 in the second chromosome heterochromatin, 11 in the third chromosome heterochromatin and, possibly, 1 in the fourth chromosome heterochromatin. In addition, there are male meiotic pairing sites in Xh and Y as well as more subtle effects associated with the deletion or duplication of specific heterochromatic regions. Clearly, heterochromatin is not *simply* a resting ground for junk or selfish DNA.

Although it is clear that discrete genetic elements are within heterochromatin, these are present in a low density relative to euchromatin and constitute only a very small fraction of the total heterochromatic DNA. Thus, a key issue concerns the role(s) of the highly repeated sequences which constitute the major portion of the heterochromatin. The selfish DNA hypothesis (Orgel and Crick, 1980) proposes that the highly repeated (or satellite) DNA sequences associated with heterochromatin are of no function to the organism.

Alternatively, the blocks of satellite DNA may have polygenic functions, subtly influencing the expression of many gene loci. The pleitropic effects associated with the distal region of Xh (Fig. 3) led to the suggestion that this heterochromatic segment affects many gene loci (Hilliker and Appels, 1982). Indeed, Mather (1944) earlier speculated that heterochromatin contributes greatly to polygenic activity. Although individual blocks of heterochromatin, such as that for the major ABO site in Xh, are dispensable in the laboratory, they may affect fitness in a natural environment. In some genotypes these ordinarily dispensable regions can dramatically affect phenotype.

Another possible function of the blocks of highly repeated DNA sequences could be to provide the necessary physical environment for the function of the major gene loci within the heterochromatin (Hilliker, 1976). The highly repeated DNA sequences may influence the chromatin structure of adjacent transcribed sequences and may also be important with respect to associations between specific chromosome regions and the nuclear membrane.

Clearly, heterochromatin has genetic functions. The genes responsible for major genetic activity have been extensively, although by no means completely, characterized. Similarly, major aspects of the DNA sequence organization have been uncovered. Progress in understanding the biological role(s) of heterochromatin will depend on further cytogenetic, genetic and molecular analyses.

REFERENCES

Appels, R. and Hilliker, A.J., 1982, The cytogenetic boundaries of the rDNA region within heterochromatin of the X chromosome of Drosophila melanogaster and their relation to male meiotic pairing sites, Genet. Res., 39:149-156.

Appels, R. and Peacock, W.J., 1978, The arrangement and evolution of highly repeated (satellite) DNA sequences with special reference to Drosophila, Int. Rev. Cytol. Suppl., 8:69-126.

Baker, W.K., 1958, Crossing over in heterochromatin, Am. Nat., 92:59-60.

Baker, W.K., 1968, Position-effect variegation, Adv. Genet., 14:133-169.

Barigozzi, C., Dolfini, S., Fraccaro, M., Rezzonico Raimondi, G., and Tiepolo, L., 1966, In vitro study of the DNA replication patterns of somatic chromosomes of Drosophila melanogaster, Exp. Cell Res., 43:231-234.

Brittnacher, J.G. and Ganetzky, B., 1983, On the components of Segregation Distortion in Drosophila melanogaster. II. Deletion mapping and dosage analysis of the Sd locus, Genetics, 103:659-673.

Brittnacher, J.G. and Ganetzky, B., 1984, On the components of

Segregation Distortion in *Drosophila melanogaster*. III. Nature of Enhancer of SD, *Genetics*, 107:423-434.

Brosseau, G.E., Jr., 1960, Genetic analysis of the male fertility factors on the Y chromosome of *Drosophila melanogaster*, *Genetics*, 45:257-274.

Brown, S.W., 1966, Heterochromatin, *Science*, 151:418-425.

Brutlag, D.L., 1980, Molecular arrangement and evolution of heterochromatic DNA, *Ann. Rev. Genet.*, 14:121-144.

Brutlag, D., Appels, R., Dennis, E.S., and Peacock, W.J., 1977, Highly repeated DNA in *Drosophila melanogaster*, *J. Mol. Biol.*, 112:31-47.

Brutlag, D. and Peacock, W.J., 1979, DNA sequences of the 1.672 g/cc satellite of *Drosophila melanogaster*, *J. Mol. Biol.*, 135: 565-580.

Carlson, M. and Brutlag, D., 1977, One of the copia genes is adjacent to satellite DNA in *Drosophila melanogaster*, *Cell*, 11:371-381.

Clark, S.H. and Kiefer, B.I., 1977, Genetic modulation of RNA metabolism in *Drosophila*. II. Coordinate rate change in 4S, 5S and poly-A associated RNA synthesis, *Genetics*, 86:801-811.

Clark, S.H., Strausbaugh, L.D. and Kiefer, B.I., 1977, Genetic modulation of RNA metabolism in *Drosophila*. I. Increased rate of ribosomal RNA synthesis, *Genetics*, 86:789-800.

Cooper, K.W., 1950, Normal spermatogenesis in *Drosophila*, *in*: "Biology of *Drosophila*", Wiley, New York. pp 1-61.

Cooper, K.W., 1959, Cytogenetic analysis of major heterochromatic elements (especially Xh and Y) in *Drosophila melanogaster*, and the theory of "heterochromatin", *Chromosoma*, 10:535-588.

Cooper, K.W., 1964, Meiotic conjunctive elements not involving chiasmata, *Proc. Natl. Acad. Sci. USA*, 52:1248-1255.

Crow, J.F., 1979, Genes that violate Mendel's rules, *Sci. Amer.*, 240:134-146.

Dawid, I.B. and Botchan, P., 1977, Sequences homologous to ribosomal insertions occur in the *Drosophila* genome outside the nucleolus organizer, *Proc. Nat. Acad. Sci. USA*, 74:4233-4237.

Dawid, I.B., Long, E.O., DiNocera, P.P., and Pardue, M.L., 1981, Ribosomal insertion-like elements in *Drosophila melanogaster* are interspersed with mobile sequences, *Cell*, 25:399-408.

Eberl, D.F., 1986, The isolation and analysis of X-linked mutations affecting embryogenesis in *Drosophila melanogaster*, M.Sc. Thesis, University of Guelph, Guelph, Ont., Canada.

Endow, S.A., 1980, On ribosomal gene compensation in *Drosophila*, *Cell*, 22:149-155.

Endow, S.A. and Glover, D.M., 1979, Differential replication of ribosomal gene repeats in polytene nuclei of *Drosophila*, *Cell*, 17:597-605.

Endow, S.A., Polan, M.L., and Gall, J.G., 1975, Satellite DNA sequences of *Drosophila melanogaster*, *J. Mol. Biol.*, 96:665-692.

Evans, W.H., 1971, Preliminary studies on frequency of autosomal nondisjunction in females of D. melanogaster, Dros. Inf. Ser., 46:123-124.

Falk, R., Baker, R. and Rahat, A., 1985, Segregation of centric Y-translocations in Drosophila melanogaster, I, Segregation determinants in males, Genet. Res., 45:51-79.

Fincham, J.R.S., 1966, "Genetic Complementation," W.A. Benjamin, New York.

Gall, J.G., 1973, Repetitive DNA in Drosophila, in: "Molecular Cytogenetics," Plenum Press, New York. pp 59-74.

Gall, J.G., Cohen, E.H. and Polan, M.L., 1971, Repetitive DNA sequences in Drosophila, Chromosoma, 33:319-344.

Ganetzky, B., 1977, On the components of Segregation Distortion in Drosophila melanogaster, Genetics, 86:321-355.

Gatti, M. and Pimpinelli, S., 1983, Cytological and genetic analysis of the Y chromosome of Drosophila melanogaster, I, Organization of the fertility factors, Chromosoma, 88:349-373.

Gatti, M., Pimpinelli, S. and Santini, G., 1976, Characterization of Drosophila heterochromatin, I, Staining and decondensation with Hoechst 33258 and quinacrine, Chromosoma, 57:351-375.

Gershenson, S., 1940, The nature of the so-called genetically inert parts of the chromosomes. Vit. Akad. Nauk. (Kiev, USSR, in Ukrainian; English translation by Eugenia Krivshenko.)

Gethmann, R.C., 1976, Meiosis in male Drosophila melanogaster, II, Non random segregation of compound second chromosomes, Genetics, 83:743-751.

Gibson, W.G., 1977, Autosomal products of meiosis arising from radiation-induced interchange in female Drosophila melanogaster, Ph.D. thesis, University of British Columbia, Vancouver, British Columbia, Canada.

Givens, J.F. and Phillips, R.L., 1976, The nucleolus organizer region of maize: Ribosomal RNA gene distribution nucleolar interactions, Chromosoma, 57:103-117.

Gustafson, J.P. and Lukaszewski, A.J., 1985, Early seed development in Triticum-Secale amphiploids, Can. J. Genet. Cytol., 27:542-548.

Hannah, A., 1951, Localization and function of heterochromatin in Drosophila melanogaster, Adv. Genet., 4:87-125.

Hardy, R.W., Lindsley, D.L., Livak, K.J., Lewis, B., Siversten, A.L., Joslyn, G.L., Edwards, J. and Bonaccorsi, S., 1984, Cytogenetic analysis of a segment of the Y Chromosome of Drosophila melanogaster, Genetics, 107:591-610.

Hartl, D.L., 1974, Genetic dissection of segregation distortion, I, Suicide combinations of SD genes, Genetics, 76:477-486.

Hartl, D.L., 1980, Genetic dissection of segregation distortion, III, Unequal recovery of reciprocal recombinants, Genetics, 96:685-696.

Hartl, D.L. and Hiraizumi, Y., 1976, Segregation Distortion, in: "The Genetics and Biology of Drosophila, Vol. 1b,"

M. Ashburner and E. Novitski, ed., Academic Press, New York. pp 616-666.

Hartl, D.L., Hiraizumi, Y. and Crow, J.F., 1967, Evidence for sperm dysfunction as a mechanism of segregation distortion in Drosophila melanogaster, Proc. Natl. Acad. Sci. USA, 58: 2240-2245.

Hauschteck-Jungen, E. and Hartl, D.L., 1982, Defective histone transition during spermiogenesis in heterozygous Segregation Distorter males of Drosophila melanogaster, Genetics, 101:57-69.

Heitz, E., 1928, Das heterochromatin der moose, I. Jb. wiss Bot., 69:762-818.

Heitz, E., 1933, Die somatische heteropyknose bei Drosophila melanogaster und ihre genetische bedeutung, Z. Zellforsch., 20:237-287.

Heitz, E., 1934, Uber α and β-heterochromatin sowie konstanz und bau der chromomeren bei Drosophila., Biol. Zbl., 45:588-609.

Henikoff, S., 1981, Position-effect variegation and chromosome structure of a heat shock puff in Drosophila, Chromosoma, 83:381-393.

Hessler, A.Y., 1958, V-type position effects at the light locus in Drosophila melanogaster, Genetics, 43:395-403.

Hickey, D.A., Loverre, A. and Carmody, G., 1986, Is the segregation distortion phenomenon in Drosophila due to recurrent active genetic transposition?, Genetics, 114:665-668.

Hilliker, A.J., 1975, Genetic analysis of the proximal heterochromatin of chromosome 2 of Drosophila melanogaster, Ph.D. thesis, University of British Columbia, Vancouver, British Columbia, Canada.

Hilliker, A.J., 1976, Genetic analysis of the centromeric heterochromatin of chromosome 2 of Drosophila melanogaster: deficiency mapping of EMS-induced lethal complementation groups, Genetics, 83:765-782.

Hilliker, A.J., 1978, The construction of ring chromosomes of autosomal heterochromatin in Drosophila melanogaster, Genetics, 90:85-91.

Hilliker, A.J., 1981, Meiotic effects of second chromosome heterochromatic deletions, Dros. Inf. Ser., 56:72-74.

Hilliker, A.J. and Appels, R., 1982, Pleiotropic effects associated with the deletion of heterochromatin surrounding rDNA on the X chromosome of Drosophila, Chromosoma, 86:469-490.

Hilliker, A.J., Appels, R. and Schalet, A., 1980, The genetic analysis of D. melanogaster heterochromatin, Cell, 21:607-619.

Hilliker, A.J., Clark, S.H., Chovnick, A. and Gelbart, W., 1980, Cytogenetic analysis of the chromosomal region immediately adjacent to the rosy locus in Drosophila melanogaster, Genetics, 95:95-110.

Hilliker, A.J. and Holm, D.G., 1975, Genetic analysis of the proximal region of chromosome 2 of Drosophila melanogaster, I,

Detachment products of compound autosomes, Genetics, 81: 705-721.
Hilliker, A.J., Holm, D.G. and Appels, R., 1982, The relationship between heterochromatic homology and meiotic segregation of compound second autosomes during spermatogenesis in Drosophila melanogaster, Genet. Res., 39:157-168.
Hilliker, A.J. and Trusis-Coulter, S.N., 1987, The functional significance of linkage group conservation: Analysis of 2;3 translocations in Drosophila melanogaster, Genetics (in press).
Hiraizumi, Y., 1981, Heterochromatic recombination in germ cells of Drosophila melanogaster, Genetics, 98:105-114.
Hiraizumi, Y., Martin, D.W. and Eckstrand, I.A., 1980, A modified model of segregation distortion in Drosophila melanogaster, Genetics, 95:693-706.
Hochman, B., 1971, Analysis of chromosome 4 in Drosophila melanogaster, II, Ethyl methanesulfonate induced lethals, Genetics, 67:235-252.
Holm, D.G., 1976, Compound autosomes, in: "The Genetics and Biology of Drosophila, Vol. 1b," E. Novitski and M. Ashburner, ed., Academic Press, New York. pp 529-561.
Holmquist, G., 1975, Hoechst 33258 fluorescent staining of Drosophila chromosomes, Chromosoma, 49:333-356.
Hsieh, T. and Brutlag, D.L., 1979, Sequence and sequence variation within the 1.688 gm/cc satellite DNA of Drosophila melanogaster, J. Mol. Biol., 135:465-481.
Kaufmann, B.P., 1934, Somatic mitoses of Drosophila melanogaster, J. Morphol., 56:125-155.
Kennison, J.A., 1981, The genetic and cytological organization of the Y chromosome of Drosophila melanogaster, Genetics, 98:529-548.
Kidd, S.J. and Glover, D.M., 1980, A DNA segment from D. melanogaster which contains five tandemly repeating units homologous to the major rDNA insertion, Cell, 19:103-119.
Lewis, E.B., 1950, The phenomenon of position effect, Adv. Genet., 3:73-115.
Lim, J.K. and L.A. Snyder, 1974, Cytogenetic and complementation analyses of recessive lethal mutations induced in the X chromosome of Drosophila by three alkylating agents, Genet. Res., 24:1-10.
Lima-de-Faria, A. and Jaworska, H., 1968, Late DNA synthesis in heterochromatin, Nature, 217:138-142.
Lindsley, D.L., Appels, R. and Hilliker, A.J., 1982, The right breakpoint of In(1)sc^{V2} subdivides the ribosomal DNA, Dros. Inf. Ser., 58:99.
Lindsley, D.L. and Sandler, L. 1958, The meiotic behaviour of grossly deleted X chromosomes in Drosophila melanogaster, Genetics, 43:547-563.
Lindsley, D.L., Sandler, L., Baker, B.S., Carpenter, A.T.C., Denell, R.E., Hall, J.C., Jacobs, P.A., Miklos, G.L., Davis, B.K.,

Gethmann, R.C., Hardy, R.W., Hessler, A., Miller, S.M., Nozawa, H., Parry, D.M. and Gould-Somero, M., 1972, Segmental aneuploidy and the genetic gross structure of the Drosophila genome, Genetics, 71:157-184.

Lohe, A.R. and Brutlag, D.L., 1986, Multiplicity of satellite DNA sequences in Drosophila melanogaster, Proc. Natl. Acad. Sci. USA, 83:696-700.

Long, E.O. and Dawid, I.B., 1980, Repeated genes in eukaryotes, Ann. Rev. Biochem., 49:727-764.

Macgregor, H.C., 1973, Amplification, polytenisation and nucleolus organisers, Nature New Biol., 246:81-82.

McKee, B., 1984, Sex chromosome meiotic drive in Drosophila melanogaster males, Genetics, 106:403-422.

Mange, A.P. and Sandler, L., 1973, A note on the maternal-effect mutants daughterless and abnormal-oocyte in Drosophila melanogaster, Genetics, 73:73-86.

Marchant, G.E., 1986, The genetic analysis of the heterochromatin of chromosome 3 of Drosophila melanogaster, Ph.D. thesis, University of British Columbia, Vancouver, British Columbia.

Martin, D.W. and Hiraizumi, Y., 1979, On the models of segregation distortion in Drosophila melanogaster, Genetics, 93:423-435.

Mather, K., 1944, The genetical activity of heterochromatin, Proc. Roy. Soc., London, B., 132:308-332.

Miklos, G.L.G. and Smith-White, S., 1971, Analysis of the instability of segregation-distorter in Drosophila melanogaster, Genetics, 67:305-317.

Muller, H.J., 1932, Some genetic aspects of sex, Am. Natur., 66:118-138.

Muller, H.J. and Painter, T., 1932, The difference of the sex chromosomes of Drosophila into genetically active and inert regions, Z. Indukt. Abstamm. Vererblehre., 62:316-365.

Nicoletti, B., Trippa, G. and DeMarco, A., 1967, Reduced fertility in SD males and its bearing on Segregation Distortion in Drosophila melanogaster, Atti. Acad. Naz. Lincei., 43:383-392.

Parry, D.M. and Sandler, L., 1974, The genetic identification of a heterochromatic segment of the X chromosome of Drosophila melanogaster, Genetics, 77:535-539.

Peacock, W.J., 1965, Nonrandom segregation of chromosomes in Drosophila males, Genetics, 51:573-583.

Peacock, W.J., Appels, R., Dunsmuir, P., Lohe, A.R. and Gerlach, W.L., 1977, Highly repeated DNA sequences: Chromosomal localization and evolutionary conservatiom, in: "International Cell Biology," B.K. Brinkely and K.R. Porter, ed., Rockefeller University Press, New York. pp 494-506.

Peacock, W.J., Appels, R., Endow, S. and Glover, D., 1981, Chromosomal distribution of the major insert in Drosophila melanogaster 28s rRNA genes, Genet. Res., 37:209-214.

Peacock, W.J., Brutlag, D.L., Goldring, E., Appels, R., Hinton, C.W. and Lindsley, D.L., 1973, The organization of highly repeated

DNA sequences in Drosophila melanogaster chromosomes, Cold Spring Harbor Symp. Quant. Biol., 38:405-416.

Peacock, W.J., Lohe, A.R., Gerlach, W.L., Dunsmuir, P., Dennis, E.S. and Appels, R., 1978, Fine structure and evolution of DNA in heterochromatin, Cold Spr. Harb. Symp. Quant. Biol., 42:1121-1135.

Peacock, W.J. and Miklos, G.L.G., 1973, Meiotic drive in Drosophila: new interpretations of the segregation distortion and sex chromosome systems, Adv. Genet., 17:361-409.

Peacock, W.J., Miklos, G.L.G. and Goodchild, D.J., 1975, Sex chromosome meiotic drive systems in Drosophila melanogaster, I, Abnormal spermatid development in males with a heterochromatin-deficient X chromosome, Genetics, 79:613-634.

Pimpinelli, S., Bonaccorsi, S., Gatti, M. and Sandler, L., 1986, The peculiar genetic organization of Drosophila heterochromatin, Trends Genet., 2:17-20.

Pimpinelli, S., Santini, G. and Gatti, M., 1976, Characterization of Drosophila heterochromatin, II, C- and N-banding, Chromosoma, 57:377-386.

Pimpinelli, S., Sullivan, W., Prout, M. and Sandler, L., 1985, On biological functions mapping to the heterochromatin of Drosophila melanogaster, Genetics, 109:701-724.

Procunier, J.D. and Tartof, K.D., 1978, A genetic locus having trans and contiguous cis functions that control the disproportionate replication of ribosomal RNA genes in Drosophila melanogaster, Genetics, 88:67-79.

Rhoades, M.M., 1978, Genetic effects of heterochromatin in maize, in: "Maize Breeding and Genetics," by D.B. Walden, ed., Wiley, New York. pp 641-671.

Ritossa, F.M., Atwood, K.C. and Spieglman, S., 1966, On the redundancy of DNA complementary to amino acid transfer RNA and its absence from the nucleolar organizer region of Drosophila melanogaster, Genetics, 54:663-676.

Rae, P.M.M., 1970, Chromosomal distribution of rapidly reannealing DNA in Drosophila melanogaster, Proc. Natl. Acad. Sci. USA, 67:1018-1025.

Roberts, P.A., 1965, Difference in the behaviour of eu- and heterochromatin: crossing-over, Nature, 205:725-726.

Rudkin, G.T., 1969, Non-replicating DNA in Drosophila, Genetics, 61:227-238.

Rushlow, C.A., Bender, W. and Chovnick, A., 1984, Studies on the mechanism of heterochromatic position effect at the rosy locus of Drosophila melanogaster, Genetics, 108:603-615.

Rushlow, C.A. and Chovnick, A., 1984, Heterochromatic position effect at the rosy locus of Drosophila melanogaster: cytological, genetic and biochemical characterization, Genetics, 108:589-602.

Sandler, L., 1970, The regulation of sex-chromosome heterochromatic activity by an autosomal gene in Drosophila melanogaster,

Genetics, 64:481-493.
Sandler, L., 1977, Evidence for a set of closely linked autosomal genes that interact with sex-chromosome heterochromatin in Drosophila melanogaster, Genetics, 86:567-582.
Sandler, L. and Braver, G., 1954, The meiotic loss of unpaired chromosomes in Drosophila melanogaster, Genetics, 39:365-377.
Sandler, L. and Golic, K., 1985, Segregation distortion in Drosophila, Trends Genet., 1:181-185.
Sandler, L. and Hiraizumi, Y., 1960, Meiotic drive in natural populations of Drosophila melanogaster, IV, Instability at the segregation distorter locus, Genetics, 45:1269-1287.
Sandler, L., Hiraizumi, Y. and Sandler, I., 1959, Meiotic drive in natural populations of Drosophila melanogaster, I, The cytogenetic basis of segregation-distortion, Genetics, 44: 233-250.
Sandler, L., Lindsley, D.L., Nicolleti, B. and Trippa, G., 1968, Mutants affecting meiosis in natural populations of Drosophila melanogaster, Genetics, 60:525-528.
Schalet, A. and Lefevre, G., Jr., 1976, The proximal region of the X chromosome, in: "The Genetics and Biology of Drosophila, Vol. 1b," M. Ashburner and E. Novitski, ed., Academic Press, London. pp 848-902.
Sederoff, R., Lowenstein, L. and Birnboim, H.C., 1975, Polypyrimidine segments in Drosophila melanogaster DNA, II, Chromosome location and nucleotide sequence, Cell, 5:183-194.
Sharp, C.B., 1977, The location and properties of the major loci affecting the segregation distortion phenomenon in Drosophila melanogaster, M.Sc. thesis, University of British Columbia, Vancouver, British Columbia.
Sharp, C.B., Hilliker, A.J. and Holm, D.G., 1985, Further characterization of genetic elements associated with the segregation distorter phenomenon in Drosophila melanogaster, Genetics, 110:671-688.
Sharp, C.B. and Hilliker, A.J., 1986, Measuring segregation distortion in Drosophila melanogaster, Can. J. Genet. Cytol., 28:395-400. (see erratum in Can. J. Genet. Cytol. 28:888.)
Sharp, C.B. and Hilliker, A.J., 1987, The k-probit transformation and segregation distortion, Dros. Inf. Ser., 65: (in press).
Spear, B.B. and Gall, J.G., 1973, Independent control of ribosomal gene replication in polytene chromosomes of Drosophila melanogaster, Proc. Natl. Acad. Sci. USA, 70-1359-1363.
Spofford, J., 1976, Position-effect variegation in Drosophila, in: "The Genetics and Biology of Drosophila, Vol. 1c," M. Ashburner and E. Novitski, ed., Academic Press, New York. pp 955-1018.
Steffensen, D.L., Appels, R. and Peacock, W.J., 1981, The distribution of two highly repeated DNA sequences within Drosophila melanogaster chromosomes, Chromosoma, 82:525-541.
Tartof, K.D., 1971, Increasing the multiplicity of ribosomal RNA genes in Drosophila melanogaster, Science, 171:294-297.

Tokuyasu, K.T., Peacock, W.J. and Hardy, R.W., 1977, Dynamics of spermiogenesis in Drosophila melanogaster, VII, Effects of segregation distorter (SD) chromosome, J. Ultrastruct. Res., 58:96-107.

Wellauer, P.K., Dawid, I.B. and Tartof, K.D., 1978, X and Y chromosomal ribosomal DNA of Drosophila: Comparison of spacers and insertions, Cell, 14:269-278.

Wieschaus, E., Nusslein-Volhard, C. and Jurgens, G., 1984, Mutations affecting the pattern of the larval cuticle in Drosophila melanogaster, III, Zygotic loci on the X-chromosome and fourth chromosome, Roux's Arch. Dev. Biol., 193:296-307.

Williamson, J.H., 1972, Allelic complementation between mutants in the fertility factors of the Y chromosome in Drosophila melanogaster, Mol. Gen. Genet., 119:43-47.

Williamson, J.H., 1976, The genetics of the Y chromosome, in: "The Genetics and Biology of Drosophila, Vol. 1b," M. Ashburner and E. Novitski, ed., Academic Press, New York. pp 667-700.

Yamamoto, M., 1979, Cytological studies of heterochromatin function in the Drosophila melanogaster male: autosomal meiotic pairing, Chromosoma, 72:293-328.

Yamamoto, M. and Miklos, G.L.G., 1977, Genetic dissection of heterochromatin in Drosophila: the role of basal X heterochromatin in meiotic sex chromosome behaviour, Chromosoma, 60:283-296.

Yedvobnick, B., Krider, H.M. and Dutton, F.L., 1980, Analysis of disproportionate replication of ribosomal DNA in Drosophila melanogaster by a micro-hybridization technique, Biochem. Genet., 18:869-877.

Zimmering, S., Sandler, L. and Nicoletti, B., 1970, Mechanisms of meiotic drive, Ann. Rev. Genet., 4:409-436.

INTERACTIVE MEIOTIC SYSTEMS

Marjorie P. Maguire

Zoology Department
University of Texas
Austin, TX

INTRODUCTION

Conundrums posed by the behavior of chromosomes in dividing cells, particularly at meiosis, have been resistant to solution. Problems remaining to be solved are large and of fundamental importance. Generally, during meiosis, each pair of homologues must align, associate closely along their length and crossover at least once; then chiasmata must be maintained until homologues can be directed to opposite poles for the first division anaphase, and sister centromeres must remain associated until they can be directed to opposite poles for correct disjunction at the second division anaphase. These maneuvers seem to call for the mobilization of structures with interactions which are not clearly visible in the context of current insight. All sexual reproduction depends upon the proper functioning of these little understood meiotic systems.

Fundamental similarities in meiosis across eukaryotes suggest that its peculiar structures and functions arose early and have been strongly conserved. Patterns also suggest that during the evolution of meiosis there may have been economical modification of structures and functions already in place for mitosis, as well as repeated use of novel meiotic structures in ways which serve the demanding requirements of its successive stages. It is the purpose of this report to summarize the current state of our understanding and ignorance with respect to the array of outstanding problems, and to emphasize promising directions for focus of future study.

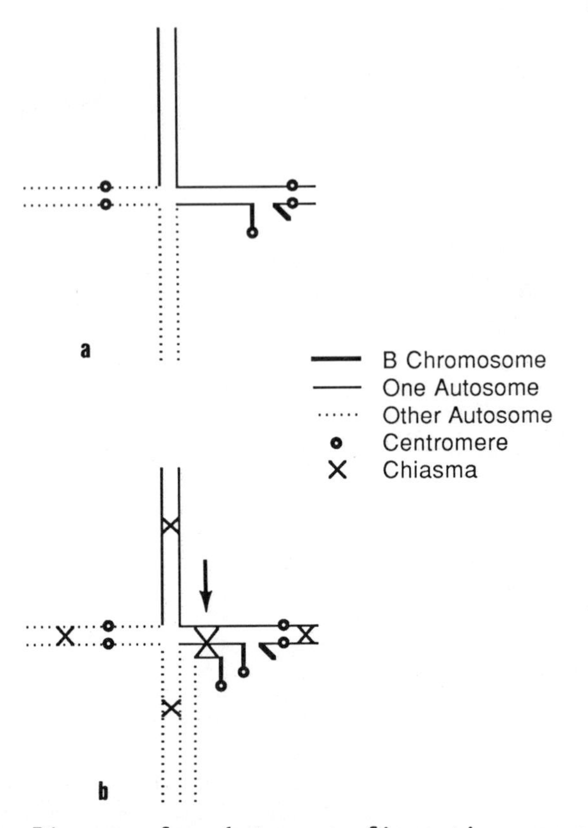

Fig. 1a. Diagram of pachytene configuration expected following complete homologous synapsis in meiocytes of heterozygotes of a complex B-A translocation of the sort derived by Rakha and Robertson (1970). These are a product of recombination in plants heterozygous for two kinds of translocation: a translocation between a B chromosome and an autosome (A chromosome) and a second translocation between two A chromosomes. b. Diagram of chromosomes of hyperploids for complex B-A translocations which carry all the chromosomes shown in a and an additional (identical) B^A chromosome, produced as a result of nondisjunction of B chromosome centromeres in the second micropore division in the course of pollen production. The chromosomes are shown homologously aligned although pachytene synapsis is assumed to be limited to two-by-two assocation. Diakinesis configurations with chiasmata located in the segments indicated are distinctive in appearance and found with substantial frequency. (For details see Maguire, 1986) It is difficult to escape the conclusion that the pairing which was effective for producing the

ALIGNMENT PAIRING OF HOMOLOGUES

Homologous chromosome pairing requires the moving together of matching chromosomes or chromosome segments across substantial distances. If the process depends upon similar underlying mechanisms throughout sexually reproducing eukaryotes, some respected ideas for alignment pairing systems seem to have serious flaws which can be briefly listed as follows: (1) It is unlikely that simple diffusion of the very large and unwieldy homologues of higher eukaryotes could by chance yield efficient meeting of matching parts since they may be widely dispersed in advance of meiosis. (2) The often suggested idea that conspicuous heterochromatic regions generally play a vital role in homologue pairing appears to be challenged by an accumulation of evidence (see Maguire 1984 for review). (3) The synaptonemal complex (SC) does not seem to function in the initial stages of pairing although it is true that once matching parts of homologues have approached to within about 300 nm at early meiotic prophase, they are generally joined by (SC) components which serve to stabilize association. (4) The probable role of telomeres is commonly overestimated. Although interactions of telomeres with the nuclear envelope (as well as the persistence of chromosome orientation established by prior mitotic anaphase poleward movement of centromere regions) may provide in many cases for closer than random positioning of some parts of homologues, the distances remaining to be traversed for association adequate for SC installation are still long range in physical-chemical terms. Homologous telomeres are usually not very closely positioned prior to synaptic intiation stage (e.g. Moens, 1973). Also studies of synapsis of complex rearrangements have suggested that first initiation of pairing tends to be distal but not terminal (Burnham et al., 1972). In addition ring chromosomes (Schwartz, 1953; Haber et al., 1984) and heterozygous, rearranged intercalary segments (Fig. 1 and Maguire, 1986) have been shown sometimes to pair very efficiently, and in ways which could not be accommodated by a two-by-two "zipping-up" of pairing initiated at homologous telomeres. (5) More complex models for homologue pairing which invoke consistent heterologous arm associations so that two genomic common sequence arrays might be conveniently appressed at synaptic stage (Lewin, 1981; Bennett, 1982) have been seriously questioned on several grounds. These include the criticism that the apparent heterologous arm associations which have been reported may result from artifact inherent in the kind of statistical analysis used (Callow, 1984); also chromosome rearrangements which alter the supposedly critical property of chromosome arms (relative mass) do

chiasma between a B^{A} chromosome in its intercalary region with the normal sequence counterpart for this segment (arrow) was not produced by zipping-up of two-by-two pairing initiated at homologous telomeres.

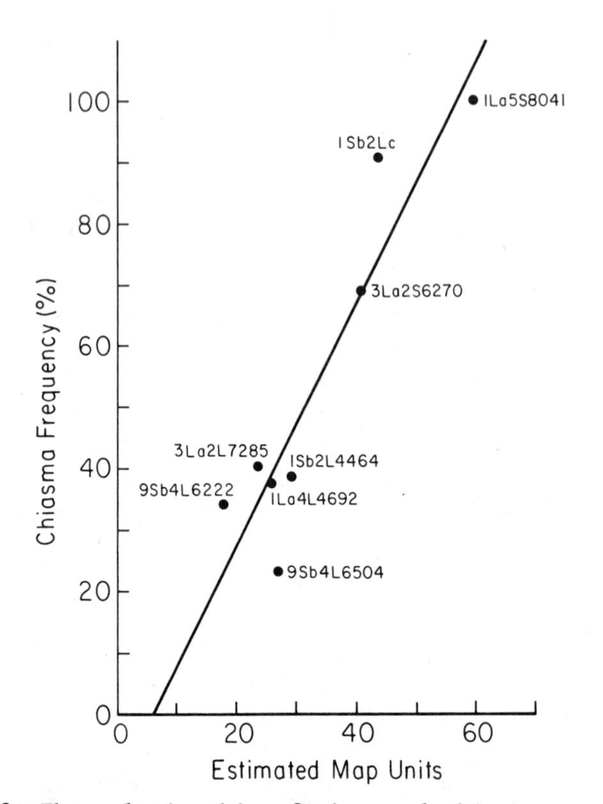

Fig. 2. The relationship of observed chiasma frequencies in intercalary translocated segments to rough estimates of their genetic length. The translocated segments are the intercalary regions of B^A chromosomes in 8 complex B-A translocation heterozygotes (see Fig. 1a. diagram). These are designated by their titles adjacent to the points and involve chromosomes 1, 2, 3, 4, 5 and 9. The chiasma frequencies found are plotted here against estimated genetic map unit contents of these segments which are based on assumption of uniform distribution of genetic map throughout the respective chromosome arms, relative lengths of maize chromosomes at pachytene as listed by Neuffer et al. (1968) and estimated translocation breakpoints tabulated by Beckett (1978). The calculated regression line indicated has a coefficient of 1.93 which is significant at the 0.01 level. (For details see Maguire, 1986). This can be viewed tentatively as suggestive of a linear relationship; heterogeneity in genetic map distribution is known to exist; it is acknowledged

not (as predicted) upset homologue pairing (Maguire, 1983a). In addition, alien chromosome substitutions (Sears, 1968) might be expected to cause meiotic chaos, at least in some cases. Consistent heterologous associations of telomeres (or approximately terminal centromeres) which produce specific intragenomic arrays in some organisms (Ashley, 1979) may in fact serve functions other than homologue pairing. (6) The notion that matching pairing sites may be bound by elastic connectors during chromosome movements in earlier divisions, so that the stage is set for homologue alignment at zygotene by way of contraction of these connectors (Maguire, 1984), is obviously burdened with severe complexity and could not apply to organisms with zygotic meiosis.

Although no clear solution to the homologue pairing problem is in sight, it seems reasonable to ask whether some observations may point toward development of a coherent model to be tested. One note of interest is that meiotic DNA replication of numerous low copy number segments (which are widely dispersed in the Lilium genome) seems to be delayed until these segments are involved in the synaptic process (Hotta et al., 1984 for review). This may be relevant since inhibition of this very late "zygotene DNA" replication also seems to inhibit synapsis itself (Roth and Ito, 1967), and it has been inferred that complementary base pairing between homologues within these segments may play an essential role in the intitial stages of the synaptic process. It is unclear, however, whether these zygotene DNA segments may somehow function in the earlier processes of alignment pairing. Nevertheless, most workers probably believe on a priori grounds that there are DNA sequences specialized to function somehow during this phase of homologue pairing: the zygomeres of Sybenga (1966), the differential pairing affinity factors of Doyle (1979) and perhaps the synaptomeres of King (1970). Hawley (1980) reported evidence which suggests there may be four specialized alignment pairing centers within the X chromosome of *Drosophila melanogaster* such that, in the presence of heterozygosity for a rearrangement breakpoint in a region bounded by two of these centers, crossover frequency is sharply reduced in that region. Results have been reported which can be interpreted as consistent with the existence of pairing centers in *Caenorhabditis elegans* (Rose et al., 1984), and also in maize (*Zea mays*) (Maguire, 1986). In maize, however, findings suggest that pairing centers may be more numerous and widely dispersed and that probably, unlike *Drosophila*, zip-up pairing which

---

that there may be errors in estimates of the relative physical lengths of the various chromosome arms as well as in estimates of translocation breakpoint positions. Capability for pairing which is effective for crossing over appears to be distributed in coarse proportion to genetic map content.

is eventually effective for crossing over can procede from putative pairing centers toward heterozygous rearrangement breakpoints (Figs. 2, 3, and Maguire, 1985a, 1986).

Direct cytological observations have indicated that organisms differ with respect to the stage at which alignment pairing occurs. In many dipterans homologues, and to some extent even rearranged homologous segments (see Puro, 1973), show a tendency to be at least loosely aligned in all cells at prophase stages (mitotic as well as as well as meiotic). In other organisms evidence has been reported that homologues tend to align first in premeiotic mitotic divisions of germ line cells, and sometimes increasingly so as meiosis is approached (Brown and Stack, 1968; Stack and Brown, 1969; McDermott, 1971; Juricek, 1975; Maguire, 1983b). In a few cells at early premeiotic mitotic prophase in maize, an extended region of a single homologue pair has been seen to to be involved in close pairing which appears very similar to synapsis at meiosis (Fig. 4 and Maguire, 1983b). On the other hand, in other organisms homologues appear to be dispersed as late as premeiotic interphase (Walters, 1970; John, 1976). Light microscope observations of bouquet stages in amphibian early meiotic prophase (Kezer and Macgregor, 1971) point to a tendency for earliest synapsis to be distally located as do a number of electron microscope observations (Gillies, 1984 for review), but sites of earliest synapsis do not necessarily indicate precise positions of chromosome centers specialized to function in alignment pairing. Homologues may become aligned for segments of considerable length before synapsis is initiated at some points within them (Donnelly and Sparrow, 1965; Therman and Sarto, 1977; Gillies, 1979; Zickler and Sage, 1981; Maguire 1983c). In polypoids the evidence for alignment of all homologues at zygotene is especially striking in some organisms (Maguire 1984, for review), and such alignment (seen in spread preparations) may even persist revealingly into pachytene, where synapsis is limited to two-by-two assocation, with changes of pairing partner (Loidl, 1986; Loidl and Jones, 1986). In the latter case in trivalent configurations, unsynapsed segments often appear to be associated to corresponding sites of the synapsed members by connections interpreted as possible remnants of alignment pairing connectors.

Of special relevance to the pairing problem are observations of organisms with zygotic meiosis (Lu and Raju, 1970; Singleton, 1953), where, immediately following karyogamy, the two parental genomes are found in two separate clusters of chromosomes. Then at a slightly later stage the two sets seem to be intermingled. Next, homologues may appear to be aligned at a substantially greater then synaptic distance, and then synapsis of homologues is initiated at sites which seem to be preferentially, but not exclusively, distally located. So far observations have been limited to fixed and stained material, but chromosomes are typically condensed at zygotene stage in these organisms and therefore may be directly observable in the

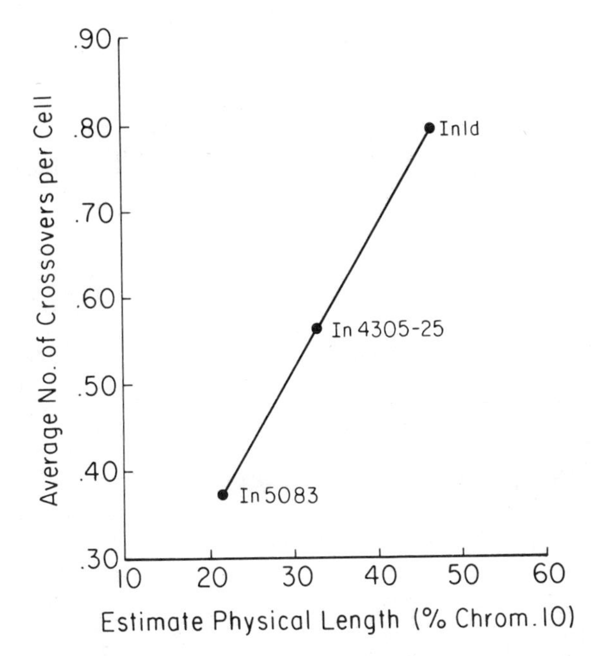

Fig. 3. The relationship of estimated crossover frequency within three overlapping maize paracentric inversions to their physical lengths. Bridge and fragment frequencies as well as possibly relevant fragment associations were used to estimate the average number of crossovers per cell. Two methods of calculation of average numbers of crossovers per cell gave similar results and were averaged. (For details see Maguire, 1985a). The clear linear relationship most simply suggests a relatively fine-grained distribution of capability for pairing which is effective for crossing over along most of the distal half of the long arm of chromosome 1, although other possibilities are not ruled out.

living state with phase or interference microscopy. What motions occur between the snapshot views seen of progressively advanced stages in fixed and stained cells could be enormously instructive. A simple guess is that the nuclear contents are mixed or stirred so that chance for meeting of matching parts is greatly enhanced, and such meetings are then somehow stablized to form lasting associations in preparation for synapsis. Rotation movements of nuclei have been reported at zygotene stage in rat spermatocytes (Parvinen and Söderström, 1976), but rotation of nuclear contents has also been observed in various cells at mitotic interphase (De

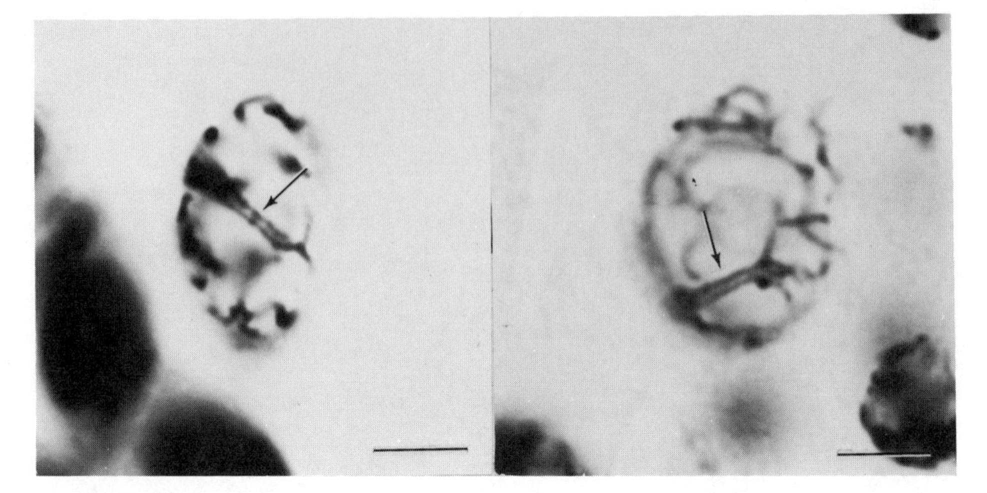

Fig. 4. Photomicrographs of maize sporogenous nuclei at early prophase of premeiotic mitosis with synapsis-like pairing association (arrows) of two matching chromosomes for a substantial part of their length. Other chromosomes within these nuclei show no comparable associations. These cells are from Feulgen stained sections of normal material. Magnification bars represent 5 μm.

Boni and Mintz, 1986), although positions of chromosomes with respect to each other generally remain stable during interphase in somatic cells (Hubert and Bourgeois, 1986 for review). Since chromosomes have no obvious means for autonomous motion and generally do not show evidence of stirring during interphase, it seems reasonable to guess that they are pushed or pulled by special interactions with other structures during the establishment of alignment pairing. Chromosome movements at mitosis seem to depend upon interactions with the spindle, but components of the cytoskeleton are usually excluded from the nucleus, and generally associate directly with chromosomes after breakdown of the nuclear envelope (Hepler and Wolniak, 1984 for review). However, recent observations of apparent nuclear perturbations prior to meiosis in maize may be relevant to homologue pairing, with possible implication of a role for the cytoskeletal structures (Maguire, unpublished). Most cytological observations of premeiotic mitosis or early meiosis have utilized either thin sections (EM), squash preparations (light microscopy) or spreads (EM). When attention was concentrated on thicker sections with light microscopy in maize it was noted that there seemed consistently to be drastic fluctuations in nuclear shape and arrangement of contents during the interval from interphase to late prophase of advanced (possibly the last) premeiotic mitoses (Figs. 5,6). At interphase in nuclei at this level of development the chromatin material was confined to the periphery of the nucleus, and most of the internal volume was occupied by a large, clear lumen which contained a single nucleolus in almost all cases. After early prophase chromosome condensation had commenced, the lumen disappeared for the most part, and the chromosomes were found compressed against and adjacent to the nucleolus. The entire nucleus generally appeared flattened into an irregular shape similar to the synizesis stage of early meiotic prophase. By late prophase (as gauged by degree of chromosome condensation) a large, clear central lumen (containing the nucleolus) was again present, and the now condensed chromosomes occupied the periphery of the nearly spherical nucleus. Similar, though less drastic, changes in shape and lumen size were found in early prophase nuclei of less advanced sporogenous cells where earlier premeiotic mitoses were in progress (Figs. 5,6), although the degree of flattening of nuclei and the amount of nuclear shape irregularity were less pronounced during prophase of these earlier divisions. Two lumens, each containing a nucleolus were occasionally found (12% of 528 at interphase) in cells at earlier premeiotic mitosis level, but only very rarely in cells at the later premeiotic mitosis level, suggesting that the pair of chromosomes which carry the nucleolus organizer tend to be more closely positioned at the later division. Nuclear shape and organization did not appear to change drastically during prophase of divisions of the surrounding somatic cells which were examined in the more advanced anthers (Figs. 5,6). A simple speculation is that cycles of compression and inflation of the nucleus (perhaps repeated over several cell generations in some organisms) may promote a special stirring of the nuclear contents that facilitates the meeting of pairing sites to yield initial association of homologues, and this in turn is followed eventually if not immediately by homologue alignment and synapsis. In many organisms the major work might be accomplished at something like a synizesis stage at early meiotic prophase, which encompasses zygotene in those organisms where it is observed. For many years synizesis was thought to represent a fixation artifact. Currently, the consensus seems to be that it is sufficiently widespread and consistent following a variety of fixation procedures to be considered meaningful. In other organisms perturbations of the nucleus which impose stirring might take some other form, or occur at earlier divisions which establish presetting for pairing. That presetting for alignment pairing can occur and be maintained (or reestablished) through division cycles is clearly demonstrated by the fact that all or most *Drosophila* cells show an apparent predisposition for such alignment of homologues, at least during prophase of mitoses.

The new observations described above, however suggestive, raise a number of questions: What is the relationship of appearances in fixed and stained material to normal conditions in living cells? If these appearances are basically realistic, what is the nature of the central lumen and its fluctuations, and what roles may be played by

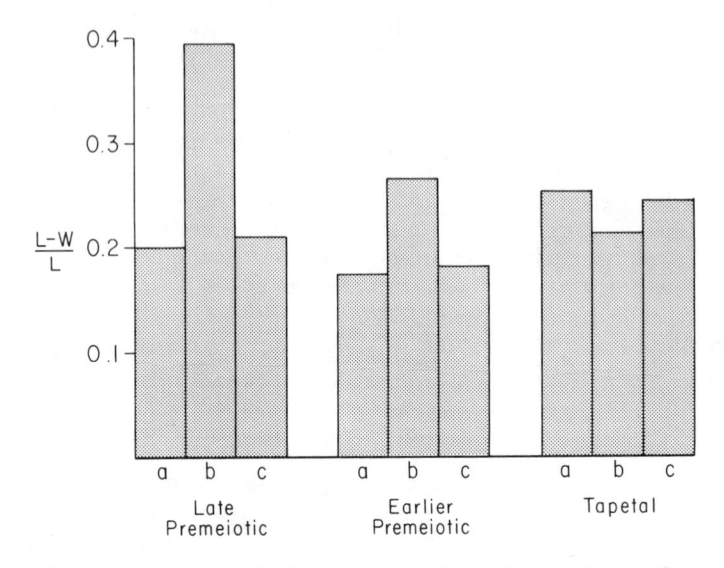

Fig.5. Estimates of departure of nuclear shape from spherical in sporogenous cells of maize. The index utilized represents the difference between nuclear length and width, per unit length. This value neglects pronounced irregularity of nuclei seen at early prophase of the late (more advanced) premeiotic mitoses and therefore underestimates the degree of possible shape changes at this stage; it is more representative of departure from spherical for the nuclei at the other stages, where shape tended to be more nearly ellipsoidal. Ocular micrometer measurements of maximum long and short axes of nuclei at the appropriate stages were taken as they were encountered during systematic scanning of slides. For late premeiotic mitoses, histogram bars represent the means of 50 nuclei at each stage, and for earlier premeiotic mitoses and tapetal mitoses, histogram bars represent the means of 25 nuclei at each stage. In each case (a) represents interphase, (b) represents early prophase and (c) represents late prophase. Means of length minus width divided by length were significantly greater in t-tests at early prophase than at interphase in both advanced premiotic mitoses and earlier premeiotic mitoses. The material used was from normal maize inbred KYS, collected from plants which had been grown in a growth chamber under controlled environmental conditions. Tassel spikelets were fixed in ethanol acetic 3:1 mixture, paraffin

nuclear matrix and cytoskeleton? Strikingly, tubulin poisons such as colchicine (Driscoll and Darvey, 1970; Hotta and Shepard, 1973; Shepard et al., 1974; Bennett et al., 1979; Salonen et al., 1982) and vinblastine (Gibson and Moses, 1986) seem to inhibit alignment pairing but do not interfere with maintenance of synapsis once it has been accomplished. In some cases this inhibition has been attributed to a possible effect of colchicine on the nuclear envelope (Shepard et al., 1974; Salonen et al., 1982). On the other hand it is well known that nuclei of the sperm of some organisms appear to be compressed and reshaped in conjunction with activity of surrounding microtubules (Dustin, 1978). Experiments indicate that dimers in the soluble tubulin pool may be able to adopt two distinct conformations: one in phosphorylated state may insert into membranes, the other may be available for polymerization into microtubules (Hargreaves et al., 1986). Is it possible that function of tubulin is relevant to homologue pairing both in membranes and in microtubules? In recent years interesting findings have emerged from studies of nuclear skeleton and spatial arrangement of chromosomes in some organisms. In general, positions of chromosomes seem to be stable during mitotic interphase; chromosomes occupy territories distributed along the inner nuclear membrane where their positions are determined by DNA-binding sites (of unknown number) in the peripheral nuclear skeleton, but the recognition system allowing the association of DNA with lamina after mitosis does not require DNA-binding sites to occupy fixed positions; tendencies for certain chromosomes to lie close together, then, seem to be due to factors specific to these chromosomes; and relative positions of chromosomes may be altered during divisions (Hubert and Bourgeois, 1986 for review). Premeiotic and meiotic nuclear and chromosome behavior now imply a structural basis which must be more precisely probed. Its essential intrinsic properties have been more clearly defined by recent work.

DETERMINATION OF CROSSOVER SITE DISTRIBUTION

The relationship of crossover site establishment to synapsis presents another meiotic puzzle; although synapsis is normally complete, crossovers are well known to be nonrandomly distributed. A number of years ago it was accidentally found in maize that for chromosome segments estimated to contain less than fifty map units which had been inverted or translocated, the estimated crossover

←

embedded, sectioned at a thickness of 15μm and Feulgen stained. Late premeiotic mitoses and tapetal mitoses were from first flower anthers, and earlier premeiotic mitoses were from second flower anthers of the same spikelets, well known to be less advanced by several days.

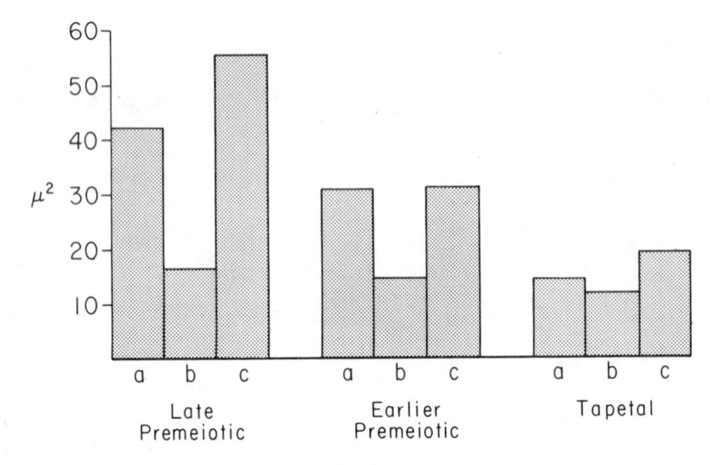

Fig. 6. Estimates of area (μm²) for maximal cross-section of nuclear lumens, (assuming ellipsoidal shape) of the same nuclei represented in Fig. 5. The cross-sectional area of the lumen in each case was calculated as the area of an ellipse with semiaxes equal to half of length and width as measured with an ocular micrometer. Lumen shape generally approached spherical at all stages, and at early prophase the lumen usually contained little more than the nucleolus. In each case histogram bars at (a) represent means at interphase, at (b) represent means at early prophase and at (c) represent means at late prophase. Means of lumen cross-sectional area were significantly greater in t-tests at interphase than at early prophase in late (advanced) premeiotic mitoses, earlier premeiotic mitoses and tapetal nuclei. These values were also significantly greater in t-tests for late prophase than interphase in advanced premeiotic mitoses and tapetal cells. Observations are suggestive of nuclear compression and inflation cycling during prophase of premeiotic mitoses.

frequency in the rearranged segements seemed to tend toward equality with their observed frequency of homologous synapsis in squash preparations at pachytene (Maguire, 1965, 1966, 1972). In many microsporocytes from inversion heterozygotes, for example, there was either nonhomologous synapsis across the inverted region, complete synaptic failure there or only partial homologous synapsis. Similar observations have been reported from other laboratories: in grasshopper (Camnula pellucida) inversions (Nur, 1968), in maize inversions (Stack, personal communication and Stack and Anderson, 1986a) and in Rhoeo spathacea translocations (Stack and Soulliere, 1984). Chovnick (1973) found conversion frequency within a

Drosophila short heterozygous paracentric inversion (where single crossover products would not be recovered) to be of approximately normal order of magnitude, even though frequency of homologous synapsis (not readily observable in this material) was expected to be low, with heterozygosity for such a small inversion. These findings are at odds with conventional thinking which envisions the positioning of crossover sites to be random along the genetic map (except for constraints on multiple exchanges usually imposed by positive chiasma interference). Under conventional rules then, short heterozygous rearranged segments should normally contain crossovers only with a frequency determined by their frequency of homologous synapsis per genetic map unit(x2). The larger values described above can be formally accounted for most readily in several ways (as suggested by Maguire, 1972): (1) Estimates of genetic map are erroneous. This is possible, but if so the varying extents of synapsis which actually occurred in the rearranged segments must have contained portions of chromosome in which normal crossover frequency substantially exceeds the average per unit cytological length, in a variety of material. (2) Change of pairing partner at synapsis increases the probability of occurrence of a crossover event (and possibly also of a conversion event). If so, this is of great interest to modelers of the mechanism of crossing over. (3) The pachytene stage observed does not faithfully represent the synapsis which may have been present at an earlier stage where it is conceivable that it was totally homologous and complete; then later losses of homologous synapsis were presumably mainly confined to cells in which crossing over had not occurred. Changes in complex synaptic configuration called "synaptic adjustment", during the interval between early and mid to late pachytene have been reported to occur, especially in mice (*Mus musculus*) (Moses, 1977; Poorman et al. 1981; Moses et al., 1982.), although searches have failed to reveal evidence for its occurrence in maize (Maguire, 1981; Gillies, 1983; Anderson and Stack, 1986). It remains possible, however, as also suggested by Rasmussen and Holm (1980), that the presence of crossing over in a heterozygous rearrangement region may prevent synaptic loss or change to nonhomologous synapsis, which would otherwise occur during the course of pachytene. This seems not to be the case in mice, where synaptic adjustment appears to procede unfailingly in the presence of substantial crossover frequency (Moses et al., 1982), and if it is the case in maize, the establishment of crossover sites must be restricted to no later than zygotene or very early pachytene. If crossover site determination (but not necessarily completion of the crossover process) occurs early, possibly even following homologous alignment but in advance of completion of SC installation, one line of reasoning proposes that the two processes might be interrelated in a way which accounts at least in part for chiasma interference. It has even been conjectured that completed SC may prevent establishment of additional crossover events following its installation (Maguire, 1968; Snow 1979; Egel-Mitani et al., 1982).

In a temperature sensitive mutant of yeast (Saccharomyces cerevisiae), however, crossover frequency has been reported to be increased by exposure to the restrictive temperature which seems to delay removal of the SC (Davidow and Beyer, 1984), implying that at least in this material additional crossover events can be initiated during the residence of the SC. More information on the distribution of crossover sites has been provided by study of recombination nodules (RNs), structures observable with EM resolution, whose distribution along the SC by mid to late pachytene generally seems to match crossover site distribution. These structures were first noted and appreciated by Carpenter (1975). Their form and number prior to mid to late pachytene often varies, and it remains unclear whether the early appearing forms are associated with potential crossover sites, possibly to be in some cases aborted or resolved as conversion only sites, and in others finally transformed to crossover sites. In some organisms RNs have been found only at pachytene and in others at zygotene as well as pachytene (Carpenter, 1979 for review). On the basis of observations of gene conversion, coconversion and crossover frequencies in several meiotic mutants of Drosophila, Carpenter (1984) has suggested that crossing over requires three distinct stages: initiation, potentiation (probably extension of heteroduplex length), and finally isomerization (see Meselson and Radding, 1975). She proposes that at or just after initiation, the presence of the early occurring RNs promotes continuation of recombination processes and finally some sites are chosen to be crossover sites in association with late RNs. Additional insight may be provided by recent work with solanaceous plants in which the numbers of recognizable RNs observed was found to depend importantly on preparation technique (Stack and Anderson, 1986b; Stack, personal communication). Observations were reported of numerous RNs during normal zygotene (after optimal preparation) in these plants, followed by the precipitous loss of many of them during the late zygotene to early pachytene period, down to a number which corresponds to the number of chiasmata found later. It was suggested that as a general rule, the RNs present earliest in a region may tend to have highest probability of being associated with a crossover event, such that new initiations of intimate pairing may be likely to be associated with a crossover. This is consistent with recent findings of Jones and Croft (1986) who have reported a general tendency for regions which usually synapse early to show relatively high chiasma frequency in Locusta and Schistocerca. Stack and Anderson (1986b) propose the following hypothesis to account for positive chiasma interference: The first successful reciprocal recombination event causes a signal to be transmitted in either direction along the SC, which tends to cause the release of nearby RNs (which have not yet established crossover intermediates). This signal would be dissipated with distance. They note that this mechanism would essentially guarantee a small number of crossovers per bivalent with only rare failures, as is usually observed. A conjecture (albeit radical) could be devised to modify and extend this proposal as follows: Incipient RNs which may be capable of modification to a form which can mediate a crossover event, may some of the time already be present just before final installation of the central component of the SC at a synaptic initiation, in closely aligned homologues. Such incipient RNs may be clearly visible only in optimally prepared favorable material. The signal transmitted from here along the bivalent, releasing less advanced RNs, might in fact be zip-up extension of the completed SC itself. Then presumably some randomly positioned incipient RNs in the path of synapsis would be sufficiently established to remain in place long enough to function in gene conversion without reciprocal exchange. Such a process could account for a tendency for a 1:1 relationship between observed homologous pairing frequencies at pachytene and crossover frequencies to occur, in regions heterozygous for rearrangements (which require synaptic initiation for homologous association). It could even be suggested that the defect in recombination deficient precondition mutants of Drosophila, such as mei 218 (Carpenter, 1984) lies in an alteration of the course or speed of synapsis itself, although at completion, synapsis appears normal. It can be imagined that the defect in the yeast mutant described above disturbs normal removal of the SC, such that initiation of removal of SC (at the restrictive temperature) tends to be followed by reinstallation with possible new insertions of functional RNs and consequent additional crossovers. These highly speculative notions rest on the assumption that incipient RNs have been more universally present than visible in zygotene preparations, but this is subject to experimental test. It has already been noted in a number of organisms that if treatments are to affect crossover frequency they must be applied by an early stage, perhaps no later than zygotene to early pachytene (see Maguire, 1972 for review). Possibly in regions heterozygous for a rearrangement, homologous synapsis tends to be initiated at a slower rate or later stage in a way that often assures the presence of an incipient RN. In preliminary comparisons of the frequency of occurrence of inversion loops at pachytene in squashed vs. spread material from a maize inversion heterozygote, Stack and Anderson (1986a) found far fewer loops in the spread material. Perhaps there is a difference in the quality of homologous synapsis in the rearranged segment.

CHIASMA MAINTENANCE UNTIL ANAPHASE I

In recent years we have come to appreciate that the process of crossing over does not inherently and of itself produce chiasmata capable of holding homologues together until they can be oriented to opposite poles at metaphase I for correct disjunction. Study of male meiosis in maize homozygous for the recessive mutant desynaptic has served to demonstrate that separate genetic controls exist for crossing over and chiasma maintenance (Maguire, 1978a). In

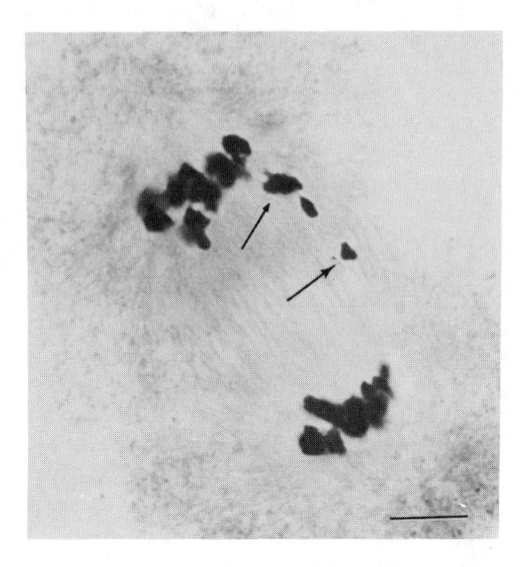

Fig. 7. Maize microsporoyte at anaphase I in material trisomic for chromosome 5 but otherwise normal. A trivalent configuration is disjoining in such a way that one chromosome 5 (not distinguishable) is progressing to the lower pole in step with the other dyads, one chromosome 5 is approaching the upper pole (small arrow) and the third has separated equationally, with one of its chromatids associated to the upper chromosome, presumably by a remnant chiasmate assocation, and the other (large arrow) apparently in progress toward the opposite pole. Magnification bar represents 10μm. With three homologues present and synapsis limited to two-by-two association at each point, separation of sister chromatids of one chromosome in the centromere region (so that they can interact separately with the spindle to give equational distribution) is consistent with expectation, if synapsis at pachytene normally contributes to later maintenance of sister chromatid association. A similar interpretation may be applicable for partial explanation of the behavior of a novel yeast chromosome which shows precocious sister centromere separation under some conditions (O'Rear and Rine, 1986). Some kinds of univalents, with varying frequency, seem to be extruded from the spindle and come to be located close to one polar region where they are often included in a telophase nucleus without separation of sister chromatids. It appears that until there is in-

desynaptic material, by diakinesis, bivalents heterozygous for a distal knob are often found to have separated to pairs of univalents, each with a knob-carrying and a knobless chromatid. From the frequency of univalents showing such equational separation which is indicative of prior exchange, it has been inferred that the crossover process is not affected by the mutant and that the genetic defect is instead responsible for lack of normal chiasma maintenance function. Similar interpretations are consistent with behavior found in a meiotic mutant of yeast (Wagstaff et al., 1982). That chiasmata (initially established by crossover events) are probably maintained by some process of sister chromatid association was implied early by Darlington (1932), but this has not often been noted by others (Maguire, 1974). Topological constraints to homologue separation during prophase I following crossing over and disintegration of the SC, are apparently not usually imposed by twisting of homologues about each other or by relational coiling of sister chromatids. Clear cytological preparations at diplotene and diakinesis show very little relational coiling of sister chromatids, but virtually universally, sisters appear to be closely associated throughout their length. Sister chromatids also often appear associated along their length at mitotic prophase and metaphase, (see Jackson, 1985; Hadlaczky et al., 1986 for examples) but more weakly so. Several differing possibilities which have been postulated for the nature of pronounced sister chromatid cohesiveness at meiotic prophase and metaphase I have not been ruled out or critically evaluated. For example, they may be associated by the hydrogen bonds of the DNA helix at positions not yet replicated (Maguire, 1978a). Even after zygotene Hotta and Stern (1976) have reported that gaps are left unreplicated until still later in meiosis. It cannot be readily estimated whether the hydrogen bonding of such very short unreplicated DNA segments could provide association of sister chromatids at prophase and metaphase I adequate for chiasma maintenance. Alternatively, sister chromatids may be associated by special binding substance in place between them at these stages (Maguire, 1974). A variety of circumstantial evidence has been reported that the SC, while it is in place, may somehow provide for the subsequent function of sister chromatid cohesiveness, during the period between pachytene and anaphase I, after its own integrity has been eliminated. This evidence consists of suggestive correlations of occurrence of synapsis through pachytene (homologous or nonhomologous) with instances of sister chromatid cohesiveness at later meiotic stages, and conversely of lack of synapsis with later failure of sister chromatid cohesiveness (Maguire, 1978a, 1978b, 1979, 1982a, 1985b). The systems studied

tervention of spindle interaction under such conditions, sister chromatids tend to remain together, even though their centromeres may be already capable of in dependent and opposite spindle orientation (Maguire, 1979).

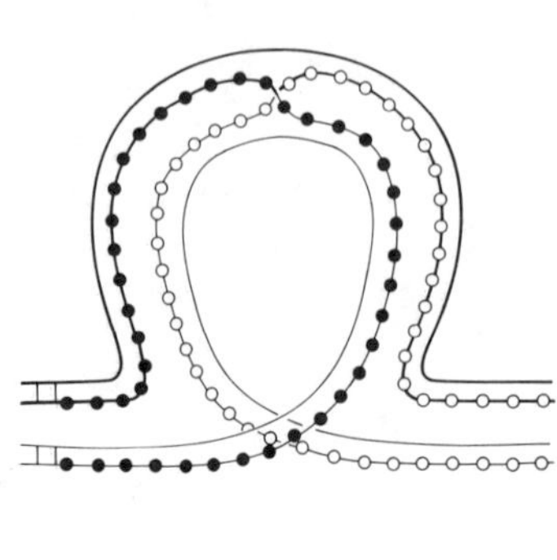

a

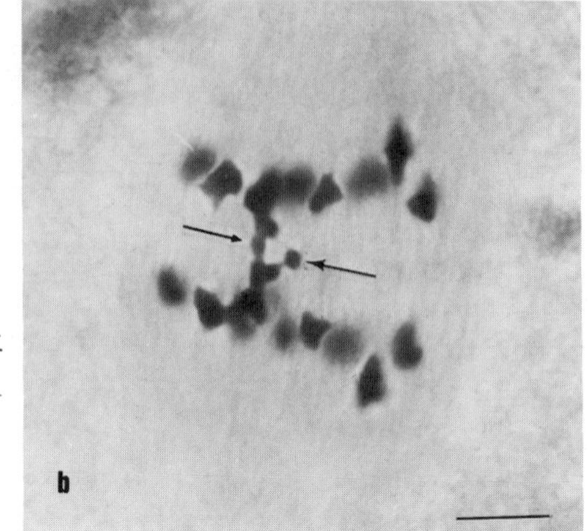

b

Fig. 8a. Diagram of configuration expected at pachytene in a chromosome arm heterozygous for a paracentric inversion, with a single crossover within the inversion. The sister chromatids of one homologue are represented by heavy lines, the sister chromatids of the other homologue by light lines. The X-shaped intersection represents a crossover event. Centromeres are represented by squares at the left in each homologue. The element destined to become a bridge when centromeres move to opposite poles at anaphase I has been marked with closed circles; the element destined to become a fragment at anaphase I has been marked with open circles. To the extent that generalized sister chromatid cohesiveness has been present and is not yet disrupted at anaphase, the fragment will tend to be associated with the normal (non-bridge) chromatids moving to opposite poles (without additional chiasmate association). b Microsporocyte in maize heterozygous for distal paracentric inversion In 5083 at early anaphase I. A crossover within the inverted region has given rise to a bridge (small arrow) and a fragment (large arrow). The fragment is clearly associated with a non-bridge chromatid moving to the lower pole and faintly associated with the corresponding non-bridge chromatid moving to the upper pole. Magnification bar represents 10μm. These associations are consistent with expectations of sister chromatid cohesiveness as indicated in a, and even though they appear chiasmate, the frequency of their occcurence at early anaphase is much higher than expectation of frequency of a second crossover event in the region distal to the inversion. Assuming that the occurrence of a crossover event in the inverted region implies that there was homologous synapsis there, these findings conform to expectations concerned chromosome behavior of trisomics (Fig. 7), meiotic mutants, translocation heterozygotes, and acentric fragments derived from crossovers within heterozygous paracentric inversions (Fig. 8). Material in which the SC has been prematurely removed by ethanol treatment also shows a tendency for separation of sister chromatids prematurely at prophase I (Maguire, 1976). Direct E.M. observations of bivalents at metaphase I in several orthopterans have revealed SC component-like remnants seemingly deployed between sister chromatids along their length in a manner which could serve to hold bivalents together at chiasmata (Moens and Church, 1979). An additional possible mechanism for sister chromatid cohesiveness has recently been suggested on the basis of reports that in the course of normal DNA replication catenation may be routinely formed at the position of meeting of replication forks (Murray and Szostak, 1985). Where such catenation is present, it may require the function of a topoisomerase activity for its resolution; sister chromatids would then tend to be held together at replication fork junctions as long as the topoisomerase activity is somehow inhibited. It is conceivable that at meiosis topoisomerase activity tends to be inhibited by chromatid-bound proteins at the crucial stages. Interestingly, it also appears that small artificial chromosomes in yeast, which consist mostly of lambda DNA together with yeast ARS and centromeres, Tetrahymena telomeres (modified by yeast) and several yeast genetic markers, do not have chiasma maintenance normal for yeast following their (relatively rare) involvement in crossing over (Dawson et al., 1986). Several explanations have been offered for this effect. It could be due to the absence in the artificial chromosomes of special DNA sequences which may be required for chiasma maintenance, or to the very short extent available for sister chromatid cohesiveness, whatever the mechanism for that cohesiveness. Crossovers near chromosomes ends in *Drosophila* seem to be less likely to ensure proper segregation than more proximal ones (Carpenter, 1973), and Fu and Sears (1973) found evidence for loss of chiasmate association in wheat (*Triticum aestivum*) where a limited extent of distal homology was present. An additional possible mechanism for sister chromatid cohesiveness has been emphasized by recent studies of behavior of mouse (*Mus musculus*) chromosomes in cultured cells, which point to possible assocations at heterochromatic regions as a mechanism for sister chromatid cohesiveness during mitotic prophase and metaphase (Lica et al.,

←

of the model that establishment of sister chromatid cohesiveness may be in part a function of the SC. More complex configurations have also been analyzed, with findings which are consistent with these postulates as well as with the supposition that chiasmata are not maintained simply as a result of installation of binding substance at crossover sites (Maguire, 1982a, 1985b).

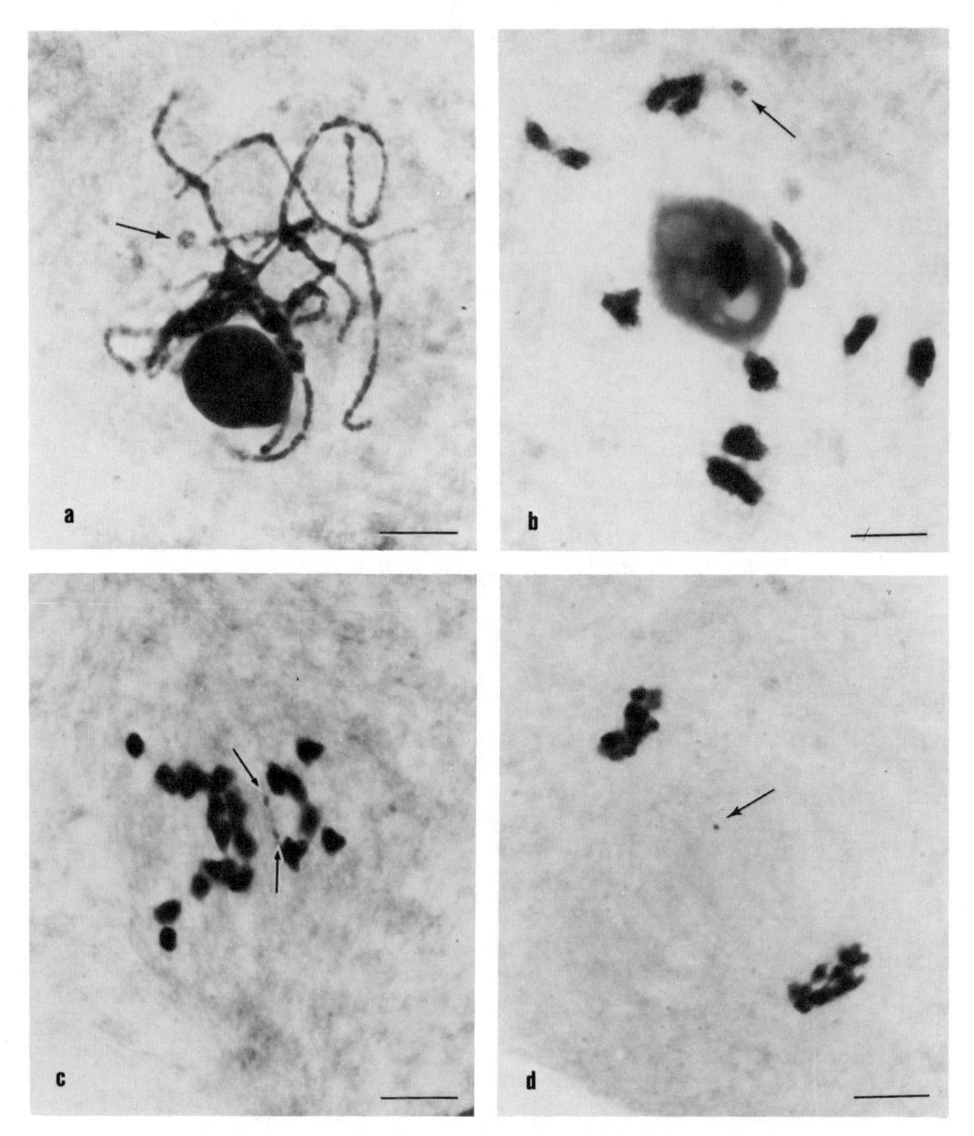

Fig. 9. Photomicrographs of maize microporocytes containing a supernumerary tiny centric fragment (arrows). Seeds of this material were kindly supplied by E.H. Coe. The fragment is known to carry dominant chromosome 9 alleles for shrunken and bronze and was initially discovered by B. McClintock in the presence of the X component (McClintock, 1978). a. Pachytene, where the fragment consistently appears decondensed and does not synapse. b. Diakinesis with the fragment still relatively decondensed. c. Anaphase I with the sister centromeres of the now condensed fragment each

1986). If such an association generally exists, it might be somehow enhanced at meiosis.

MAINTENANCE OF SISTER CENTROMERE ASSOCIATION UNTIL ANAPHASE II

If normal disjunction is to occur at anaphase of meiosis II, sister centromeres must somehow be held together until this stage, a fact which reflects one of the crucial differences between meiosis I and ordinary mitotic division. It is unlikely that centromeres are normally not replicated until this stage, as is sometimes suggested. Possibly the most compelling evidence against that suggestion is the fact that sister centromeres of univalent chromosomes have often been observed to interact independently with the spindle at anaphase I. Frequently, these are seen as laggards which separate equationally and move to opposite poles at anaphase I, but the sister chromatids of univalents may also move to opposite poles in step with normal dyads where they are less conspicuous. A special study was performed to check whether absence of SC in the centromere region at pachytene is correlated with early separation of sister centromeres and failure of normal anaphase II disjunction (Maguire, 1982b). This involved observation of meiotic behavior in heterozygotes for each of three different maize pericentric inversions which differed with respect to the usual mode of pairing at pachytene. In two of these, synapsis of the centromere region was usually found, while in the third there was substantial frequency of cells with apparent synaptic failure of the centromere containing inverted region at pachytene. Synapsis and crossing over generally occurred in the uninverted distal chromosome regions so that bivalents generally oriented normally at metaphase I even in the inversion heterozygote with frequent pachytene synaptic failure of the centromere region. But at anaphase II in that case only there was substantial frequency of cells which contained two lagging monads in the plate region of the spindle, and where cells could be identified as sisters, sister cells showed identical behavior at anaphase II. This is interpreted as suggestive evidence that synapsis of the centromere region may indeed be important to provision for sister centromere association until anaphase II. There is some additional support for this notion in recent

---

interacting with the spindle so that its sister chromatids are moving to opposite poles, in step with the normal dyads. _d_. Anaphase II with monad of fragment lagging in the plate region. Magnification bar represents 10μm. These observations are consistent with the interpretation that capability for separate interaction of sister centromeres with the spindle at anaphase I is correlated with prior failure of synapsis of the centromere region at pachytene.

observations (Maguire, unpublished) of the meiotic behavior of a tiny centric fragment in maize which has not been seen to synapse at pachytene, frequently separates equationally at meiosis I, moving for the most part in step with normal dyads, and then is left lagging in the metaphase II plate region (Fig. 9).

CONCLUDING REMARKS

The relatively unobtrusive techniques of genetic manipulation and cytological observation of chromosome behavior can contribute importantly to the definition of ultrastructural and molecular properties to be sought in ultimate explanations. Findings have suggested some general shapes of possible solutions for long standing problems. With respect to the most intractable of these, the mechanisms of alignment pairing, chiasma interference, and chiasma and dyad maintenance until correct disjunction is established, the following ideas seem promising as working hypotheses: (1) Alignment pairing of homologues may be initiated by chromosome movements inspired by their relatively passive interactions with components of the cytoskeleton as well as with nuclear skeleton and matrix; these motions may facilitate the meeting of specialized chromosomal pairing centers. (2) The determination of crossover sites (with consequent nonrandom positioning and chiasma interference) may result from the mode of deployment of the SC in conjunction with timely presence of functioning RNs. (3) In addition to its crossover related function it may be a function of the SC to contribute to provision for strong meiotic sister chromatid cohesiveness which seems to be responsible for chiasma maintenance until anaphase I, and possibly also for dyad maintenance until anaphase II (both of which are necessary for normal disjunction). These sketchy models are simplistic and undoubtedly naive, but they are consistent with a battery of information from a wide variety of sources; they are set forth here with the hope of inspiring further experimentation and observation. It is worth remembering that Mendel first deduced the pattern of meiotic division on the basis of genetic experiments.

ACKNOWLEDGEMENTS

This work has been supported by PHS grant GM19582. Seeds have been very generously supplied by G.G. Doyle, D.S. Robertson, E.H. Coe and the Maize Genetic Cooperation Stock Center.

LITERATURE CITED

ANDERSON, L. and STACK, S., 1986, Synaptic adjustment of inversion loops does not occur in _Zea mays_., _J. Cell Biol_., 103: 493a.

ASHLEY, T., 1979, Specific end-to-end attachment of chromosomes in Ornithogalum virens, J. Cell Sci., 38: 357-367.

BECKETT, J. B., 1978, B-A translocations in maize. I. Use in locating genes by chromosome arms, J. Hered., 69:27-36.

BENNETT, M. D., 1982, Nucleotypic basis of the spatial ordering of chromosomes in eukaryotes and the implications of the order for genomic evolution and phenotypic variation, in: "Systematics Association Special Volume No. 20: Genome Evolution," ed. G.A. Dover and R.B. Flavell, Academic Press, New York. pp. 239-261.

BENNETT, M.D., TOLEDO, L.A., STERN, H., 1979, The effect of colchicine on meiosis in Lilium speciosum cv. 'Rosemede', Chromosoma, 72: 175- 189.

BROWN, W.V., and STACK, S.M., 1968, Somatic pairing as a regular preliminary to meiosis, Bull. Torrey Bot. Club, 95: 369-378.

BURNHAM, C.R., STOUT, J.T., WEINHEIMER, W.H., KOWLES, R.V., and PHILLIPS, R.L., 1972, Chromosome pairing in maize, Genetics, 71: 111-126.

CALLOW, R.S., 1984, Comments on Bennett's model of somatic chromosome disposition, Heredity, 54: 171-177.

CARPENTER, A.T.C., 1973, A meiotic mutant defective in distributive disjunction in Drosophila melanogaster, Genetics, 73: 393-428.

CARPENTER, A.T.C., 1975, Electron microscopy of meiosis in Drosophila melanogaster, females. II. The recombination nodule-A recombination-associated structure at pachytene? Proc. Natl. Acad. Sci. USA, 72: 3186-3189.

CARPENTER, A.T.C., 1979, Recombination nodules and synaptonemal complex in recombination-defective females of Drosophila melanogaster, Chromosoma, 75: 259-292.

CARPENTER, A.T.C., 1984, Meiotic roles of crossing-over and gene conversion, Cold Spring Harb. Symp. Quant. Biol., 49: 23-29.

CHOVNICK, A., 1973, Gene conversion and transfer of genetic information within the inverted region of inversion heterozygotes, Genetics, 75: 123-131.

DARLINGTON, C.D., 1932, Recent advances in Cytology, P. Blakiston's Son and Co., Philadelphia.

DAVIDOW, L.S., and BEYERS, B., 1984, Enhanced gene conversion and post-meiotic segregation in pachytene arrested Saccharomyces cerevisae, Genetics, 106: 165-183.

DAWSON, D.S., MURRAY, A.W., and SZOSTAK, J.W., 1986 An alternative pathway for meiotic chromosome segregation in yeast, Science, 234: 713-717.

DE BONI, U., and MINTZ, A.H., 1986, Curvilinear three-dimensional motion of chromatin domains and nucleoli in neuronal interphase nuclei, Science, 234: 863-866.

DONNELLY, G.M. and SPARROW, A.H., 1965, Mitotic and meiotic chromosomes of Amphiuma, J. Hered, 56: 90-98.

DOYLE, G.G., 1979, The allotetraploidization of maize. Part 1. The physical basis - differential of pairing affinity, Theoret. Appl. Genet., 54: 103-112.

DRISCOLL, C.J., and DARVEY, N.L., 1970, Chromosome pairing: Effect

of colchicine on an isochromosome, Science, 169: 290-291.
DUSTIN, P., 1978, Nuclear and cytoplasmic shaping in spermatogenesis, in: "Microtubules", Springer-Verlag. Berlin, Heidelberg, New York, pp 232-238.
EGEL-MITANI, M., OLSON, L.W., and EGEL, R., 1982, Meiosis in Asperigillus nidulans: Another example for lacking synaptonemal complexes in the absense of crossover interference, Hereditas, 97: 179-187.
FU, T.K. and SEARS, E.R., 1973, The relationship between chiasmata and crossing over in Triticum aestivum, Genetics, 75: 231-246.
GIBSON, J. B. and MOSES, M.J., 1986, Effects of vinblastine sulfate on the synaptonemal complex in Mus musculus, Genetics, 113: s66.
GILLIES, C.B., 1979, The relationship between synaptonomal complexes, recombination nodules and crossing over in Neurospora crassa bivalents and translocation quadrivalents, Genetics, 91: 1-17.
GILLIES, C.B., 1983, Ultrastructural studies of the association of homologous and non-homologous parts of chromosomes in the mid-prophase of meiosis in Zea mays, Maydica, 28: 265-287.
GILLIES, C.B., 1984, The synaptonemal complex in higher plants. CRC Critical Reviews in Plant Sciences 2: 81-116.
HABER, J.E., THORNBURN, P.C., and ROGERS, D., 1984, Meiotic and mitotic behavior of dicentric chromosomes in Saccharomyces cerevisiae, Genetics, 106: 185-205.
HADLACZKY, G., WENT, M., and RINGERZ, N.R., 1986, Direct evidence for non-random localization of mammalian chromosomes in interphase nuclei, Exp. Cell Res., 167: 1-15.
HARGREAVES, A.J., WANDOSEL, F., and AVILA, J., 1986, Phosphorylation of tubulin enhances its interactions with membranes, Nature, 323: 827- 828.
HAWLEY, R.S., 1980, Chromosomal sites necessary for normal levels of meiotic recombination in Drosophila melanogaster. I. Evidence for and mapping of the sites, Genetics, 94: 625-646.
HEPLER, P.K. and WOLNIAK, S.M., 1984, Membranes in the mitotic apparatus: their structure and function, Int. Rev. Cytol., 90: 169-238.
HOTTA, Y. and SHEPARD, J., 1973, Biochemical aspects of colchicine action on meiotic cells, Molec. Gen. Genet, 122: 243-260.
HOTTA, Y. and STERN, H., 1976, Persistent discontinuities in late replicating DNA during meiosis in Lilium, Chromosoma, 55: 171-182.
HOTTA, Y., TABATA, S., and STERN, H., 1984, Replication and nicking of zygotene DNA sequences. Control by a meiosis-specific protein, Chromosoma, 90: 243-253.
HUBERT, J., and BOURGEOIS, C.A., 1986, The nuclear skeleton and the spatial arrangement of chromosomes in the interphase nucleus of vertebrate somatic cells, Human Genetics, 74: 1-15.
JACKSON, R.C., 1985, Mitotic instability in Haplopappus: Structural

and genetic causes, Amer. J. Bot., 72: 1452-1457.
JOHN, B., 1976, Myths and mechanisms of meiosis, Chromosoma, 54: 295-325.
JONES, G.H., and CROFT, J.A., 1986, Surface spreading of synaptonemal complexes in locusts II. Zygotene pairing behavior, Chromosoma, 93: 489-495.
JURICEK, D.K., 1975, Non-random chromosome distribution in radial metaphases from the Chinese hamster, Chromosoma, 50: 313-326.
KEZER, J., and MACGREGOR, H.C., 1971 A fresh look at meiosis and centromeric heterochromatin in the red-backed salamander, Plethodon cinereus cinereus (Green), Chromosoma, 33: 146-166.
KING, R.C., 1970, The meiotic behavior of the Drosophila oocyte, Int. Rev. Cytol., 28: 125-168.
LEWIN, R., 1981, Do chromosomes cross talk? Science, 214: 1334-1335.
LICA, L.M., NARAYANSWAMI, S., and HAMKALO, B.A., 1986, Mouse satellite DNA, centromere structure and sister chromatid pairing, J. Cell Biol., 103: 1145-1151.
LOIDL, J., 1986, Synaptonemal complex spreading in Allium. II Tetraploid A. vineale., Can. J. Genet. Cytol., 28: 754-761.
LOIDL, J., and JONES, G.H., 1986, Synaptonemal complex spreading in Allium I. Triploid A. sphaerocephalon, Chromosoma, 93: 420-428.
LU, B.C. and RAJU, N.B., 1970, Meiosis in Coprinus II. Chromosome pairing and the lampbrush diplotene stage of meiotic prophase, Chromosoma, 29: 305-316.
MAGUIRE, M.P., 1965, The relationship of crossover frequency to synaptic extent at pachytene in maize, Genetics, 51: 23-40.
MAGUIRE, M.P., 1966, The relationship of crossing over to chromosome synapsis in a short paracentric inversion, Genetics, 53: 1071-1077.
MAGUIRE, M.P., 1968, The effect of synaptic partner change on crossover frequency in adjacent regions of a trivalent, Genetics, 59: 381-390.
MAGUIRE, M.P., 1972, The temporal sequence of synaptic initiation, crossing over and synaptic completion, Genetics, 70: 353-370.
MAGUIRE, M.P., 1974, The need for a chiasma binder, J. Theoret. Biol., 45:485-487.
MAGUIRE, M.P. 1976, The effect of ethanol on meiotic chromosome behavior in maize, Caryologia, 29: 41-47.
MAGUIRE, M.P., 1978a, Evidence for separate genetic control of crossing over and chiasma maintenance in maize, Chromosoma, 65: 173-183.
MAGUIRE, M.P., 1978b, A possible role for the synaptonemal complex in chiasma maintenance, Exp. Cell Res., 112: 297-308.
MAGUIRE, M.P., 1979, An indirect test for a role of the synaptonemal complex in the establishment of sister chromatid cohesiveness, Chromosoma, 70: 313-321.
MAGUIRE, M.P., 1981, A search for the synaptic adjustment phenomenon in maize, Chromosoma, 81: 717-725.
MAGUIRE, M.P., 1982a, The mechanism of chiasma maintenance. A study based upon behavior of acentric fragments produced by

crossovers in heterozygous paracentric inversions, Cytologia, 47: 699-711.
MAGURIE, M.P., 1982b, Evidence for a role of the synaptonemal complex in provision for normal chromosome disjunction at meiosis II in maize, Chromosoma, 84: 675-686.
MAGUIRE, M.P., 1983a, Homologous chromosome pairing remains an unsolved problem: A test of a popular hypothesis, Genetics, 104: 173-179.
MAGUIRE, M.P., 1983b, Chromosome behavior at premeiotic mitosis in maize, J. Hered., 74: 93-96.
MAGUIRE, M.P., 1983c, Homologue pairing and synaptic behavior at zygotene in maize, Cytologia, 48: 811-818.
MAGUIRE, M.P., 1984, The mechanism of meiotic homologue pairing, J. Theoret. Biol., 106: 605-615.
MAGUIRE, M.P., 1985a, Crossover frequencies within paracentric inversions in maize: the implications for homologue pairing models, Genet.Res., 46: 273-278.
MAGUIRE, M.P., 1985b, Evidence on the nature and complexity of the mechanism of chiasma maintenance in maize,. Genet. Res., 45: 37-49.
MAGUIRE, M.P., 1986, The pattern of pairing that is effective for crossing over in complex B-A chromosome rearrangements in maize. III. Possible evidence for pairing centers, Chromosoma, 94: 71-85.
MCCLINTOCK, B., 1978, Mechanisms that rapidly reorganize the genome, Stadler Genet. Symp., 10: 25-47.
MCDERMOTT, A., 1971, Human male meiosis: Chromosome behaviour at pre-meiotic and meiotic stages of spermatogenesis, Can. J. Genet. Cytol., 13: 536-549.
MESELSON, M.S., and RADDING, C.M., 1975, A general model for genetic recombination, Proc. Natl. Acad. Sci., USA, 72: 358-361.
MOENS, P.B., 1973, Mechanisms of chromosome synapsis at meiotic prophase, Int. Rev. Cytol., 35: 117-134.
MOENS, P.B., and CHURCH, K., 1979, The distribution of synaptonemal complex material in metaphase I bivalents of Locusta and Chloealtis (Orthoptera: Acrididae), Chromosoma, 73: 247-254.
MOSES, M.J., 1977, Microspreading and the synaptonomal complex in cytogenetic studies, Chromosomes Today, 6: 71-82.
MOSES, M.J., POORMAN, P.A., RODERICK, T.U., and DAVISSON, M.T., 1982, Synaptonemal complex analysis of mouse chromosomal rearrangements. IV. Synapsis and synaptic adjustment in two paracentric inversions. Chromosoma, 84: 457-474.
MURRAY, A.W., and SZOSTAK, J.W., 1985, Chromosome segregation in mitosis and meiosis, Ann. Rev. Cell Biol., 1: 289-315.
NEUFFER, M.D., JONES, L., and ZUBER, M.S., 1968, The mutants of maize, Crop Science Society of America. Madison, Wisconsin.
NUR, U., 1968, Synapsis and crossing over within a paracentric inversion in the grasshopper, Camnula pellucida, Chromosoma, 25: 198-214.
O'REAR, J., and RINE, J., 1986, Precocious meiotic centromere

separation of a novel yeast chromosome, Genetics, 113: 517-529.
PARVINEN, M. and SÖDERSTRÖM, K.O., 1976, Chromosome rotation and formation of synapsis, Nature, 269: 534-535.
POORMAN, P.A., MOSES, M.J., DAVISSON, M.T., and RODERICK, T.H., 1981, Synaptonemal complex analysis of mouse chromosomal rearrangments. III. Cytogenetic observations on two paracentric inversions, Chromosoma, 83: 419-429.
PURO, J., 1973, Tricomplex, a new type of autosome complement in Drosophila melanogaster, Hereditas, 75: 140-143.
RAKHA, F.A., and ROBERTSON, D.S., 1970, A new technique for the production of A-B translocations and their use in genetic analysis, Genetics, 65: 233-240.
RASMUSSEN, S.W., and HOLM, P.B., 1980, Mechanics of meiosis, Hereditas, 93: 187-216.
ROSE, A.M., BAILLIE, D.L., and CURRAN, J., 1984, Meiotic pairing behavior of two free duplications of linkage group I in Caenorhabditis elegans, Mol. Gen. Genet., 195: 52-56.
ROTH, T.F. and ITO, M., 1967, DNA dependent formation of the synaptonemal complex at meiotic prophase, J. Cell Biol., 35: 247-255.
SALONEN, K., PARANKO, J., and PARVINEN, M., 1982 A colcemid-sensitive mechanism involved in regulation of chromosome movements during meiotic pairing, Chromosoma, 85: 611-618.
SCHWARTZ, D., 1953, The behavior of an X-ray-induced ring chromosome in maize, Am. Nat., 87: 19-28.
SEARS, E.R., 1968, Relationships of chromosomes 2A, 2B and 2D with their rye homoeologue, in Proc. 3rd. Int. Wheat Genet. Symposia, Aust. Acad. Sci., Canberra, pp 53-61.
SHEPARD, J., BOOTHROYD, E.R., and STERN, H., 1974, The effect of colchicine on synapsis and chiasma formation in microsporoytes of Lilium, Chromosoma, 44: 423-437.
SINGLETON, J.R., 1953, Chromosome morphology and the chromosome cycle in the ascus of Neurospora crassa, Am. J. Bot., 40: 124-144.
SNOW, R., 1979, Maximum likelihood estimation of linkage and interference from tetrad data, Genetics, 92: 231-245.
STACK, S.M., and ANDERSON, L.K., 1986a, The relation between synapsis, the synaptonemal complex, and crossing over in Zea mays, Am. J. Bot., 73: 687-688.
STACK, S., and ANDERSON, L., 1986b, Two-dimensional spreads of synaptonemal complexes from solanaceous plants III. Recombination nodules and crossing over in Lycopersicum esculentum (tomato), Chromosoma, 95: 253-258.
STACK, S.M., and BROWN, W.V., 1969, Somatic and premeiotic pairing of homologues in Plantago ovata, Bull. Torrey Bot. Club, 96: 143-149.
STACK, S.M., and SOULLIERE, D.L., 1984, The relation between synapsis and chiasma formation in Rhoeo spathacea, Chromosoma, 90: 72-83.
SYBENGA, J., 1966, Reciprocal translocation and preferential pairing

in auto-tetrploid rye, Chromosomes Today, 1: 66-70.
THERMAN, E. and SARTO, D.E., 1977, Premeiotic and early meiotic stages in the pollen mother cells of Eremurus and in human embryonic oocytes, Hum. Genet., 35: 137-151.
WAGSTAFF, J.E., KLAPHOLZ, S., and ESPOSITO, R.E., 1982, Meiosis in haploid yeast, Proc. Nat. Acad. Sci. USA, 79: 2986-2990.
WALTERS, M.S., 1970, Evidence on the time of chromosome pairing from the preleptotene spiral stage in Lilium longiflorum "Croft", Chromosoma, 29: 375-418.
ZICKLER, D., and SAGE, J., 1981, Synaptonomal complexes with modified lateral elements in Sordaria humana: Development of and relationship to the "recombination nodules", Chromosoma, 84:305-318.

DISTRIBUTION PATTERNS OF NONHISTONE CHROMOSOMAL PROTEINS ON POLYTENE CHROMOSOMES: FUNCTIONAL CORRELATIONS

S.C.R. Elgin, S.A. Amero, J.C. Eissenberg[°], G. Fleischmann[*], D.S. Gilmour, T.C. James[+]

Department of Biology
Washington University
St. Louis, MO 63130

INTRODUCTION

The polytene chromosomes of *Drosophila melanogaster* offer a unique opportunity to examine the structural organization of the genome of an organism for which an extensive genetic map has been established. In many of the larval tissues of Drosophila the cells become quite large; the DNA replicates, but rather than separating, the sister chromatids remain synapsed through several rounds (up to ten) of replication, resulting in large polytene chromosomes which are easily visible in the light microscope. The pattern of bands and interbands, indicative of the variation in packaging that occurs along the chromatin fiber, is highly reproducible and allows one to identify specific loci. This pattern has been correlated with the extensive genetic map which has been delineated over the last several decades through the analysis of deletions, insertions, and rearrangements. With the development of *in situ* hybridization techniques for polytene chromsomes by Pardue et al. (1970) and the advent of gene cloning through recombinant DNA techniques, it has become possible to locate the positions of specific genes and to survey the distribution of repetitious elements with considerable accuracy. As a consequence, we probably know more about the organization of the *Drosophila melanogaster* genome than about that of any other eukaryote.

[°]Present address: Department of Biochemistry, St. Louis University Medical Center, St. Louis, MO 63104, USA.
[*]Present address: Ruhr-Universitat Biochem; Lehrstuhl für Biochemie; D-4630 Bochum 1, Germany
[+]Present address: Department of Molecular Biology; Wesleyan University; Middletown, CT 06457, USA

This system, then, is most attractive for attempts to sort out the functions of the nonhistone chromosomal proteins (NHC proteins). While there are only five types of histones (with some variants) required for the packaging of DNA into the basic 100 Å chromatin fiber, several hundred NHC proteins can be detected by two-dimensional gel electrophoresis (Peterson and McConkey, 1976). Some of these are enzymes of chromosomal metabolism, some are specific effectors of gene expression (analogous to those seen in prokaryotic cells), and some are thought to play a role in differential packaging of the eukaryotic genome. Assigning functions to the individual proteins is a difficult task, particularly for those that play a structural role. By using specific antibodies to detect a given protein, one can determine the distribution pattern of an NHC protein in the polytene chromosomes, and draw inferences regarding its role by correlation with the known characteristics of the antibody-associated loci.

The basic procedure we have used is diagrammed in Fig. 1. Salivary glands are dissected from third instar larvae and extraneous material removed. The glands are placed on a clean slide in a drop of 45% acetic acid, and the cells and nuclei are broken open by gentle movement of the coverslip. The chromosome arms are then spread by tapping on the cover slip with the eraser end of a pencil. Finally, the spread chromosomes are flattened by firm pressure on the cover slip. The slide is quickly frozen in liquid nitrogen, the cover slip is flipped off, and the slide is placed in Tris-buffered saline. Slides can be stored for several weeks in 67% glycerin-33% phosphate-buffered saline. Slides are "stained" by first incubating the chromosomes with a suitable dilution of an antibody directed against a chromosomal protein and then, after washing, with a secondary, fluorescein-conjugated antibody. Chromosomes are examined and photographed using phase contrast and fluorescence optics sequentially. (See Silver and Elgin, 1976, 1978 for a more detailed description of the procedure.)

As with any cytological technique, one must be aware of the potential problems of a given fixation protocol. 45% acetic acid, which is necessary to obtain the distinctive banding pattern needed for mapping, will extract some of the histones and other proteins from the polytene chromosomes. To avoid this problem, we carry out a fixation of the intact salivary gland with 2% formaldehyde prior to squashing when studying proteins of this type. Tests using antibodies to histone H1 (one of the most readily extracted of the basic chromosomal proteins) show that adequate fixation is achieved by this procedure. However, one must now be concerned that some antigenic determinants may be obscured or denatured by the fixation. Greatest confidence is possible when similar results are obtained using both of the above of fixation protocols. Despite these caveats, a cytological approach can be very efficient and very useful. It has recently been possible to confirm the cytological results by biochemical and genetic studies in several instances.

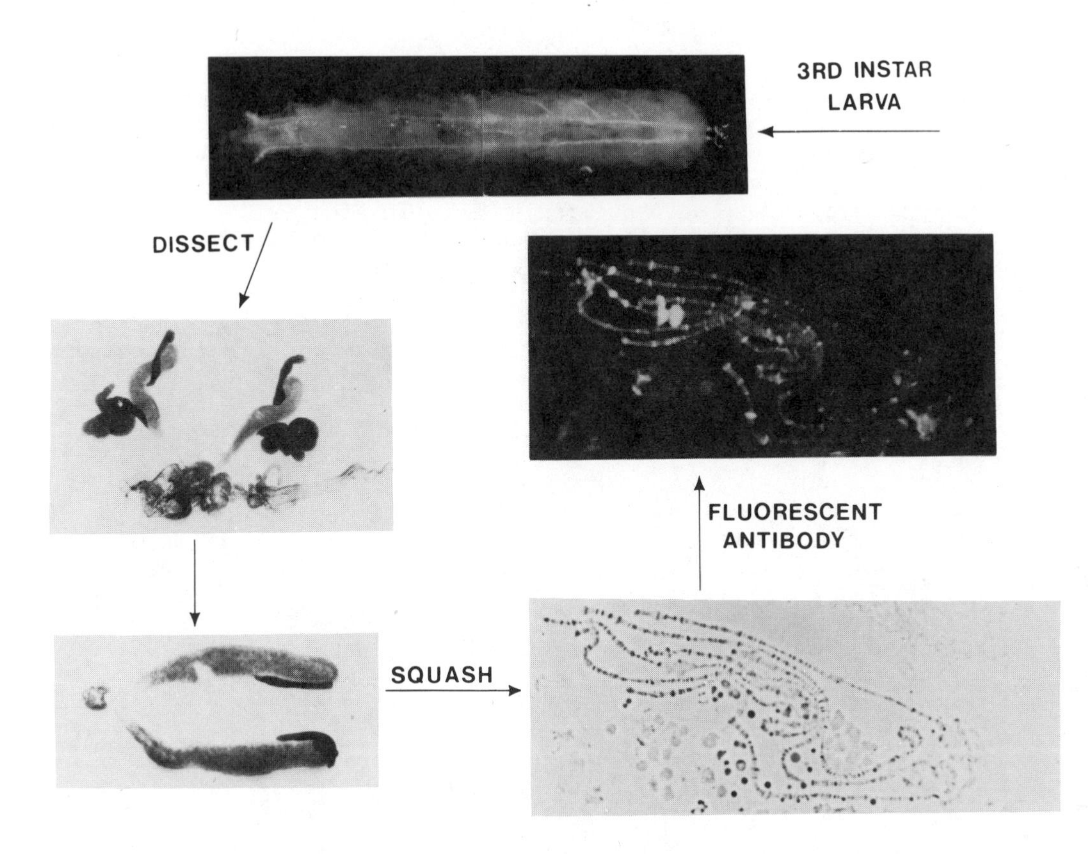

Fig. 1. Immunofluorescent Staining of Polytene Chromosomes of Drosophila. Illustration of steps in the dissection of salivary glands and preparation of stained chromosomes. See text for details.

In this brief review we will examine three cases that together provide a good example of the type of results one can currently obtain using this approach to the study of NHC proteins. First, we will consider the results obtained for topoisomerase I, an enzyme whose interactions with the genome have also been mapped by biochemical techniques; second, we will examine the results for a heterochromatin-specific protein; and third, we will consider the results obtained for several proteins which appear to be associated with active loci.

DNA TOPOISOMERASE I

Topoisomerase I (topo I), the nicking-closing enzyme, can relax negatively supercoiled DNA. The enzyme can serve as a "swivel" for double-stranded DNA, and has been postulated to play such a role in both transcription and replication of the genome. (See Wang, 1985, for a detailed review.) The distribution pattern of topo I on the Drosophila polytene chromosomes is shown in Fig. 2. Topo I is prominently associated with a subset of loci, including the nucleolus and all of the visible puffs. On heat shock, one observes a diminution of staining at the normally active sites, and a very intense staining at the (now highly active) heat shock loci (Fleischmann et al., 1984). This shift mimics the known changes in transcriptional activity and the previously observed shifts in the distribution pattern of RNA polymerase II. The nucleolus remains stained with topo I before and after heat shock; rRNA synthesis is active under both conditions.

The results indicate that the distribution pattern of topo I reflects the pattern of gene activity in the polytene chromosomes, and that changes in the topo I pattern can occur very quickly as the pattern of gene expression changes. Two subsequent biochemical studies have demonstrated that the shift in topo I distribution is accompanied by a shift in the pattern of topo-DNA interaction. First, Gilmour et al. (1986) have used a UV-photocrosslinking technique to examine the association of topo I with the Drosophila heat shock genes. In this experiment, normal and heat-shocked cells are briefly irradiated with UV light; this results in the crosslinking of a small percentage of the chromosomal proteins to the DNA. Noncovalent protein-DNA interactions are disrupted with detergent, and the protein-DNA conjugates are purified by CsCl centrifugation. After the DNA is cut with a restriction enzyme, the complexes are immunoprecipitated (using in this case the antibodies against Drosophila topo I) and the purified pellet and supernatant DNA fractions are size-separated on agarose gels. The partitioning of specific DNA fragments can be determined by hybridizing cloned DNA probes to Southern blots of the gels. Such an analysis has shown that topo I is associated with the heat shock genes after, but not before, activation. By use of appropriate restriction fragments the authors demonstrated that the association occurs with the DNA which is transcribed, and not with the surrounding non-transcribed spacer DNA (Gilmour et al., 1986).

A more detailed analysis can be obtained by making use of the drug camptothecin. This drug specifically inhibits topo I, increasing the stability of the transient covalent protein-DNA intermediate that forms during the relaxation reaction (Hsiang et al. 1985). Isolation of this covalent intermediate and subsequent purification of the DNA allows one to map the site of interaction by mapping the position of the nicks left in the DNA. Such a mapping experiment is shown in Fig. 3. An indirect end-labelling procedure has been used with strand-specific probes, allowing one to determine the pattern of nicks along each strand of the DNA.

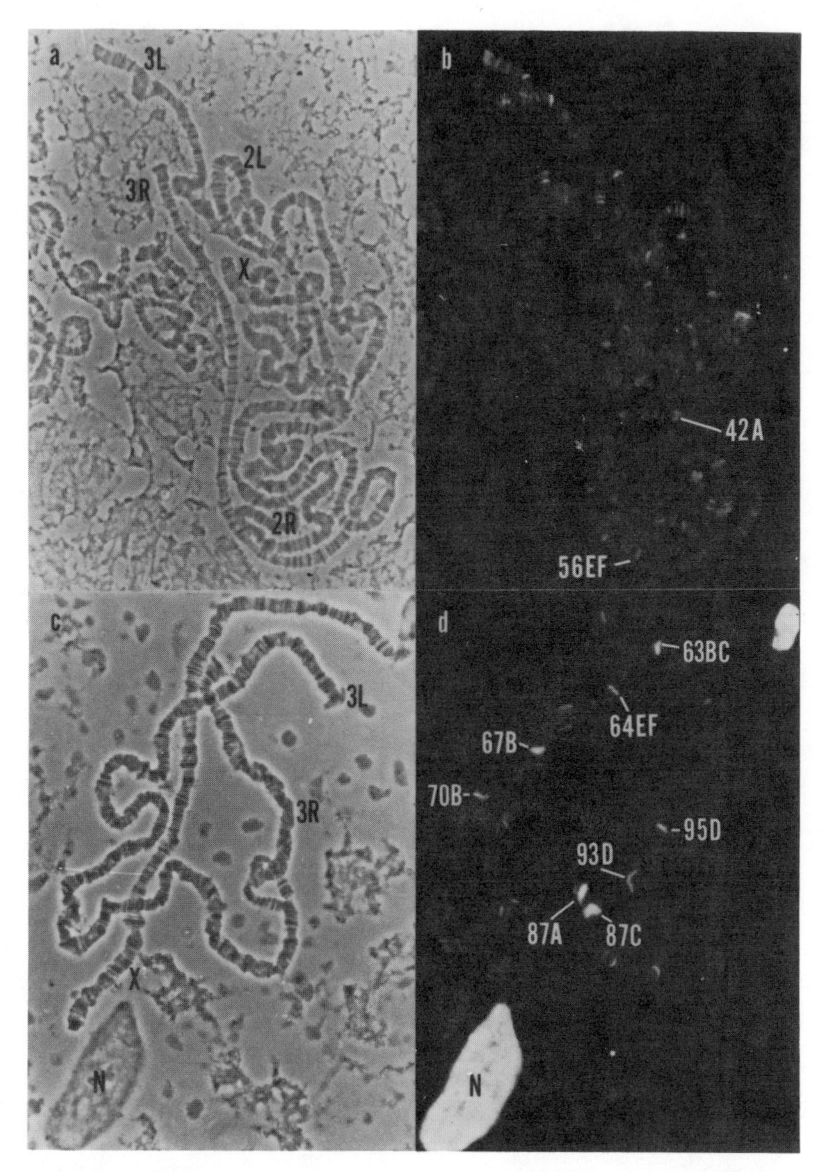

Fig. 2. Distribution Pattern of Topoisomerase I on Polytene Chromosomes. Chromosomes were obtained from third instar larvae grown at 25°C (a and b) and from larvae heat shocked at 37°C for 20 minutes prior to dissection (c and d). (a and c) Phase contrast images; (b and d) fluorescent images. N, nucleolus; X, 2L, 2R, 3L and 3R, chromosome arms; major heat shock loci are indicated in panel d. See Fleischmann et al. (1984) for further details.

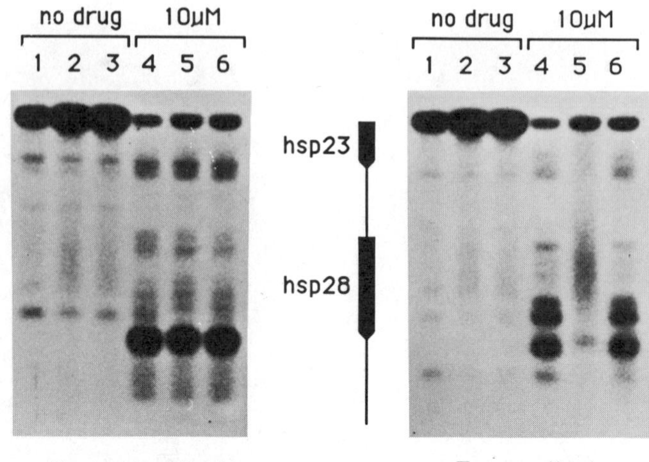

Fig. 3. Association of Topoisomerase I with the *hsp28* and *hsp23* Genes Mapped by Camptothecin-induced DNA Cleavage. Heat-shocked cells were treated with either 10 μM camptothecin or an equivalent volume of dimethyl sulfoxide for 3 min. at 36°C and then collected by a 5-min spin in a clinical centrifuge. Cells were lysed in 1% SDS, and a portion of the lysate was treated immediately with proteinase K; the DNA was phenol-chloroform extracted and ethanol precipitated. The remaining DNA was not treated with protease but extracted several times with chloroform followed by ethanol precipitation. The DNA was cut with *Eco*RI in the presence of 0.5 mM phenylmethylsulfonyl fluoride and then ethanol precipitated. Precipitates were dissolved in 0.5% SDS-50 mM Tris (pH 8.0)-10 mM EDTA, and half was treated with proteinase K. Both samples were again ethanol precipitated and then run on an alkaline gel, along with the sample treated immediately with proteinase K. Cuts on specific strands were detected with strand-specific probes. (Left panel) Cuts in the nontranscribed strand. Lanes: 1, 2, and 3, cells not treated with camptothecin; 4,5, and 6, cells treated with camptothecin. Lanes 1 and 4 are samples that were treated with proteinase K before the organic extraction. Lanes 2 and 5 are samples that were never treated with proteinase K. Lanes 3 and 6 are samples that were treated with proteinase K after restriction digestion, just before electrophoresis. The probe detects cleavage fragments whose 5′ ends were generated with camptothecin. (Right panel) The same blot as in panel A hybridized with a probe specific for the transcribed strand. This probe detects cleavage fragments whose 3′ ends were generated with camptothecin. From Gilmour and Elgin (1987) with permission.

Extensive nicking of the DNA of the heat shock genes is observed after, but not before, their activation. As shown, the nicking coincides with the region of transcription, and occurs on both the transcribed and nontranscribed strands. If the DNA is prepared without protease treatment, protein is found associated with the 3′, but not the 5′, end of the nick, confirming that the cleavage is a consequence of topo I activity. Interestingly, treatment with the inhibitor reduces the level of transcription, but does not appear to block the production of full-length transcripts (data not shown). This suggests that the topo I interactions with the DNA, being relatively transient, do not destabilize the polymerase or otherwise interfere with its continued progression along the template (Gilmour and Elgin, 1987). However, other interpretations are possible, and further work is needed to clarify this point.

Thus three independent experimental analyses support the conclusion that topo I is associated with active genes. Two of the techniques make use of antibodies; one does not. Each allows a different level of resolution. We conclude that topo I is a significant constituent of the transcriptionally active chromatin complex, available to relieve unconstrained torsional stress as needed. It is possible that topo I plays a role in "unfolding" the chromosome, perhaps in a disassembly and reassembly of the nucleosome array to facilitate polymerase passage. The results confirm the utility of immunofluorescence staining of polytene chromosomes to provide an assessment of a protein distribution pattern for the genome as a whole.

A HETEROCHROMATIN-ASSOCIATED PROTEIN

Given the obvious differences in the packaging of euchromatin and heterochromatin, one can anticipate differences in the population of chromosomal proteins present in these two fractions of the genome. We have recently prepared antibodies against a set of proteins which bind very tightly to DNA (extracted only at 1-2 M potassium thiocyanate). Four monoclonal lines have been identified which secrete antibodies recognizing a protein of 18 kD. Staining of the polytene chromosomes has shown this protein to be associated primarily with the α- and β-heterochromatin (James and Elgin, 1986) (Fig. 4). The few sites observed to be stained in the euchromatic arms are primarily sites previously classified as "intercalary heterochromatin" based on cytogenetic behavior. Staining of the fourth chromosome is also observed, an interesting observation in that this chromosome shows some heterochromatic behavior (see review by Hilliker et al., 1982).

In this case the antibodies obtained have been used to screen a λgt11 expression library constructed from Drosophila embryo cDNA. Recombinant DNA clones have been obtained, and a full-length cDNA clone has been sequenced. *In situ* hybridization indicates that the gene is encoded at locus 29A. This is of particular interest, as Griglatti and his colleagues (Sinclair et al., 1983) have mapped

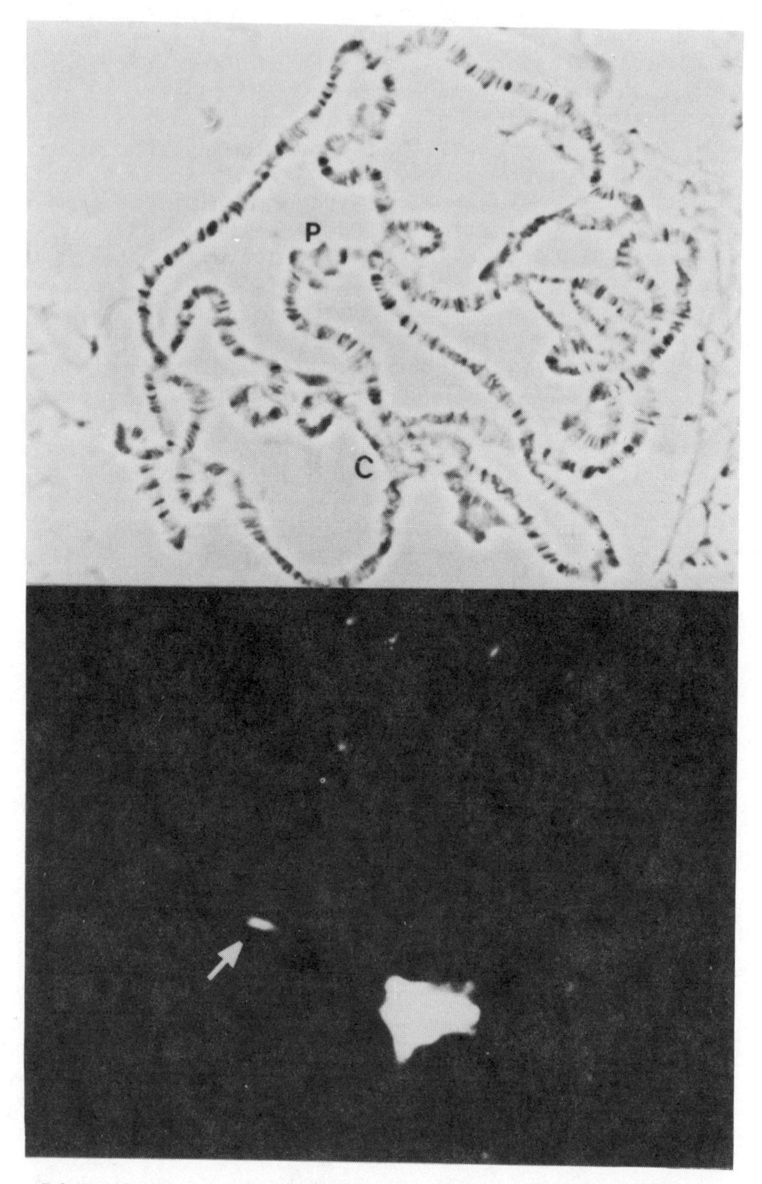

Fig. 4. Distribution Pattern of the 18kD Heterochromatin-associated Protein on Polytene Chromosomes. Staining is primarily of the chromocenter (C) and a few loci along the chromosome arms. The arrow points to a region of intercalary heterochromatin on chromosome arm 2L. There is no staining of puffs (p). Upper panel-phase contrast; lower panel-fluorescent images. See James and Elgin (1986) for further details.

the *Su(var)205* mutation to this region. This mutation, a homozygous lethal, is a suppressor of position effect variegation when heterozygous. Position effect variegation involves somatic inactivation of genes relocated to heterochromatin by chromosomal rearrangements or by transformation (reviewed by Spofford, 1976). Since the suppression effect appears to be mediated by heterochromatin formation, suppression of position effect variegation is a phenotype which might be anticipated for a hypomorphic or null mutation of a heterochromatin-associated protein gene. Interestingly, a cDNA clone of the 18kD protein detects an aberrant RNA in a northern blot of RNA from *Su(var)205* (Eissenberg, James and Elgin, unpublished). Using the recombinant DNA clones, we are in the process of obtaining a molecular map of the gene encoding the protein; we plan to characterize mutants of the gene and to manipulate the gene copies using P-element mediated transformation. This access to genetic techniques for the study of a structural chromosomal protein is essential to develop an understanding of its biological function.

PROTEINS ASSOCIATED WITH ACTIVE LOCI

In addition to the enzymes necessary for transcription, one could speculate that the proteins necessary for generating and maintaining appropriate (and different) chromatin structures and for the packaging and processing of the transcribed RNA might be found to be associated with active sites. In several instances we have observed that antibodies prepared against specific chromosomal proteins show a pattern of selective association with the set of loci catalogued by Ashburner (1972) as puffing (showing high levels of RNA transcription) at some time during the developmental period in which the chromosomes can be observed. The correlation is good, but not perfect; generally on the order of 80%-90% of the active/inducible loci are observed to be stained. In some instances the pattern is constant, suggesting that the protein under study is associated with the loci before, during, and after puffing (Silver and Elgin, 1977; Mayfield et al., 1979; Howard et al., 1981); in other cases the pattern appears to shift during development, suggesting an association limited to the actual period of transcription (S.A. Amero, V. Dietrich, and S.C.R. Elgin, unpublished observations). Interestingly, the patterns observed also differ in their generality. While in most of the above cited cases the antigen appears to be associated with the heat shock loci when these loci are active, there are exceptions. In two cases we have observed patterns of this general type in which the antigen is not associated with the heat shock loci, either before or after their activation by stress (Amero et al., unpublished observations; Elgin et al., 1987). This suggests that there are prominent differences in the mode of activation that can be identified using this cytological approach. Recent studies using monoclonal antibodies prepared against proteins extracted from *Drosophila melanogaster* embryo nuclei after a brief treatment with nucleases will illustrate this case. Two of these antibodies identified a

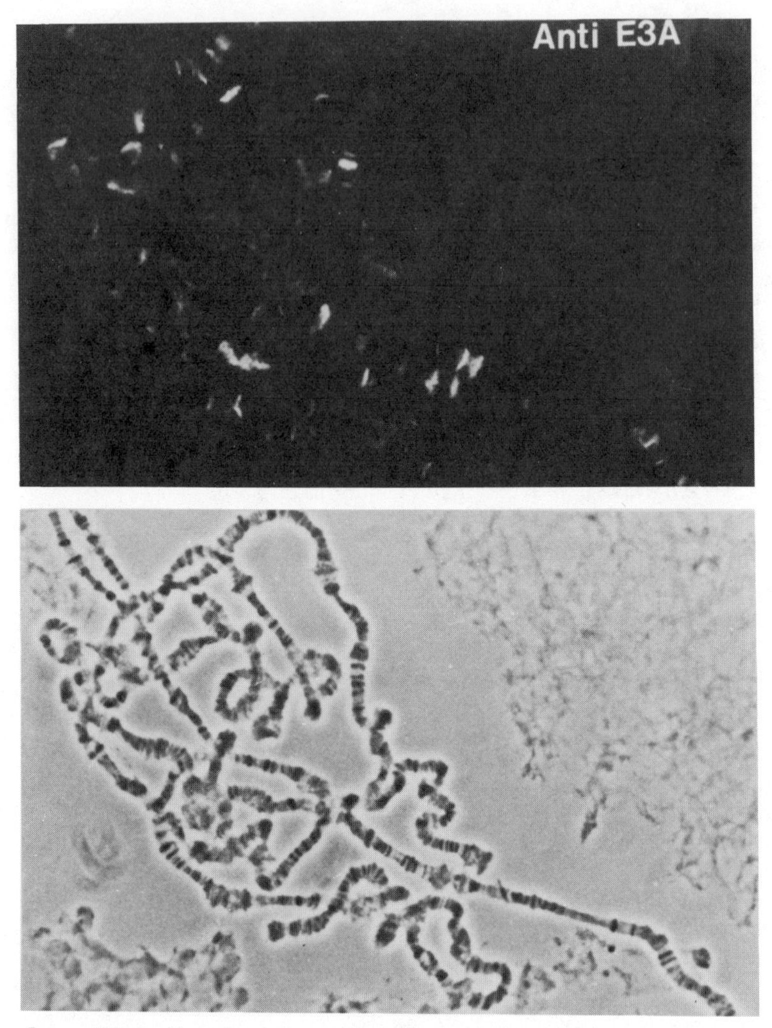

Fig. 5. Distribution Pattern of Antigen E3A on the Polytene Chromosomes. Major puffs are prominently stained. Upper panel-fluorescent image; lower panel-phase contrast image.

specific antigen of ca. 100 kD. These antibodies give the pattern of chromosome staining seen in Fig. 5. The developmentally regulated loci are stained when they are active (puffed); the staining appears closely correlated with activity, in that examination of some of the prominent sites in chromosomes from larvae of different stages indicates that these loci are not stained before puffing. In this case, the major heat shock loci are not stained, with the exception of 93D, which is also expressed under developmental regulation (S.A. Amero et al., unpublished observations). These antibodies are now being used to screen a λgt11 expression library as a next step in the characterization of this chromosomal protein.

CONCLUSIONS

These three examples illustrate how a cytological analysis can be used in conjunction with biochemical and genetic techniques to learn about the role of different NHC proteins. Since the cytological approach can be applied to structural proteins as well as those having specific DNA-binding or enzymatic activities, it should be very useful in our continuing effort to analyze this complex group of proteins.

REFERENCES

Ashburner, M., 1972, Puffing patterns in Drosophila melanogaster and related species, in : "Developmental Studies on Giant Chromosomes," W. Beerman, ed., Springer-Verlag, New York, pp. 101-151.

Elgin, S.C.R., Dietrich, V., Steiner, E.K., Mita, K., Olmsted, J.B., Allis, C.D., and Gorovsky, M.A., 1987, Selective staining of polytene chromosomes by antibodies specific for a Tetrahymena histone H2A variant, manuscript submitted.

Fleischmann, G., Pflugfelder, G., Steiner, E.K., Javaherian, K., Howard, G.C., Wang, J.C., and Elgin, S.C.R., 1984, Drosophila DNA topoisomerase I is associated with transcriptionally active regions of the genome, Proc. Natl. Acad. Sci. USA, 81: 6958-6962.

Gilmour, D.S. and Elgin, S.C.R., 1987, Localization of specific topoisomerase I interactions within the transcribed region of active heat shock genes by using the inhibitor camptothecin, Molec. Cell Biol., 7: 141-148.

Gilmour, D.S., Pflugfelder, G., Wang, J.C., and Lis, J.T., 1986, Topoisomerase I interacts with transcribed regions in Drosophila cells, Cell, 44: 401-407.

Hilliker, A.J., Appels, R., and Schalet, A., 1980, The genetic analysis of D. melanogaster heterochromatin, Cell, 21: 607-619.

Howard, G.C., Abmayr, S.M., Shinefeld, L.A., Sato, V.L., and Elgin, S.C.R., 1981, Monoclonal antibodies against a specific nonhistone chromosomal protein of Drosophila associated with active genes, J. Cell Biol., 88: 219-225.

Hsiang, Y.-H., Hertzberg, R., Hecht, S., and Lin, L.F., 1985, Camptothecin induces protein-linked DNA breaks via mammalian DNA topoisomerase I, J. Biol. Chem. 260: 14873-14878.

James, T.C., and Elgin, S.C.R., 1986, Identification of a nonhistone chromosomal protein associated with heterochromatin in Drosophila melanogaster and its gene, Molec. Cell Biol., 6: 3862-3872.

Mayfield, J.E., Serunian, L.A., Silver, L.M., and Elgin, S.C.R., 1978, A protein released by DNase I digestion of Drosophila nuclei is preferentially associated with puffs, Cell, 14: 539-544.

Pardue, M.L., Gerbi, S.A., Eckhardt, R.A., and Gall, J.G., 1970, Cytological localization of DNA complimentary to ribosomal

RNA in polytene chromosomes of Diptera, Chromosoma, 29: 268-290.

Peterson, I.L., and McConkey, E.H., 1976, Non-histone chromosomal proteins from HeLa cells: a survey by high resolution, two-dimensional electrophoresis, J. Biol. Chem., 251: 548-554.

Silver, L.M., and Elgin, S.C.R., 1976, A method for determination of the in situ distribution of chromosomal proteins, Proc. Natl. Acad. Sci. USA, 73: 423-427.

Silver, L.M., and Elgin, S.C.R., 1977, Distribution patterns of three subfractions of Drosophila nonhistone chromosomal proteins: possible correlations with gene activity, Cell, 11: 971-983.

Silver, L.M., and Elgin, S.C.R., 1978, Immunological analysis of protein distribution in Drosophila polytene chromosomes, in: "The Cell Nucleus," H. Busch, ed., Academic Press, New York. pp. 216-263.

Sinclair, D.A.R., Mothus, R.C., and Grigliatti, T.A., 1983, Genes which suppress position effect variegation in D. melanogaster are clustered, Mol. Gen. Genet., 191: 326-333.

Spofford, J.B., 1976, Position effect variegation in Drosophila, in: "The Genetics and Biology of Drosophila 1c," M. Ashburner and E. Novitski, eds., Academic Press, New York, pp. 955-1018.

Wang, J.C., 1985, Topoisomerase, Ann. Rev. Genet. 54: 665-698.

MOLECULAR MAPPING OF PLANT CHROMOSOMES

Steven D. Tanksley
Joyce Miller
Andrew Paterson
Robert Bernatzky

Department of Plant Breeding
252 Emerson Hall
Cornell University
Ithaca, NY 14853

Genetic maps have been constructed for a number of plant species. For maize and tomato, two of the better studied species, the maps are comprehensive and include several hundred markers each (O'Brien, 1986). It is an arduous task to construct a genetic map and their existence is a tribute to the dedication of the geneticists and breeders who have made it their labor. The techniques of molecular biology have recently offered a more rapid approach to generating genetic maps and in some ways these maps may prove to be more useful than their classical counterparts. The technique, often referred to as restriction fragment length polymorphism or "RFLP" analysis, involves cloning unique sequences of DNA from the nuclear genome. These clones are then used as radioactive probes to detect homologous sequences in plant DNA which has been cut with various restriction enzymes, separated on agarose gels and blotted onto nylon filters. Alleles are identified by differences in the size of the restriction fragments to which the probes hybridize (Fig. 1). The segregation of RFLP markers can be monitored in progeny from controlled genetic crosses and, by using standard genetic analysis, linkage groups can be constructed.

There are several considerations that must be taken into account when contemplating the construction of a genetic map using RFLP technology. Over the past several years, our laboratory has constructed a molecular linkage map of tomato

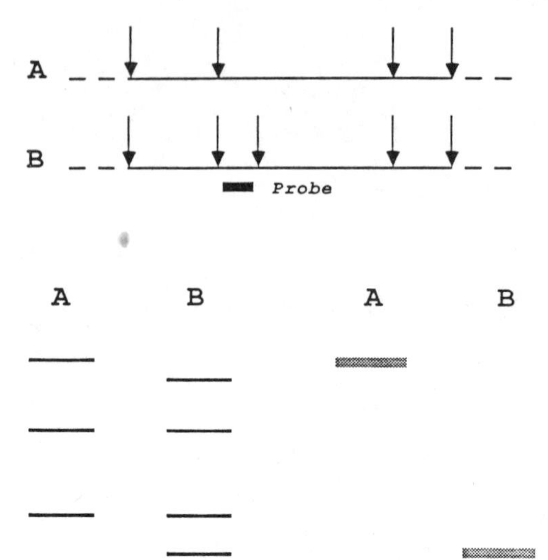

Fig. 1. Detection of restriction fragment length polymorphism (RFLP) using a cloned probe. DNA from two inbred lines (A and B) are cleaved (arrows) with a restriction endonuclease. Sequence differences between lines result in five cleavage sites in line B but only four in line A. Restriction fragments are separated according to size on an agarose gel (bottom left) and transferred to nylon membrane for probing. When nick-translated probe is added to filter it will hybridize to homologus sequences which can be detected by autoradiography (bottom right). In this case, the homologous sequence is found on different sized restriction fragments in the two inbred lines. The restriction fragment length polymorphism can then be monitored in segregating progeny to confirm allelism and locus copy number.

chromosomes. The lessons we have learned may be helpful to others conducting similar projects or for those who may be contemplating such action. Our research also provides additional insight into organization of genes in the tomato genome.

CLONING SINGLE COPY DNA

In most eukaryotes, the nuclear genome harbors a large proportion of DNA sequences which are repeated many times. Often these repeats are interspersed with unique sequence DNA in a manner that makes it difficult to isolate clones consisting entirely of single copy DNA. The proportion of repeated DNA and the extent to which it is interspersed with single copy DNA is generally a function of the overall DNA content of the species being studied (Flavell, 1980). For species with small genomes like Arabidopsis (haploid DNA content = 0.07 pg), the majority of nuclear DNA is single copy, uninterrupted by repeats (Pruitt and Meyerowitz, 1986). However, for species with large genomes such as wheat, unique sequences longer than a few hundred base pairs are rare (Flavell, 1980).

In order to do RFLP mapping, one must isolate clones free of repeated DNA. Two methods have been used to accomplish this task.

cDNA clones

Classical genetic studies and research at the molecular level both indicate that most plant genes are single or low copy (Tanksley and Pichersky, 1987). cDNA clones, derived from gene transcripts, are therefore a good source of single copy clones. In tomato, we have shown that approximately 60% of random cDNA clones correspond to single loci and can be used effectively as probes for RFLP mapping (Bernatzky and Tanksley, 1986a, Fig. 2).

Genomic clones

cDNA clones are limited by their length (they tend to be less than 1 kb) and, in some cases, their lack of perfect homology due to introns (cDNA's derive from processed transcripts and thus lack introns). In practical terms, the first limitation is more serious than the second. Autoradiography is one of the rate limiting steps in detecting single copy DNA with radioactive probes. The shorter the clone, the longer the required exposure time during autoradiography. Using standard nick-translation and a clone of 500 bp, exposure times range from 1 to 5 days. Ideally, one would like to have the results the same day. The amount of radioactivity that can be incorporated into a cloned sequence is proportional to its length; thus, longer clones yield shorter exposure times and are therefore preferable. To obtain a large supply of long clones (>2 kb), it is necessary to move from cDNA clones to genomic clones. However, for most plant species, the majority of random genomic clones are likely to carry repeated sequences, making them unsuitable for RFLP mapping. To get around this problem, one must devise a way to select for clones free of repeated DNA. In our laboratory, two methods have been used to accomplish this goal--one employed before cloning and the other after.

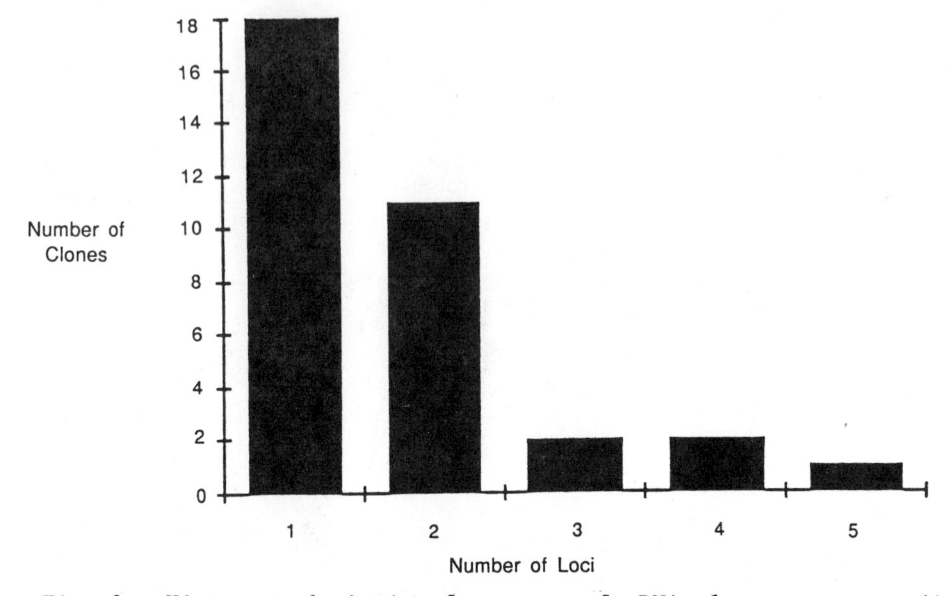

Fig. 2. Histogram depicting frequency of cDNA clones corresponding to various number of loci in tomato genome. Number of loci per clone was determined by Southern analysis in F2 population (from Bernatzky and Tanksley, 1986a).

Post-cloning selection. Large numbers of bacterial colonies can be grown directly on nylon hybridization filters -- each of the colonies harboring a plasmid into which plant DNA has been cloned. The colonies are lysed on the filter, the denatured plasmid DNA bound to the filter and the filter is then probed with nick-translated total nuclear DNA for 12-24 hours. The degree to which bound plasmids hybridize with their homologous counterpart in the radioactive nuclear DNA is a function of the copy number of that sequence in the genome. The concentration of highly repetitive DNA sequences is high enough to drive the hybridization reaction to near completion, but low copy sequences are at such a low concentration that little hybridization occurs during that period. Autoradiographic exposure reveals those colonies carrying plasmids with highly repetitive DNA (Fig. 3). In our hands, colony hybridization has been successful in eliminating cloned sequences present in more than 50 copies in the genome. The remaining clones must be checked via probing onto genomic DNA to determine which ones are actually single copy.

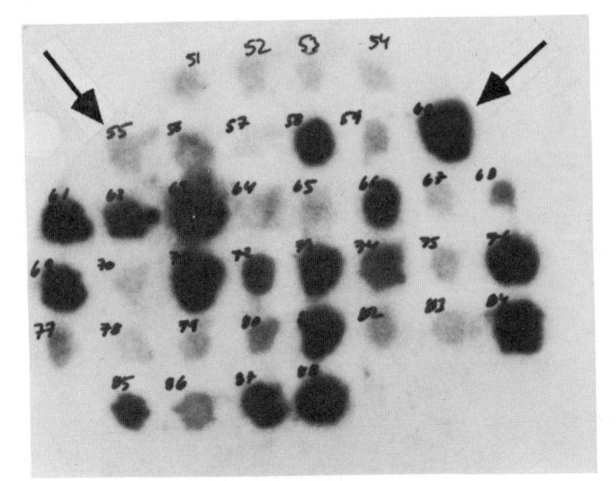

Fig. 3. Autoradiogram derived from colony hybridization experiment. Each spot represents a different tomato genomic clone. Colony filter was probed with nicktranslated total DNA. Areas of light signal (e.g. left arrow) indicate clones of sequences present in low copy in the nucleus. Areas of heavy signal (e.g. right arrow) represent clones of highly repeated sequences.

Pre-cloning selection. Much of the nuclear DNA in eukaryotes is C-methylated (Doerfler, 1983). Data from maize suggests that repeated DNA sequences are methylated to a greater degree than single copy DNA and by using restriction enzymes that are C-methylation sensitive for cloning, it is possible to produce libraries that are greatly enriched for single copy nuclear DNA (B. Burr, personal comm.). In tomato, we have screened libraries constructed with the C-methylation sensitive enzyme *PstI* and the methylation insensitive enzyme *EcoRI*. The fragments cloned were of the same size range (1.5-3.0 kb). Several hundred clones from each library were screened with nick-translated chloroplast DNA and mitochondrial DNA to remove any organellar clones. Subsequently they were screened with total nuclear DNA to identify high copy repeats. The data suggested that the *EcoRI*-generated library contains many more clones with repeated DNA (Miller and Tanksley, unpubl.). To verify these results, approximately 50 random clones from each library were nick-translated and hybridized to filters containing restricted nuclear DNA. The results verify the colony hybridization experiments and indicate that approximately 92% of the *PstI* clones correspond to sequences present only once in the genome whereas the majority of the *EcoRI* clones contained repeated elements (Fig. 4). Many of these single copy genomic clones have now been mapped onto the tomato linkage map (Fig. 5).

The ability to efficiently select single copy genomic clones allows one to work with clones much larger than could be obtained from cDNA libraries. We are presently using single copy genomic clones that range in size from 2 - 5 kb. Using clones of this length, the time required for autoradiography is reduced from several days to several hours.

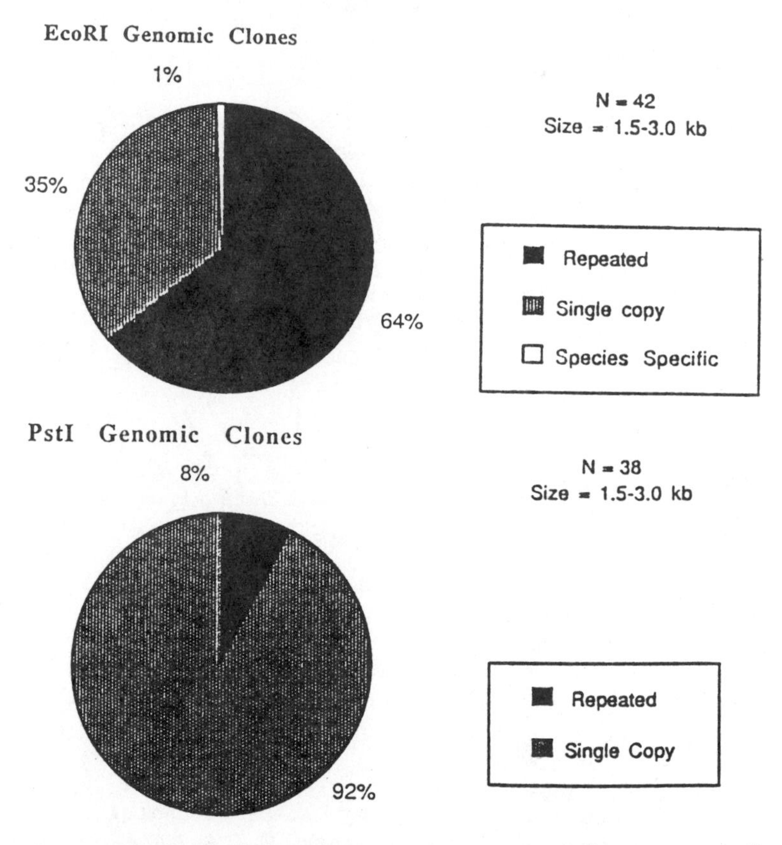

Fig. 4. Proportion of clones possessing repeated sequences from genomic libraries constructed with restriction endonucleases EcoRI and PstI. Statistics were generated by probing randomly selected clones from each library onto restricted, nuclear DNA (N = number of clones tested; species specificity derived by comparing DNA from *L. esculentum* and *L. pennellii*).

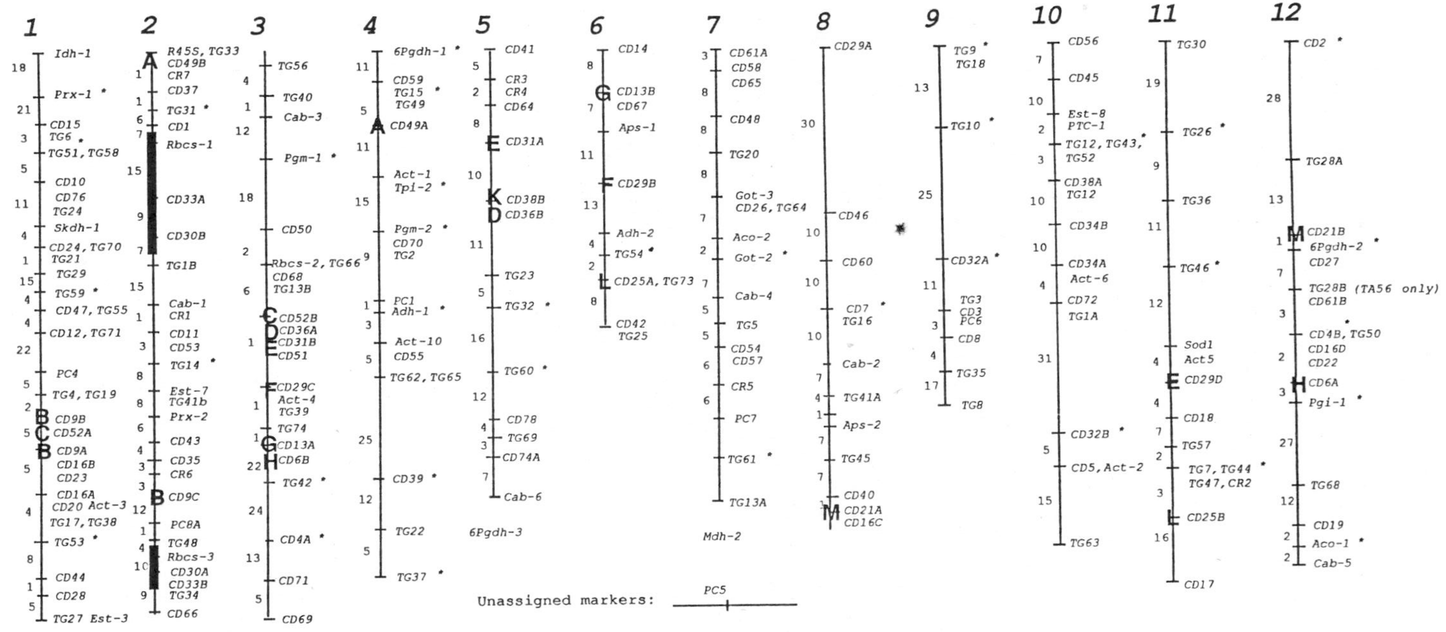

* Verified by dosage effects with primary trisomics

Fig. 5. Molecular linkage map of tomato chromosomes.
CD's = loci detected by Southern analysis with random cDNA clones. Letters after each locus indicate different loci possessing gene sequences homologous to the same cDNA. For example cDNA clone pCD49 was found to have homologous loci on chromosome 2 (CD49B) and chromosome 4 (CD49A). TG = loci detected by hybridization with single copy genomic sequences (not necessarily genes). Large letter on chromosomes (e.g. A,B,C etc) shows chromosomal sites of some of the known gene duplications in tomato detected by hybridization with cDNA clones. Note that most gene duplications are unlinked. Bold line on chromosome 2 indicates a possible duplication of a chromosomal fragment containing an RBCS gene and two gene detected analysis with random cDNA's (CD33 and CD30).

MAPPING RFLP'S INTO LINKAGE GROUPS

One of the attractive features of RFLP mapping is that it can be accomplished in any plant species that undergoes sexual reproduction. Genetic maps are based on meiotic recombination among homologous chromosomes carrying alternative alleles. To determine linkage relationships among genetic loci, one must generate crosses that segregate for the genes of interest. Since most of the traditional genetic markers affect some aspect of the whole plant phenotype (such as anthocyanin production, height, leaf shape, etc.) and are hence prone to epistatic effects with one another, it has been difficult to score more than a few segregating markers simultaneously in a single cross. The generation of classical genetic maps have thus required hundreds of crosses and the analysis of thousands of segregating progeny.

The level of allelic variation detectable with DNA probes is much higher than that for morphological markers and there are no epistatic effects. By making crosses between parents that are genetically dissimilar, and thus likely to possess different alleles, it is possible to score segregation at hundreds of different loci in the same segregating population. Thus, a detailed genetic map can be constructed by analyzing less than 100 plants with many different DNA probes. In maize, where very high levels of DNA polymorphism are found, it is sufficient to make crosses betwen unrelated inbreds (Helentjaris et al., 1986). In tomato, lettuce and rice, where levels of polymorphism are less, wider crosses (interracial or interspecific) have been a more appropriate choice (Bernatzky and Tanksley, 1986b; R. Michelmore, person. comm.; unpublished data, this laboratory).

Classical genetic maps have been constructed largely through the use of backcross generations. Morphological markers are usually recessive and thus only in the testcross can all genotypes be identified. DNA markers behave in a codominant fashion allowing identification of all genotypes in any generation. Under these conditions, the F2 generation is in many ways superior to the testcross. We base this assertion on estimation of two parameters: 1) the <u>maximum</u> map distance which can be distinguished from random assortment, and 2) the <u>minimum</u> map distance at which recombination can be detected between two markers for given population size (n) and confidence level (Fig. 6). Allard (1956) provides maximum likelihood equations for estimating recombination values (r) and attached standard errors (SEr) for most common mating designs and segregation patterns. Confidence intervals for estimation of a particular recombination value may be constructed by multiplying the SEr by the standard normal deviate (z) or Student's t statistic corresponding to the number of individuals and desired confidence level. Such confidence intervals suffer the limitations (Silver, 1985; Green, 1981) of being symmetrical and often including undefined negative values. Silver (1985) uses the binomial distribution to obtain confidence intervals for r from recombinant inbred lines. This can be done simply for backcross populations as all recombinants (except double crossover gametes) can be detected. In F2 populations, not all recombination events are detectable since double heterozygotes may derive from either two parental or two recombinant gametes. The proportion of cryptic recombinants declines with the square of r, important for loose linkages but negligible as r approaches zero. Hence, we have chosen the following estimators for the two parameters above, presented in Fig. 6.

1) The maximum map distance at which linkage can be detected represents the greatest distance at which $r + t_{0.01,n-2}(SE_r) < 50cM$ where n is numbered individuals in mapping population, t is Student's t parameter for significance probability 0.01 (or otherwise specified) and n-2 degrees of freedom, and r is a point estimate of recombination value. This will overestimate detectable linkage at small n, but provides a good estimate for populations larger than 30.

2) The minimum map distance at which recombination can be detected represents the smallest distance at which the binomial probability of no recombinant is <0.01 (see below). F2 cryptic recombinants have been lumped into 'nonrecombinant' class because they are not informative by this approach, but for the values of r associated with this parameter they will be rare.

F2 populations permit detection of more distant linkage and resolution of tighter linkage because recombination in both mega- and microgametophyte is detectable. In a testcross, recombination in only one gametophyte can be detected. Implicit in these estimates and confidence intervals is that each individual locus exhibits Mendelian segregation, and that linkage is the only cause of nonindependence among loci. Nonrandom

Probability of recombinants (q)

Bc; $r = q$ .. r = recomb distance between loci

F2; $[1 - [(1-r)^2 + \frac{1}{2} r^2]] = q$

Binomial probability of zero recombinants among n individuals

$$\frac{n!}{0!(n-0)!} p^o q^n \;\text{-->}\; q^n$$

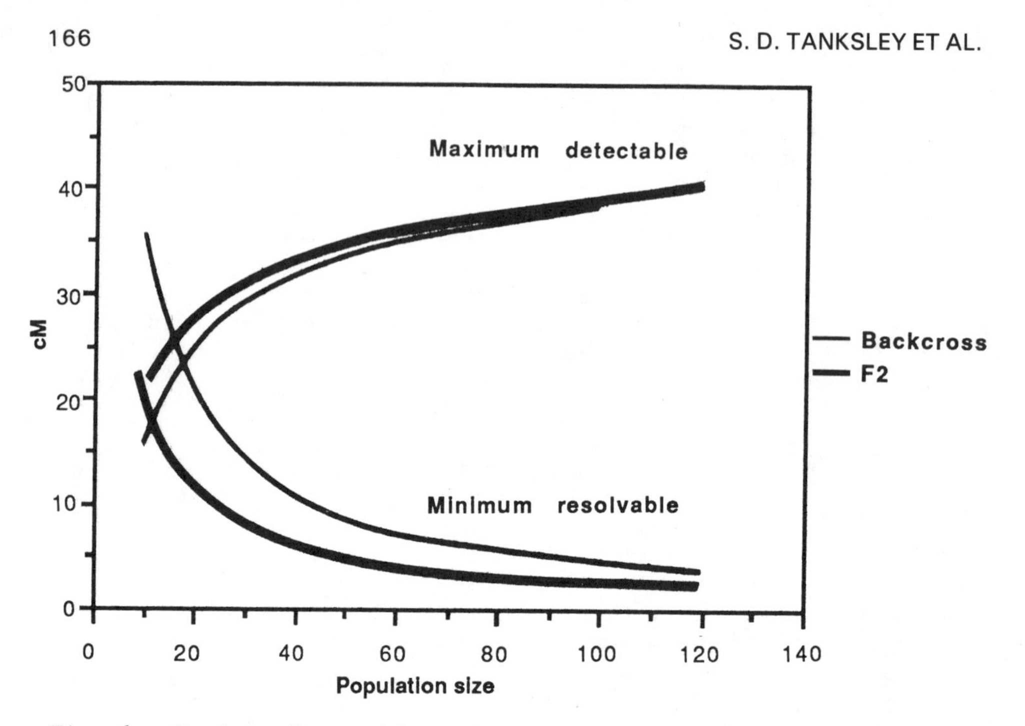

Fig. 6. Maximum detectable and minimum resolvable map distances between markers utilizing backcross and F2 populations. Curves are for 99% confidence level (see text for details).

gamete sampling reduces the validity of these estimates, and is justification for conservatism in situations where this is likely (e.g. wide crosses).

An alternative to F2 populations for RFLP mapping in plants has recently been raised by Burr (person. comm.). This involves the use of recombinant inbreds (independent progeny lines from F2 or backcross generations selfed down to homozygosity). The advantage of recombinant inbreds is that they can be propagated from seeds and thus maintained indefinitely. Segregating generations, like the F2, are ephemeral. The genotypes cannot be propagated by seed. Bulk harvesting DNA from F3 plants derived from individual F2 plants allows continuation of the mapping population for an extended period of time but not indefinitely as with the recombinant inbreds. The most serious limitation to recombinant inbreds is the time required for their synthesis. Even for plants with short generation times, the time investment is likely to extend into several years.

ASSIGNING LINKAGE GROUPS TO CHROMOSOMES--USE OF PRIMARY TRISOMICS

Aneuploid stocks have played a key role in plants in assigning linkage groups to chromosomes. In tomato, we have relied heavily on the use of primary trisomics. Primary trisomics possess an additional copy of one chromosome. Tomato has a haploid complement of twelve chromosomes and primary trisomics are available for each of these chromosomes.

Primary trisomics have been used to assign isozyme-coding genes to chromosomes based on protein dosage (Nielsen and Scandalios, 1974; Tanksley, 1979). It follows that it should be feasible to use trisomics to assign RFLP markers to chromosomes via DNA dosage effects. Our first attempts at this involved isolating DNA from each of the trisomics, along with diploid controls, and probing equal quantities of DNA from each with a clone of unknown chromosmal assignment using the slot blotting technique. If the clone did not reside on the extra chromosome in the trisomic, the signal from autoradiography should be equal or slightly less than the diploid control. However, if the clone did reside on the extra chromosome, the signal would be increased (up to 150%) since DNA from those plants would possess three copies instead of two in the diploid. This technique involves between sample comparisons and in our experience it is difficult to reliably detect the change in hybridization signal on the positive trisomic versus the controls. As an alternative to this approach, we adopted a multiple probing technique, whereby DNA from the trisomics is digested with a restriction enzyme, separated on an agarose gel and blotted onto a nylon membrane. This filter is then simultaneously probed with several clones. The relative intensity of the fragments corresponding to each clone can then be compared with one another to detect dosage (Fig. 7). With this approach, the comparisons are within samples (lanes), significantly reducing the problems inherent in making between sample comparisons. Using this technique, we have been able to assign DNA markers to each of the tomato chromosmes (Fig. 5).

COMPLETING LINKAGE MAPS

Ideally, one would desire a linkage map with markers at 1 cM distances or less -- especially when contemplating the possibility of trying to walk or hop from a known clone to a gene of interest. To accomplish this would likely involve analyzing thousands of clones and many years. In this regard we have generated estimates of the number of clones needed to construct maps of varying degrees of saturation (Fig. 8). Two parameters have been estimated as a function of the number of clones mapped: 1) The average distance between markers. This

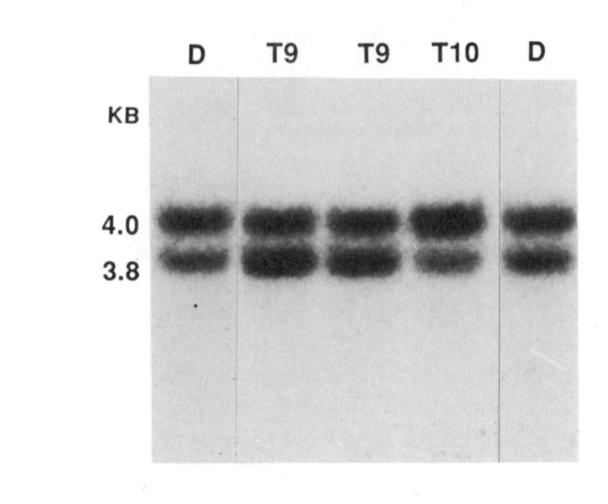

Fig. 7. Autoradiogram demonstrating DNA dosage effects with primary trisomics. D = diploid, T9 = triplo 9, T10 = triplo 10. Hybridizing fragments (4.0 kb and 3.8 kb) correspond to loci on different chromosomes as determined by changes in relative intensity of two fragments as compared to the diploid control. For example, in triplo 9 DNA, the 3.8 kb fragment increases relative to the 4.0 fragment indicating the locus corresponding to the 3.8 kb fragment is on chromosome 9. The 4.0 kb fragment shows a relative dosage in triplo 10 DNA indicating the location of the corresponding locus on chromosome 10.

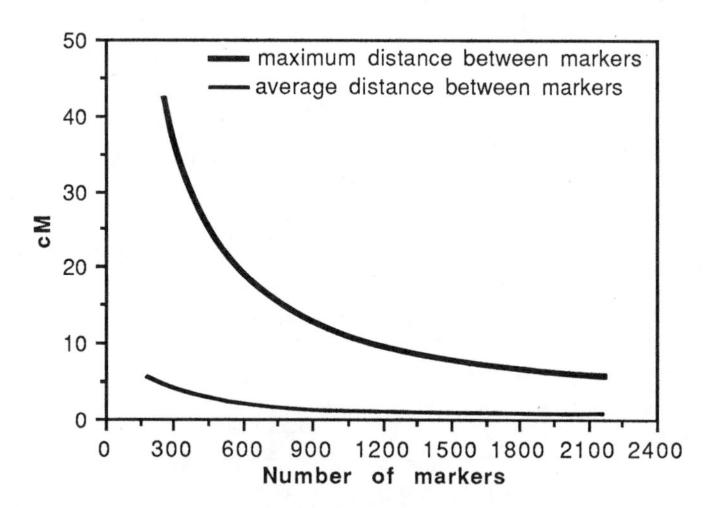

Fig. 8. Average and maximum distance expected between markers on a linkage map depending on number of random clones mapped. Curves are based on a genome with 1200 cM (e.g. 12 chromosomes each of 100 cM). Maximum distance curve is for 99% confidence level.

is the estimated total map units in a genome divided by the number of clones mapped. For instance in a species with 12 chromosomes of 100 cM each there would be a total of 1200 cM. If 120 clones were mapped, the average distance between clones would be 10 cM. 2) Maximum distance between any two markers on the map. This variable is perhaps more important than the first since it determines the degree to which all portions of the genome are covered. Our estimates for this variable (Fig. 8) are based on the Poisson distribution (99% probability level) and the assumption that recombination values are additive. From Fig. 8, it can be seen that if 1200 clones are mapped, the average distance between markers would be 1 cM, but the maximum distance between any two markers would be approximately 10 cM. So while most areas would have markers at distances of 1 cM or less, one would still expect to find gaps of up to 10 cM. The probability of larger gaps would be <0.01. To reduce the gaps to less than 5 cM would require mapping more than 1000 additional clones. In reality, it would be impractical to produce maps with uniform coverage of 1 cM or less via this random mapping approach. To accomplish this, it seems inevitable that methods must be developed to allow one to select clones which fill in the gaps.

For most practical purposes in plant breeding, maps with markers every 10-20 cM are sufficient for developing tags for genes of interest or for detecting quantitative trait loci (QTL's). This degree of saturation is feasible with the random mapping approach. The molecular maps currently available in tomato (Fig. 5) and maize (Helentjaris et al., 1986) are in this range of saturation.

Another problem concerning the completeness of any linkage map is determining the ends of the linkage groups. Despite the numbers of markers on any chromosome, the possibility always exists that additional loci reside beyond the most distal loci on the map. The solution to the problem will almost certainly involve a combination of genetic mapping and *in situ* hybridization to chromosomes to determine the actual physical positions of loci on the genetic and cytological maps.

Another consideration about linkage maps is the lack of uniformity in distribution of single copy sequences and/or heterogeneity in recombination along the chromosomes. Either of these phenomena could result in gaps in the linkage maps. RFLP maps are based on single copy DNA sequences. There is much data in the literature to suggest that single copy DNA and repeated DNA are not necessarily uniformly interspersed throughout the chromosomes (Jones, 1970; Flavell, 1980). Telomeres and centromeres tend to possess higher amounts of repeated DNA and would thus be underrepresented in collections of single copy clones. Recombination is also known not be strictly a function of the physical length of a piece of chromosomal DNA. Where genetic and physical maps have been compared, there is clear evidence that certain regions of the chromosomes experience higher rates of crossing over (Lindsley and Grell, 1968; Rick, 1971). Highly recombinogenic areas of chromosomes may be perceived as large intervals on the genetic map.

GENE DUPLICATIONS IN THE TOMATO GENOME

By studying the genetics of a relatively large number of different cDNA clones, it is possible to estimate the proportion of genes which are duplicated in a species. In tomato, we have determined the inheritance of nuclear sequences homologous to approximately 100 unique random cDNA clones as well as several cloned genes of known function (Bernatzky and Tanksley, 1986a,b; Vallejos et al., 1986; Pichersky et al., 1985). Several conclusions can be drawn from these studies: 1) Most genes in this diploid species are represented by a single copy at one locus. 2) Abundant proteins (e.g. RBCS -- small subunit ribulose-bis-phosphate carboxylase and CAB -- chlorophyll a/b binding polypeptide) are normally encoded by gene families consisting of as many as 10-15 copies. 3) When there is more than one copy of a gene, the different copies are frequently found on different chromosomes.

Fig. 5 shows the map position of some of the duplicate loci in tomato detected by genetic analysis of random cDNA's. It is interesting to note that most of the unlinked, duplicate loci appear in solitary arrangement and are not part of larger segments of duplicated chromosomes. The exception to this is a short piece of chromosome 2 which possibly resulted from an intra-chromosomal duplication (Fig. 5). If the duplications of tomato genes were a result of a past polyploidy event in the lineage leading to the tomato genus, one would expect duplications to appear as linked sets. The fact that we do not find this situation in tomato casts doubts on polyploidization as the mechanism that led to the duplication of these loci. Unequal crossing over is known to lead to tandem duplications, but does not readily account for unlinked duplications. While one cannot rule out the possibility that some of the unlinked duplications in tomato resulted from separation of tandem loci, it seems improbable that the high proportion of duplicate genes could have arrived at their present independent locations through such a mechanism.

There is good evidence that transposons can generate duplications in bacteria and indirect evidence for this in plants (Kleckner, 1981; Courage-Tebbe, 1983). The possibility thus exists that the duplications in tomato may have been generated, at least in part, by the activity of transposons.

APPLICATIONS OF MOLECULAR MAPS TO PLANT BREEDING

The applications of molecular markers and their corresponding maps to plant breeding have been adequately covered in other publications (Burr et al., 1983; Tanksley, 1983; Helentjaris et al., 1985). These maps provide geneticists and breeders a ready method of tagging any gene of interest with a selectable, codominant marker. In cases where the gene of interest is recessive or difficult to score accurately in segregating generations, the utility of a linked molecular marker is obvious. The molecular map also provides an approach for detecting and tagging genes controlling quantitative variation. Quantitative genetics has long been under the purview of biometry and seldom has it been possible to measure effects of single quantitative trait loci. Perhaps it is this aspect of molecular maps and RFLP markers, the application to quantitative genetics, which has been the most promulgated and promoted. At this point in time it is an area with more speculation than critical experimentation. Only after controlled experiments are conducted in a number of plants with molecular linkage maps can we really assess the potential utility of molecular markers in the genetic improvement of quantitative characters.

ACKNOWLEDGEMENTS

Research support by grants from USDA-CRGO, NSF and Agrigenetics Corp. Thanks to Steffie David for typing of the manuscript.

REFERENCES

Bernatzky, R., and Tanksley, S. D., 1986a, Majority of random cDNA clones correspond to single loci in the tomato genome, Mol. Gen. Genet, 203:8-14.

Bernatzky, R., and Tanksley, S. D., 1986b, Toward a saturated linkage map in tomato based on isozymes and random cDNA sequences, Genetics, 112:887-898.

Burr, B., Evola, S., Burr, F. A., and Beckmann, J. S., 1983, The application of restriction fragment polymorphisms to plant breeding, in: "Genetic Engineering: Principles and Methods", Vol. 5, J. K. Setlow and A. Hollaender, (eds), Plenum Press, New York, pp. 45-59.

Courage-Tebbe, U., Doring, H.-P., Federoff, N., and Starlinger, P, 1983, The controlling element Ds at the Shrunken locus in *Zea mays*: structure of the unstable sh-m5933 allele and several revertants, *Cell*, 34:383-393.

Doerfler, W, 1983, DNA methylation and gene activity, *Ann. Rev. Biochem.*, 52:93-124.

Flavell, R. B., 1980, The molecular characterization and organization of plant chromosomal DNA sequences, *Ann. Rev. Plant Physiol.*, 31:569.

Green, E. L., 1981, "Genetics and Probability in Animal Breeding Experiments", Oxford University Press,

Helentjaris, T., King, G., Slocum, M., Siedenstrang, C., and Wegman, S., 1985, Restriction fragment polymorphism as probes for plant diversity and their development as tools for applied plant breeding, *Pl. Mol. Biol.*, 5:109-118.

Helentjaris, T., Slocum, M., Wright, S., Schaefer, A., and Nienhuis, J., 1986, Construction of genetic linkage maps in maize and tomato using restriction fragment length polymorphisms, *Theor. Appl. Genet.*, 72:761-769.

Jones, K. W., 1970, Chromosomal and nuclear location of mouse satellite DNA in individual cells, *Nature*, 225:912-915.

Kleckner, N., 1981, Transposable elements in prokaryotes, *Ann. Rev. Genet.*, 15:341-404.

Lindsley, D. L., and Grell, E. H., 1968, Genetic variations of *Drosophila melanogaster*, *Carnegie Inst. Wash.*, Publ. No. 627.

Nielsen, G., and Scandalios, J. G., 1974, Chromosomal location by use of trisomics and new alleles of an endopeptidase in *Zea mays*, *Genetics*, 77:679.

O'Brien, S. J., 1986, *Genetic Maps*. Cold Spring Harbor Laboratory.

Pichersky, E., Bernatzky, R., Tanksley, S. D., Breidenbach, R. B., Kausch, A. P., and Cashmore, A. R., 1985, Molecular characterization and genetic mapping of two clusters of genes encoding chlorophyll a/b-binding proteins in *Lycopersicon esculentum* (tomato), *Gene*, 40:247-258.

Pruitt, R. E., and Meyerowitz, E. M., 1986, Characterization of the genome of *Arabidopsis thaliana*, *J. Mol. Biol.*, 187:169-183.

Rick, C. M., 1971, Some cytogenetic features of the genome in diploid plant species, *Stadler Symposia*, 1,2:153-174.

Silver, J., 1985, Confidence limits for estimates of gene linkage based on analysis of recombinant inbred strains, *J. Hered.*, 76:436-440.

Tanksley, S. D., 1979, Linkage, chromosomal association and expression of *Adh-1* and *Pgm-2* in tomato, *Biochem. Genet.*, 17:1159.

Tanksley, S. D., 1983, Molecular markers in plant breeding, *Plant Mol. Biol. Rep.*, 1:3-8.

Tanksley, S. D., and Pichersky, E., 1987, Organization and evolution of sequences in the plant nuclear genome, *in*: "Plant Evolutionary Biology", L. D. Gottlieb and S. K. Jain, (eds), Chapman and Hall, London.

Vallejos, C. E., Tanksley, S. D., and Bernatzky, R., 1986, Localization in the tomato genome of DNA restriction fragments containing sequences homologous to the rRNA (45s), the major chlorophyll a/b binding polypeptide and the ribulose bisphosphate carboxylase genes, Genetics 112:93-105.

TRANSGENIC ARABIDOPSIS

G. P. Rédei, Csaba Koncz and Jeff Schell

University of Missouri, 117 Curtis Hall, Columbia, MO 65211, and Max-Planck-Institut, Köln 30, D-5000

INTRODUCTION

A nuclear gene locus of (*Arabidopsis thaliana* L.) (Rédei, 1973; Rédei and Plurad, 1973) causes hereditary alterations in the genetic material of the plastids. Its effectiveness is quite remarkable in as much as the rate mutation when either one of the three known recessive alleles become homozygous, increases by a

Fig 1. Plant homozygous for the *chm*^{2} allele displays sectoring.

factor of about 10^6 over the spontaneous level. The mutator activity is revealed by the numerous green, yellow and white sectors on the leaves and stems of the plants (Figs. 1 and 2). Some of the mutations induced have pleiotropic effect: in addition to alteration of the plastids the shape of the leaves is also affected. Since the sorting out of the mutant plastids is clearly a non-random process, leaves or entire plants may become homo- or near homoplastidic within a single generation (Rédei, 1974). The homoplastidic condition generally cannot be stabilized, however, unless the recessive inducer is blocked by rendering the plants heterozygous or by the removal of the chm alleles from the nucleus. Details of these procedures were worked out and their effectiveness has been proven (Rédei, unpublished).

The mutator alleles are only slightly different in effectiveness, and all can induce both forward and apparent reverse mutation. In some of the mutant tissues the frequency of reversion is high (Fig. 3), in others it does not take place at

Fig. 2. Plants homozygous for the chm^1 allele illustrating some of the phenotypic effects of the mutator. Plant on the left shows uniform deformity of the leaves and sectoring, the individual in the middle has a large albina sector which -- albeit it grows as well as the parts with green pigments--does not display reversions, plant on the right indicates high rate of mutation and sorting out. Aseptic cultures like these permit better visualization of the pigment-free or low-pigment leaves.

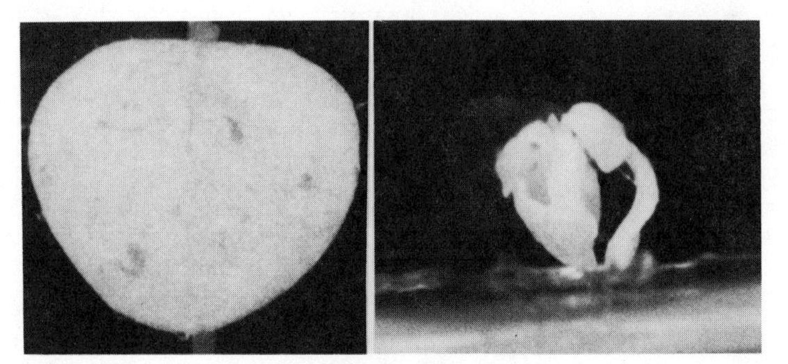

Fig. 3. Green sectors in an albina leaf (left) and in a small semi-lethal seedling (right) grown in vitro, indicating phenotypic reversions.

all (Fig. 2). The ability of back mutation does not seem to be related to the viability of the original genetic alteration. It appears that the mutator is capable of inducing both point mutations and losses, and it may be responsible also for rearrangements in the plastome.

At the time when the first such mutator was discovered (Li and Rédei, unpublished data), only classical genetic analysis and electron microscopic studies were feasible on the nuclear gene and its effects on the ultrastructure of the photosynthetic apparatus (Rédei, 1973; Rédei and Plurad, 1973). Recent advances in molecular genetics open new approaches to the analysis of the plastid DNA, and to the tagging and isolation of the nuclear gene itself. This study will be concerned with the development of techniques which eventually may lead to a physical characterization of the chm nuclear gene and its various mutant alleles.

TRANSFORMATION OF ARABIDOPSIS

The early studies of Barbara McClintock at the University of Missouri on chromosome breakage (McClintock, 1931; 1938) lead her to the discovery of the Ac - Ds system of maize (Zea mays L.) (McClintock, 1951) and to the genetic exploration of the controlling elements (McClintock, 1956). These mobile genetic elements could be identified in a variety of prokaryotic and eukaryotic organisms (Shapiro, 1983) and permitted the tagging of a number of genes with appropriate probes.

Arabidopsis does not have verified transposable elements albeit some mutations such as im (Rédei, 1963; Rédei et al., 1974) and chm bear some resemblance to them regarding the pattern of variegation. For gene tagging transformation provides a better alternative because once an insertion takes place that is expected to stay put without additional manipulations. For transformation different procedures may be available, including various agrobacterial vectors (Márton et al., 1979; Otten et al., 1981; De Greeve et al., 1982), electroporation (Potter et al., 1984), direct gene transfer by plasmids into protoplasts (Paszkowski et al., 1984), injection of DNA into the shoot apices (de la Peña et al., 1987), and DNA-coated tungsten microprojectiles (Klein et al., 1987). Some of these procedures and applications have been briefly summarized recently (Cocking and Davey, 1987).

For our studies the techniques developed by Horsch et al. (1985) and applied to Arabidopsis by Lloyd et al. (1986) were adapted. This procedure, with modifications, have been successfully used by other laboratories (see 3rd Int. Arabidopsis Meeting, 1987) although Feldmann and Marks (1987) developed an entirely different method involving infection of germinating seeds by Agrobacterium carrying kanamycin resistance genes.

Vectors

The experiments discussed below used 6 different vectors developed by Csaba Koncz. Five of the vectors were free of the oncogenes of the Ti plasmids. The oncogenic vector was used only for testing the effectiveness of the system. For their overall features see Fig. 4. Vector pPCV 311, containing the oncogenic region, was the largest, approximately 22.4 Kb. It contained the neomycinphosphotransferase II gene (Km^R) attached to the nopaline synthase promoter (pNOS) and termination signals represented by the open boxes. It carried also the octopine synthase gene. The arrows indicate the directon of the transcription. B black boxes stand for the left and right border sequences of the T-DNA. pPCV 310 (approx. 14.2 Kb) is identical to pPCV 311, except it is free of genes, 5, 2, 1 and 4 of the tumor inciting plasmid. pPCV 631 has a hygromycin phosphotransferase gene of bacterial origin (Hyg^R) hooked to a nopaline synthase (pNOS) promoter and terminations signals and thus conveying resistance to the antibiotic hygromycin B. This vector does not contain the octopine synthase gene but it has the NPT II gene albeit without a promoter. Thus successful transformants become usually resistant to hygromycin but they can express kanamycin resistance only if the insertion takes place in such a way that the prokaryotic NPT II gene comes under the control of a plant promoter.

pPCV 601 is a vector with a promoter-free Km^R gene. Since under our experimental conditions transformants are screened on the basis of antibiotic resistance, isolation of transformants is very difficult because only those rare cases will be found which are integrated behind a plant promoter. Nevertheless a few

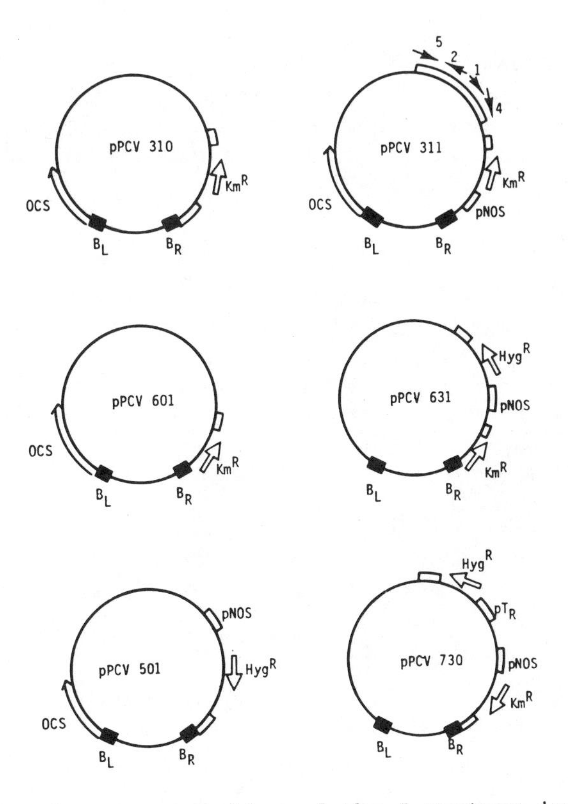

Fig. 4. Schematic representation of the 6 vectors involved. The diagrams are not on scale.

transformed cell colonies were obtained with this vector too. Transformants can be further identified by octopine synthase activity.

pPCV 730 carries both hygromycin and kanamycin resistance genes, equipped by a T_R and NOS promoter, respectively. Both genes can be freely expressed after integration irrespective of the site of insertion. This vector does not have an octopine synthase gene. pPCV 501 contains the hygromycin phosphotransferase gene in an opposite orientation; it contains the NOS promoter, and it has a functional octopine synthase gene. All these are binary vectors, and the VIR functions are provided in trans orientation.

In all experiments transformants were selected by either kanamycin or hygromycin resistance. Both of these antibiotics affect primarily the 16S ribosomal RNA (Moazed and Noller, 1987) yet their effect on Arabidopsis is quite distinct. Hygromycin seems to be a better selective agent than kanamycin although transformants have been obtained with both in our laboratories.

Hygromycin can entirely prevent growth of the plants without interfering with the synthesis or maintenance of leaf pigments. Kanamycin, at appropriate concentration, prevents the formation of all pigments and causes the bleaching of the tissues exposed. Their effects are also complicated by cefotaxime (claforan) when it is present in the culture medium.

Bacterial Cultures

Agrobacterium was grown on YEB medium (Lichtenstein and Draper, 1985) at 20°C with vigorous agitation. In about a day, high cell density was obtained. The cells were collected by centrifugation (Sorvall GS-3 rotor, 5,000 rpm, 10 min., 150 ml Corex bottle) and washed with tissue culture medium containing sucrose but no hormone or agar. For infection, the cell density was adjusted to about O.D. 0.8 at 550 nm.

The bacteria were stored on YEB masterplates containing 100 μg/mL rifampicin and carbenicillin, each, and they were kept in the refrigerator for about two months before reculturing.

Preparation of the Plant Material and Infection

Arabidopsis, Columbia wild type, was grown aseptically on E medium (Rédei, 1965) in cotton-plugged 16 x 150 mm test tubes under 12 hrs light cycles in a growth chamber at approximately 24°C or in the greenhouse. Illumination was provided by daylight fluorescent tubes at about 500 foot candle. The plants were used at the stage when the flowering stem reached to near half height of the inner space between the 5 mL medium and the enclosure Fig. 5.

From both rosette and stem leaves the base was cut off, and the stems were sliced into 10 to 20 mm long segments. The leaves were wounded with several cuts using sharp scalpels and the stem pieces were also pricked. The explants were then bathed for about 20 minutes in the bacterial suspension. After blotting the pieces, they were placed on tissue culture medium without hormones and incubated for two days at approximately 24°C.

Plant Culture Medium

As mineral salt solution, the nutrients suggested by Murashige and Skoog (1962) were used in some of the experiments. Much better regeneration was obtained by a new medium developed during the course of these studies, and this is now being used routinely for all phases of the transformation experiments.

The new nutrient solution (designated R4) contained the following ingredients in mg/L: NH_4NO_3 1,800, KNO_3 800, $MgSO_4$·

Fig. 5. Test tube-grown plants ready for bacterial infection.

$\cdot 7H_2O$ 100, $CaH_4(PO_4)_2 \cdot H_2O$ 100, KH_2PO_4 100, K_2HPO_4 90, Fe-pentetic 67.1. This latter component was prepared as a 0.02 M solution of $FeSO_4 \cdot 7H_2O$ and the chelating agent diethylenetriamine pentaacetic acid, and 5 mL was added to each L of the final medium.

The minor salt mixture was identical to that in the Murashige and Skoog solution. Vitamins were supplied as suggested by

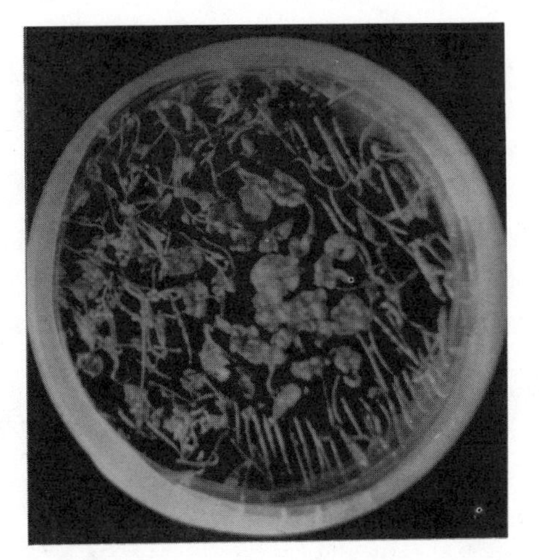

Fig. 6. The infected explants were laid on the surface of 30 mL nutrient medium in 100 x 10 mm sterile polystyrene plates and sealed with parafilm.

by Gamborg et al. (1968), except that in place of nicotinic acid nicotinic acid amide was employed. The medium contained also 3% sucrose, and it was solidified by 0.18% Gellan Gum (Gel-Gro, ICN Biochemicals, Cleveland, OH).

Proliferation of Transformed Cells

Two days after infection the explants were removed from the first solid medium where considerable growth of the bacteria has taken place and time was thus allowed for the integration of the T-DNA. The second solid medium contained cefotaxime (500 mg/L) to eliminate further bacterial growth which would seriously damage the plant tissue. Cefotaxime sodium is a semisynthetic cephalosporin antibiotic stops bacterial growth immediately but does not eliminate Agrobacterium cells until several passages have taken place. It is noticeably decomposed by two weeks. This antibacterial agent is relatively harmless to Arabidopsis cells, however, it interacts undesirably with aminoglycoside antibiotics (e.g. kanamycin). We have used a water-soluble powder obtained either as a gift from the Hoechst Company or purchased their product from Calbiochem, La Jolla, CA.

This second medium contained 2,4-dichlorophenoxyacetic acid (0.025 mg/L) and isopentenyl adenosine [6(γ,γ-dimethylallylamino) purine riboside, abbr. 9RiP] generally 1.5 mg/L. Depending on the selectability of the transformed cells, we generally used either 100 mg/L kanamycin sulfate or 15 mg/L hygromycin B. Calli developed to excisable size within about three weeks (Fig. 7).

Fig. 7. Five-week old calli isolated on 100 mg/L kanamycin sulfate (pPCV 310).

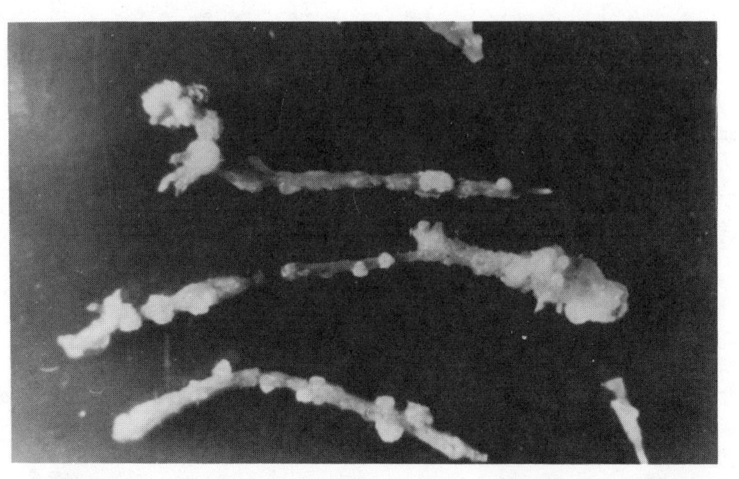

Fig. 8. Development of transformed cell colonies on infected stem segments (pPCV 310) placed on a medium containing 100 mg/L kanamycin sulfate.

The efficiency of the transformation was quite variable. In the experiment shown above from 583 explants 528 calli were obtained (ca. 0.91). In other experiments only a few percent successful transformation was observed. Generally, stem segments displayed considerably more antibiotic resistant calli than leaves (Fig. 8). The stem leaves were much more susceptible to transformation than rosette leaves. The age and vigor of the tissues were decisive factors. Some leaves developed several calli whereas others died shortly after infection.

The calli excised were subcultured every two to three weeks on shoot-inducing medium containing 2 mg/L 9RiP and 0.1 mg/L α-naphthalene acetic acid (NAA). Some of the calli developed leaves within a few weeks, for others several months were required to show leaf initials. After leaf initials appeared the dose of 9RiP was reduced to 1.5 mg/L. Within 5 months approximately 1/3 of the calli displayed leaves but the laggards still continued with leaf differentiation indefinitely. Nearly all the calli represented in Table 1 formed leaves and flowering stems after 8 months.

After a few weeks the level of cefotaxim in the culture medium was reduced to 200 mg/L and maintained this way until the bacteria were eliminated. After bacterial infection of the cultures ceased cefotaxim was omitted from the media. Following a few subcultures kanamycin and hygromycin, respectively were

Table 1. A typical transformation experiment (pPCV 501) yielded the results shown below.

		%
Total of leaves and stems infected	1977	100
Surviving calli by 5 months	701	35.5
Calli differentiating only leaves by 5 months	128	6.5
Plants with flowering stem obtained	190	9.6

discontinued in the media to assure more vigorous growth. Although the transformed calli tolerated these antibiotics, they did better without them.

Differentiation of Complete Plants from Transformed Calli

The plantlets seen on Fig. 9 eventually form a flowering shoot on 1.5 mg/L 9RiP and 0.1 mg/mL NAA. In some experiments better shoot formation resulted on 0.5 mg/mL 6-benzylaminopurine (BAP) and 0.05 mg/mL NAA. At this stage some of the plantlets we transferred into test tubes of various sizes to allow better

Fig. 9. When leaf differentiation begins, the majority of the calli form multiple plantlets. Because of the high level of cytokinin, the leaves are generally purplish or deep purple but soon after this hormone is reduced they turn green again and the leaf blades assume wider normal shape.

Fig. 10. Transgenic plants at the shooting stage in test tubes.

growth (Fig. 10). Particularly favorable were the 20 x 150 mm test tubes. Some of the plantlets formed viable seeds within the test tubes, obviating the need for further transplantation.

Unfortunately, very few of the plants developed roots under these culture conditions. On aseptic media, the presence of roots was not absolutely necessary for seed production. Rooting was attempted by treatments with various auxin and cytokinin combinations and single or mixed treatments of indoleacetic acid and indolebutyric acid but no really satisfactory solution is available to this problem.

Many initially rootless plants thrived well when transplanted from the axenic culture to an artificial potting mix (Pro-Mix BX), and produced seeds. Some of them fed only through the calli still attached to the stem, others rooted spontaneously. Attempts to root the plants by dipping the base into "Rootone F" (Union Carbide) commercial rooting agent controlled to some extent fungus infection but showed no other beneficial effect. Actually, many transplants bleached as a consequence of this treatment and perished.

Some transformants produced well over 1000 seeds while others yielded only a couple. The percentage of viable seed per silique varied a great deal but the majority of seed was of good quality and germinated well. The appearance of the regenerated transformants was quite different from seed-grown ones or from the much faster regenerated individuals which were not involved

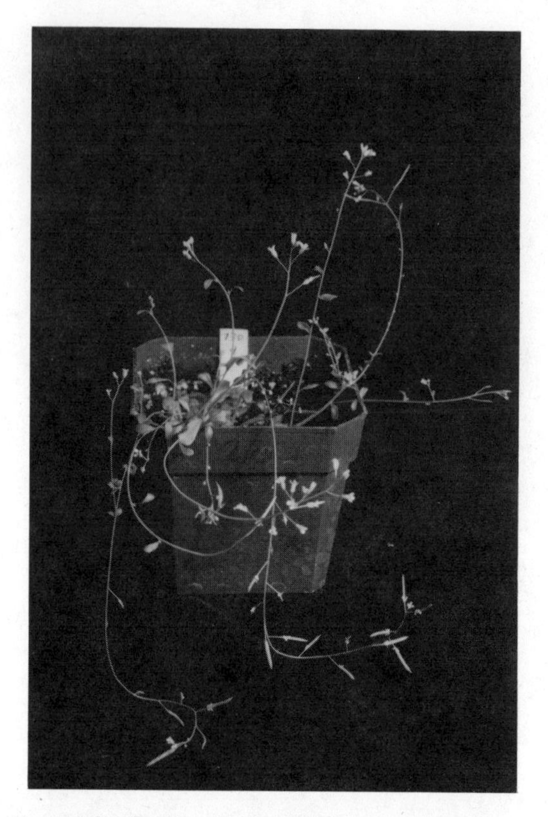

Fig. 11. Regenerated plant of typical appearance displaying fruits and flowers. Frequently, these type of plants continued developing new shoots after this stage and produced many more fertile fruits.

with bacterial infection and the accompanied treatments with antibiotics. Most of the transformed plants displayed a bushy habit (Fig. 11) with a larger number of stems than the regularly regenerated ones and certainly many more than those originating from seed.

VERIFICATION OF TRANSFORMATION

Although all the calli came through several passages of selective media, some regenerated individuals turned out to be susceptible to the antibiotics when the progeny was germinated on hygromycin or kanamycin media, respectively. So far only a small number of plants were tested, therefore the percentage of escapees cannot be accurately or meaningfully determined. Generally, the seedlings are more sensitive to the selective antibiotics than the coresponding calli.

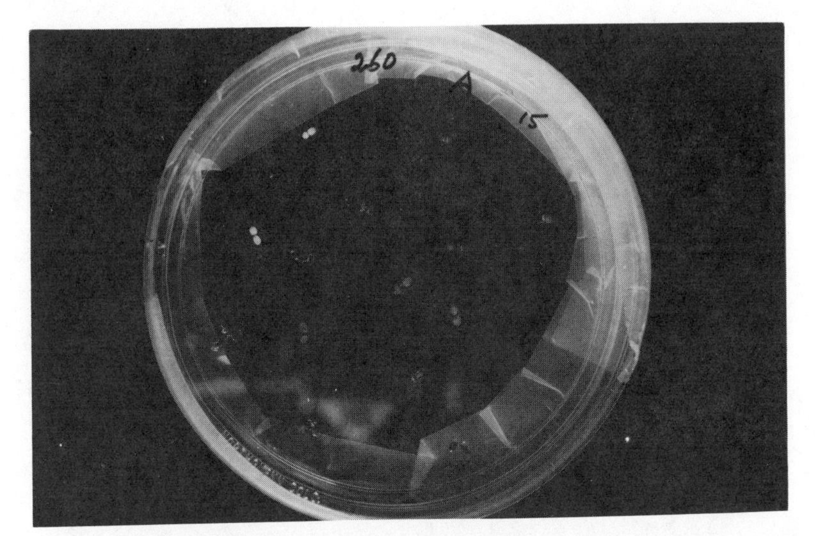

Fig. 12. T_2 progeny of plant 310-A tested on 60 µg/mL kanamycin sulfate; 14/15 seeds planted germinated and 12 (dark green cotyledons) appeared resistant and 2 (pale) was found to be susceptible to the antibiotic.

Progenies of Transgenic Plants

So far seed was obtained on transformants infected by pPCV 310, pPCV 501 and pPCV 730. Only small progenies were tested for segregation (Fig. 12 and 13). The proportion of the resistant seedlings indicated that the most common occurrence was where more than one copy of the T-DNA was inserted into the chromosomes of the plants. In all T_2 populations where less than 20 individuals were tested there was at least one susceptible plant, suggesting that no more than two nonhomologous chromosomes were involved in the transformation of the original cells. The segregation ratios did not provide precise information on the number of inserts because it was impossible to tell why in some progenies a large proportion of the seeds failed to germinate. The non-germinating seeds might have been the result of physiological conditions prevailing during the differentiation and development of the reproductive structures. It was also conceivable that the dead seed contained insertions into vital genes and thus suffering from dominant lethal mutations.

Only 13 seeds were planted of plant 501-13 (pPCV 501) from which 2 failed to germinate, 2 were obviously resistant to hygromycin B, 2 were pale, and 7 albina showed up. This indicated that insertional mutations may have occurred. These mutations may have had different phenotypic consequences in the heterozygous and homozygous condition and they may have epistatic interactions.

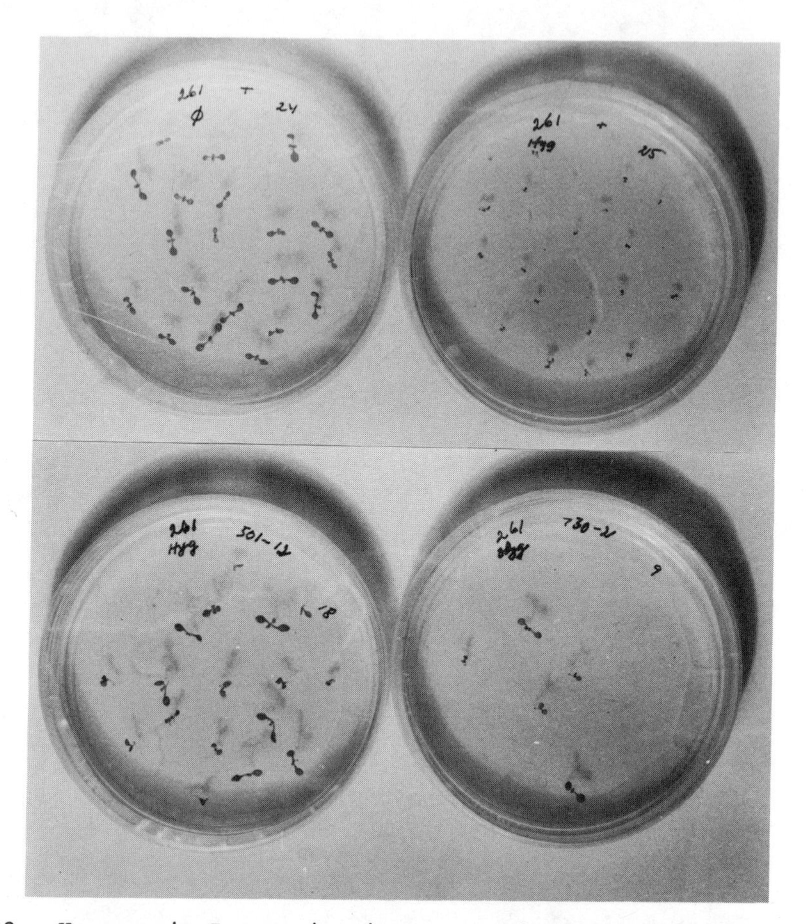

Fig. 13. Hygromycin B even in minute quantity (2.5 μg/mL) has clear effect on Arabidopsis at the cotyledonous stage. Top: wild type plants germinated without hygromycin (left), and on hygromycin containing medium (right). Bottom: T_2 progenies of transgenic plants showing segregation for hygromycin resistance. Left: transformed by pPCV 501, right: by pPCV 730. Note the difference in the size of the cotyledons. Even more conspicuous was the difference in the elongation of the roots (not visible on the picture because of the white background).

Biochemical Evidence of Transformation

Octopine tests. Of the six vectors used, four contained information for octopine synthesis. Octopine is an arginine analog which can be produced only by plant cells which were transformed by appropriate Agrobacterium strains carrying the specific T-DNA gene (OCS). The analysis was carried out by the procedure of Otten and Schilperoort (1978) as modified by Murphy and Otten (1985). Small callus pieces are sufficient for the detection of this opine. From each plant assays were made both with and without the complete reaction mixture. The material examined revealed octopine from the complete system only. After four hours of incubation aliquots were separated by paper electrophoresis and the resulting fluorescent spots on phenanthrenequinone-treated paper were photographed under ultraviolet light by a polaroid camera (Fig. 14). Some of the octopine spots were quite faint yet other line of evidence (neomycinphosphotransferase II activity, Southern blots) indicated that the same plant was transformed indeed.

Neomycinphosphotransferase assays. The aminoglycoside, kanamycin may have several different effects on the ribosomes (Moazed and Noller, 1987), among them the most important is probably the inhibition of amino acid translocation by attaching

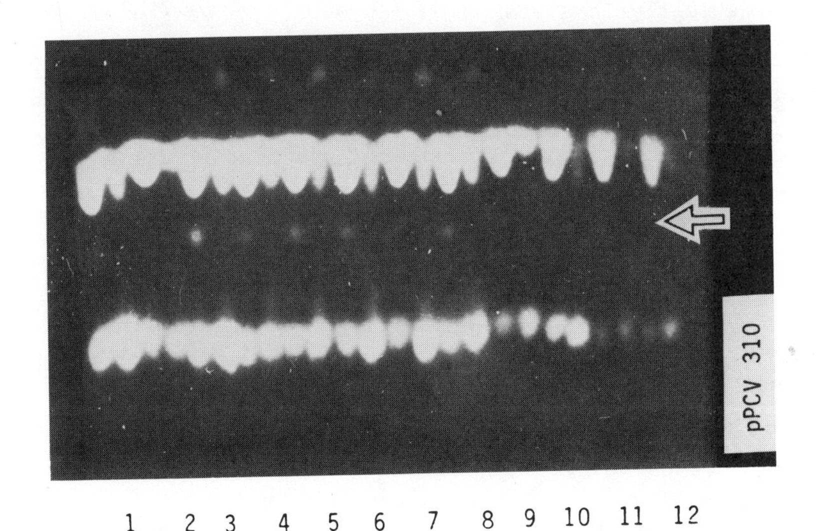

Fig. 14. The majority of the transformed tissues provided extracts with detectable octopine synthase activity. Octopine appeared as small fluorescent spots (arrow). The large spot above octopine is arginine, and below NADH, both ingredients of the reaction mixture.

to the A site and also by interfering with decoding by sequestering critical macromolecules and distorting conformation of structural domains necessary for protein synthesis. Umezawa and coworkers (1967) showed 20 years ago that a gene carried on an R plasmid of *Escherichia coli* was capable of inactivating several aminoglycosides by virtue of phosphorylation at critical and specific hydroxyl groups. A single phosphorylation may suffice to eliminate most of the toxicity of this group of antibiotics. The phosphokinase with a narrower spectrum of specificity was called neomycinphosphotransferase I, and the other capable of adding phosphates to an additional class of molecules was named neomycinphosphotransferase II (see Haas and Dowding, 1975).

The vectors used in these studies conveyed kanamycin resistance to the transgenic plants because they contained the structural gene of NPT II derived from Tn5 of *E. coli*. NPT II activity was assayed according to the procedure developed by Reiss et al. (1984) with modifications by Schreier et al. (1985). Plant extracts were subjected to discontinuous polyacrylamide gel electrophoresis. Subsequently, the gel was overlayered--on a sealed glass plate--with a 1% agarose containing 1 mg/mL kanamycinsulfate and 100 μCi ATP in an incubation buffer. The solidified agarose was then overlayed by 1 sheet of P81 phosphocellulose paper and 3 sheets of 3MM filter paper and on top of them a pile of blotting material was weighted down. Then a 3 hr incubation took place. After this the P81 paper was deproteinized enzymatically in a sealed bag at 60 ^{o}C (1 mg/mL protease in 1% SDS). The phosphocellulose paper was then washed several times by warm 10mM phosphate buffer, pH 7.5. After drying it was autoradiographed. The bottom spots represent the phosphorylated kanamycin molecules, and thus neomycinphosphotransferase activity. The spots above show some other unspecific phosphokinase activities of the plant extracts (Fig. 15).

Presence of Foreign DNA in the Genetic Material of the Plants. From Ti plasmid vectors only the T-DNA, bounded by 25-base pair direct repeats, is integrated into the plant chromosome(s) (Wang et al., 1987). The presence of intact left and right border sequences (Fig. 4) were essential for the facile insertion of the passenger DNA. The integrated T-DNA retains most of the bases of the left border sequences frequently whereas generally there is no junction beyond the third base of the right border (Wang et al., 1987).

The integration is mediated by the virulence loci (*vir*) located on the bacterial plasmid and chromosome (*chv*). The genes were activated in response to plant wound signal molecules of acetosyringone.

For the detection of the physical presence of the foreign

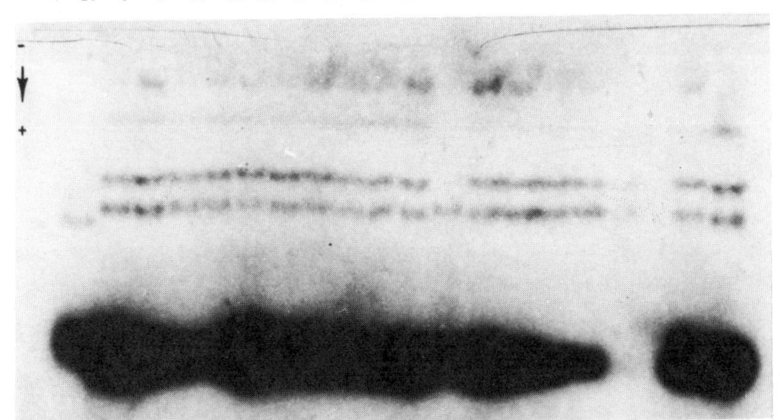

Fig. 15. Autoradiogram of the products of a neomycinphosphotransferase assay. The first lane (T) at the left shows NPT II activity from transgenic tobacco tissue. All other lanes are from Arabidopsis plants transformed by pPCV 310 (except the xx control lane). Altogether 37 plants were assayed and all appeared the same. Note the difference in mobility of the unspecific kinases in tobacco and Arabidopsis (in the middle of the photo).

DNA, Southern analysis was used (Southern, 1975). As probes both the left and right ends of the T-DNA were employed, including the octopine synthase (OCS) and the neomycinphosphotransferase (NPT II) genes, respectively (Fig. 16).

All five plants--unexpectedly--have the same fragment pattern. This may be a fortuitious coincidence or it may be that they were derived from the same original transformed cell. The fast growing calli can be quite friable at the early stages and they all may represent the same tissue clone. These plants originated from the very first attempts at transformation of Arabidopsis in the laboratory.

All five plants appeared to have two inserts. This information was supported by the genetic observations, that none of the T_2 progenies tested segregated for 3 resistant: 1 susceptible to kanamycin or hygromycin, respectively. Rather, approximately digenic ratios were observed (Fig. 13). Several more plants must be analyzed and additional restriction enzymes must be used before further conclusions are made. It is obvious that the OCS and the NPT II genes were actually incorporated into the plant chromosomes.

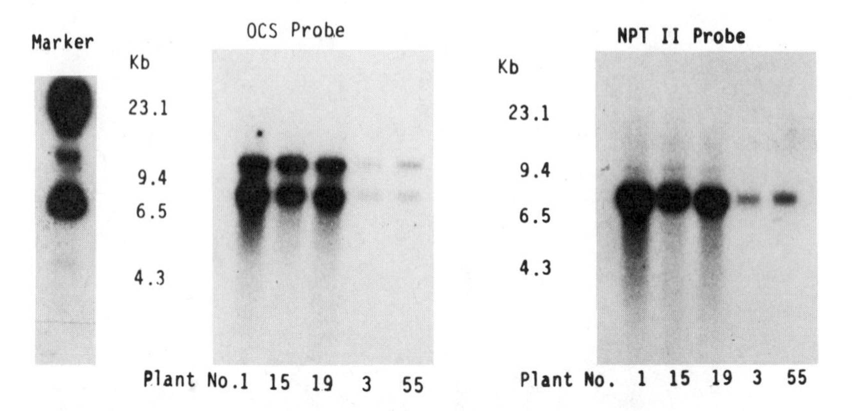

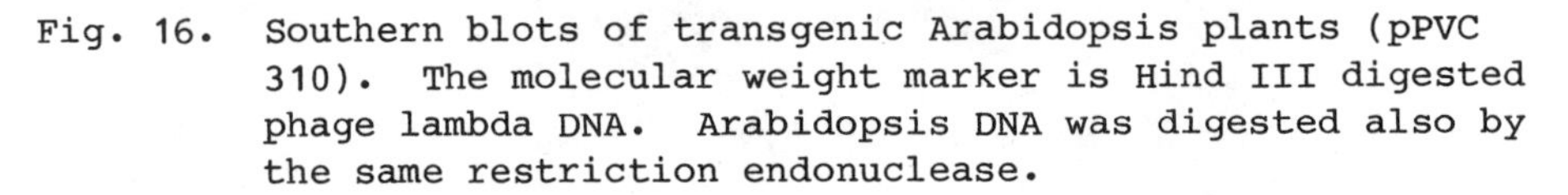
Fig. 16. Southern blots of transgenic Arabidopsis plants (pPVC 310). The molecular weight marker is Hind III digested phage lambda DNA. Arabidopsis DNA was digested also by the same restriction endonuclease.

GENETIC ANALYSIS OF THE TRANSGENES

At the moment the locations of the transgenes in the Arabidopsis genome are not known. The chromosomal assignment of the inserts can be made most economically by trisomic analysis. Fortunately primary trisomics are available for all five chromosomes of the plant genome, in addition one telotrisomic can be used (Lee-Chen and Steinitz-Sears, 1967). The primary trisomics are phenotypically distinguishable from the disomic plants and from each other, and they can be identified without chromosome counts (Fig. 17).

Assignment of Inserts to Chromosomes

The transgenes do not have counterparts in the plant genome, and they are expressed as dominant mutations. In the case where more than one insert is within a particular chromosome, trisomic analysis can not determine their location unless they were situated on different arms and telotrisomics would be available for at least one of the two arms concerned. The advantage of using trisomic analysis over test crosses or the product method (Rédei, 1982) is the lack of ambiguity regarding the chromosome involved. Actually, making only four different crosses suffices; if an insert is not in one of the four, it must be in the fifth. In the case where conventional crosses are used, and linkage is sought with strategically positioned markers along the chromosome, the clarity of the results are influenced by the frequency of recombination between the markers involved and the insert(s). The analysis may be particularly cumbersome when plant color markers

Fig. 17. The primary trisomics of Arabidopsis on Columbia wild type background grown under natural summer day length in Missouri. At the upper left corner is a normal disomic plant. The appearance of the telotrisomics available is very similar to the plant shown in the upper right corner, and it is trisomic for only one arm of the same chromosome.

are used that may mimic the effects of the selective media on the susceptible segregants.

The simpliest procedure is to cross the trisomics by transgenic plants and study only the progeny of the trisomic individuals from the first generation. The transmission of the extra chromosome is generally limited to the egg, and even through the egg, transmission is less than the 50% theoretically expected. Thus, among the offspring of trisomics no more than 1/5 or 1/10 of the individuals are trisomic. Obviously trisomics can produce two main types of gametes: monosomics and disomics. The proportion of the chromosome markers is also affected by the distance of the markers (in this consideration the insert) from the syntenic centromeres. Longer distances permit more frequent recombination between the insert and the centromere and favors the higher frequency of identical segments in the same gamete (Table 2).

The theoretical expectations may not be realized among the gametes actually formed because the inserts may affect the transmission of the chromosome involved. Heterozygosity for the insert is not a problem in the initial cross but in such a case only half of the trisomic offspring will be useful for the

Table 2. Gametic output of trisomics (Iii) with one insert (I); i stands for lack of an insert.

Insert is absolutely linked to the centromere				
II	Ii	ii	I	i
0	2	1	1	2
Insert segregates independently from centromere				
II	Ii	ii	I	i
1	6	5	4	8

analysis. Since the male transmission of disomic gametes is always much reduced or nil, the trisomics must be used as the pistillate parent.

Since the genetic constitution of the gametes cannot be directly determined, self-pollinated progeny of the trisomics must be used for the analysis. Knowledge of the gametic expectations is necessary to predict the sporophytic proportions (Table 3).

The expected segregation ratios can be easily distinguished from the 3 I : 1 i, expected when the insertion does not involve

Table 3. Phenotypic proportions among the progeny of trisomics heterozygous for an insertion (Iii). The numbers refer to the phenotypes only and disomic and trisomic individuals are pooled on the basis of presence or absence of the insert. Albeit six different possibilities were considered, under practical experimental conditions intermediate situation between the last two rows of the table are expected to occur most commonly.

Transmission of disomic gametes through	Absolute linkage			Independent segreg.		
	I	i		I	i	
Male and female	3	1	(0.25)	2.41	1	(0.29)
Female only	2	1	(0.33)	1.77	1	(0.36)
Neither male nor female	5	4	(0.44)	1.25	1	(0.44)

Table 4. Gametic output of recessive insertional mutations (AAa).

The mutation is absolutely linked to a centromere				
AA	Aa	aa	A	a
1	2	0	2	1

The mutation segregates independently from centromere				
AA	Aa	aa	A	a
5	6	1	8	4

the chromosome present in three doses. Even the worst scenario of 2 : 1 segregation of the trisomic's offspring is distinguishable from the 3 : 1 at the 0.95 probability level in a population of 133. Once the synteny is determined, the more precise chromosomal position can be sought with the aid of the available chromosomal markers. Wallroth et al. (1986) used for transformation *Petunia hybrida* with at least one genetic marker on each of the seven chromosomes.

Chromosomal Assignment of Insertional Mutations

Insertion of the T-DNA within the boundaries of plant genes may result in inactivation of that gene detectable as recessive mutations. These recessive mutations can also be assigned to chromosomes by trisomic analysis even easier than the inserts as long as the insertional mutations have clear stable phenotype.

The combined disomic and trisomic offspring of heterozygous trisomics reveals the chromosomal location of the insertional recessive mutations. When the recessive mutation does not involve the chromosome triplicate in the trisomic, the phenotypic proportions are 3 normal wild type : 1 recessive mutant. When the trisome carries the mutation the segregation ratio varies from between 17 : 1 to 8 : 1, depending on its vicinity to the centromere, and even more importantly on the degree of transmission (Table 5).

When synteny is known, the situation of the new mutation within the linkage group can be determined by standard procedures. In cases where there are no markers within less than 20% recombination of the mutations vicinity, Haldane's mapping function may be useful for more precise mapping (Rédei, 1982). Mapping inserts and insertional mutations is of considerable interest because it may reveal much about the organization of chromosomes. Arabidopsis has only 1 - 3% as much DNA as the common angiosperms, and it is particulary low in redundant sequences (Pruitt and Meyerowitz, 1986).

Table 5. Phenotypic proportions in the progeny of trisomics of AAa genetic constitution.

Transmission of disomic gametes through	Absolute linkage A	a		Independent segreg. A	a	
Male and female	35	1	(0.03)	22.04	1	(0.04)
Female only	17	1	(0.06)	13.40	1	(0.07)
Neither male nor female	8	1	(0.11)	8	1	(0.11)

DISCUSSION

Transformation of Arabidopsis by the use of various agrobacterial vectors does not pose any more problems. At this time over 16,000 calli selected on hygromycin and kanamycin media, respectively are available. More than 12,000 differentiated rosette leaves or they are at the flowering or seed-bearing stages.

Hygromycin was found to be an effective selective agent (Lloyd et al., 1986), but it appears now that transformation to kanamycin resistance is also practical with the leaf or stem infection methods. During the course of this study much simpler procedures than previously used (such as feeder cells) proved to be quite effective. The mineral media and hormone regimes reported in this paper eventually result in almost 100% shoot regeneration on the surviving calli. Shoot formation occasionally is detectable on the 9th day of the cultures.

One serious bottleneck in the process still remains, namely the lack of a reliable procedure for rooting the shoots. Frequently, high frequency root differentiation adversely affects the formation of normal fruits and viable seed. However, successfully rooted plants produce more and better quality seed. If this last hurdle can be overcome, the procedure of transformation of Arabidopsis may be suitable to targeting specific genes with inserts.

According to an estimate the number of gene loci of Arabidopsis capable of yielding visible mutations is approximately 28,000 (Rédei et al., 1984). In case insertional mutations occur at random, and uniformly along the entire genome of Arabidopsis, each of these loci have about 1/28.000 (ca. 4×10^{-5}) chance for being hit by the T-DNA. Since the size of individual genes, just as mutability, varies over a wide range, the probability of specific insertions can be expected to differ as well. Ignoring the extremes, and assuming average chances one can estimate the probability of inflicting a particular gene with an insertion. In order to find at least 1 inactivation by insertion at a chosen level of probability one must solve the following equation:

$$f^n = P$$

or it can be rewritten in a more meaningful way because one wishes to determine the size of the population (n) required where among 28,000 potentially hit genes there would be no more than 5% chance of missing one desired type of alteration: $(27,999/28,000)^n = 0.05$ hence $n = \log 0.05/(\log 27,999) - (\log 28,000) = 83,883$. In other words, if the chances for all genes are the same for being hit by an insertion, among ca. 84,000 insertional mutation cases at least 1 must be the desired type with a probability of 0.95. Since the transformation experiments are carried out with diploid tissues, 42,000 cells may suffice.

In our transformation experiments the number of transgenic calli obtained approached this figure. In addition, there is an indication for the presence of insertional mutations among the transgenic plants in our as well as in the experiments carried out by Alan Lloyd (personal communication). Thus tagging even specific selected genes may be within reach in extended experiments.

There is also another possibility by gene tagging through the use selectable structural genes introduced into the plants without appropriate promoters. The fusion of the selectable (reporter) gene with other genes permits the identification of the other component as well as their genetic and developmental regulation (André et al., 1987). Plasmids pPCV 601 and 631 used in these studies can be employed for such analyses.

SUMMARY

Although Arabidopsis was recalcitrant to transformation for a number of years, the production of transgenic indivdiuals became quite practical within the past year. Transformation was accomplished with six different binary vectors conveying one or two of the following functions to the plant cells: octopine synthesis, neomycinphosphotransferase II and hygromycinphosphotransferase. Over 16,000 transgenic calli were selected and 12,000 of them could be regenerated into shooty plants. The success of transformation was verified by selective growth, octopine assays, neomycinphosphotransferase II activity, Southern hybridization and analysis of the seedling progenies. The procedures used in infection, for tissue culture and regeneration are reported. The procedures for assigning inserts and insertional mutations to chromosomes are outlined and the problems involved in gene tagging by T-DNA are discussed.

Acknowledgements

The conscientious technical assistance of Magdi Rédei is much appreciated. We are indebted to Joyce Reinbott for the preparation of the typescript. This research was supported by the Max-Planck-Institut, Köln, Federal Republic of Germany, The United States Department of Agriculture grant 86-CRCR-1-2015, the Graduate School of the University of Missouri-Columbia, the Missouri Agricultural Experiment Station. Missouri Agricultural Experiment Station Journal Series No. 10253.

LITERATURE

André, D., Colau, D., Schell, J., Van Montagu, M., and Hernalsteens, J-P. 1987, Gene tagging in plants by a T-DNA insertion mutagen that generates ARH(3')II-plant gene fusions, Mol. Gen. Genet., 204:512-518.

Cocking, E. C., and Davey, M. R., 1987, Gene transfer in cereals, Science, 236:1259-1262.

De Greeve, H., Leemans, J., Hernalsteens, J. P., Thia-Toong, L., De Beuckeleer, M., Willmitzer, L., Otten, L., Van Montagu, M., and Schell, J., 1982, Regeneration of normal and fertile plants that express octopine synthase, from tobacco grown galls after deletion of tumour-controlling functions, Nature, 300:752-755.

Feldmann, K. A., and Marks M. D., 1987, Agrobacterium-mediated transformation of germinating seeds: A non-tissue culture approach, Mol. Gen. Genet., 207:in press.

Gamborg, O. L., Miller, R. A., and Ojima, K., 1968, Nutrient requirements of suspension cultures of soybean root cells, Expt. Cell Res., 50:151-158.

Haas, M. J., and Dowding, J. E., 1975, Aminoglycoside-modifying enzymes, Methods of Enzymology, 43:611-627.

Horsch, R. B., Frey, J. E., Hoffmann, N. L., Wallroth, M., Eichholtz, D. A., Rogers, S. G., and Fraley, R. T., 1985, A simple and general method for transferring genes into plants, Science, 227:1229-1231.

Klein, T. M., Wolf, E. D., Wu, R., and Sanford, J. C., 1987, High-velocity microprojectiles for delivering nucleic acids into living cells, Nature, 327:70-73.

Lee-Chen, S., and Steinitz-Sears, L. M., 1976, The location of linkage groups in Arabidopsis thaliana, Can. J. Genet. Cytol., 9:381-384.

Lichtenstein, C. P., and Draper, J., 1985, Genetic engineering of plants, in: "DNA Cloning: A Practical Approach", Gover, D. M., Ed., IRL Press, Washington, DC, 2:67-119.

Lloyd, A., Barnason, A. R., Rogers, S. G., Byrne, M. C., Fraley, R. T., and Horsch, R. B., 1986, Transformation of Arabidopsis thaliana with Agrobacterium tumefaciens, Science, 234:464-466.

Márton, L., Wullems, G. J., Molendijk, L., and Schilperoort, R. A., 1979, In vitro transformation of cultured cells from Nicotiana tabacum by Agrobacterium tumefaciens, Nature, 277:129-131.

McClintock, B., 1931, Cytological observations of deficiencies involving known genes, translocations and an inversion in Zea mays, Missouri Agricultural Experiment Station, Research Bull. 163.

McClintock, B., 1938, The fusion of broken ends of sister half-chromatids following chromatid breakage at meiotic anaphases, Missouri Agricultural Experiment Station Research Bull. 290.

McClintock, B., 1951, Chromosome organization and genetic expression, Cold Spring Harbor Symp. Quant. Biol., 16:13-47.

McClintock, B., 1956, Controlling elements and the gene, Cold Spring Harbor Symp. Quant. Biol., 21:197-216.

Moazed, D., and Noller, H. F., 1987, Interaction of antibiotics with functional sites in the 16S ribosomal RNA, Nature, 327:389-394.

Murashige, T., and Skoog, F., 1962, A revised, medium for rapid growth and bio assays with tobacco tissue culture, Physiol. Plant., 15:473-497.

Murphy, P., and Otten, L., 1985, Detection of opines in transformed plant tissue, in: "EMBO Course on Transfer and Expression of Genes in Higher Plants", Max-Planck-Institut für Züchtungsforsch., Köln, FRG., pp. 41-46, Czernilofsky, P., Schell, J., and Willmitzer, L., eds.

Otten, L., De Greve, Hernalsteens, J. P., Van Montague, M., Schieder, O., Straub, J., and Schell, J, 1981, Mendelian transmission of genes introduced into plants by the Ti plasmids of Agrobacterium tumefacients, Mol. Gen. Genet., 183:209-213.

Otten, L., and Schilperoort, R. A., 1978, A rapid microscale method for the detection of lysopine and nopaline dehydrogenase activities, Biochem. Biophys. Acta., 517:497-500.

Paszkowski, J., Shillito, R. D., Saul, M., Mandak, V., Hohn, T., Hohn, B., and Potrykus, I., 1984, Direct gene transfer to plants, EMBO J., 3:2717-2722.

Potter, H., Weir, L., and Leder, P., 1984, Enhancer-dependent expression of human immunoglobulin genes intoduced into mouse pre-B lymphocytes by electroporation, Proc. Natl.

Acad. Sci., USA, 81:7161-7165.
Pruitt, R. E., and Meyerowitz, E. M., 1986, Characterization of the genome of Arabidopsis thaliana, J. Mol. Biol., 187:169-183.
Rédei, G. P., 1963, Somatic instability caused by a cysteine-sensitive gene in Arabidopsis, Science, 139:767-769.
Rédei, G. P., 1965, Genetic blocks in the thiamine synthesis of the angiosperm Arabidopsis, Amer. J. Bot., 52:834-841.
Rédei, G. P., 1973, Extra-chromosomal mutability determined by a nuclear gene locus in Arabidopsis, Mutation Res., 18:149-162.
Rédei, G. P., 1974, Genetic mechanisms in differentiation and development. Pp. 183-209, in: "Genetic Manipulation with Plant Material", Ledoux, L., ed., Plenum, New York.
Rédei, G. P., 1982, "Genetics", Macmillan, New York.
Rédei, G. P., Acedo, G. N., and Sandhu, S. S., 1984, Mutation induction and detection in Arabidopsis, in: "Mutation, Cancer and Malformation", Chu, E. Y. and Generoso, W. M., eds., Plenum, New York, pp. 285-313.
Rédei, G. P., Chung, S. C., and White, J. A., 1974, Mutants, antimetabolites and differentiation, Brookhaven Symp. Biol., 25:281-296.
Rédei, G. P., and Plurad, S. B., 1973, Hereditary structural alterations of plastids induced by a nuclear mutator gene in Arabidopsis, Protoplasma, 77:361-380.
Reiss, B., Sprengel, R., Will, H., and Scaller, H., 1984, A new sensitive method for qualitative and quantitative assay of neomycin phosphotransferase in crude cell extracts, Gene, 30:211-218.
Schreier, P., Seftor, E. A., Schell, J., and Bohnert, B. J., 1985, The use of nuclear encoded sequences to direct the light-regulated synthesis and transport of a foreign protein into plant chloroplasts, EMBO J., 4:25-32.
Shapiro, J. A., ed., 1983, "Mobile Genetic elements", Acad. Press, New York.
Southern, E., 1975, Detection of specific sequences among DNA fragments separated by gel electrophoresis, J. Mol. Biol., 98:503-517.
Umezawa, H., Okanishi, M., Kondo, S., Hamana, K., Utahara, R., and Maeda, K., 1967, Phosphorylative inactivation of aminoglycosidic antibiotics by Escherichia coli carrying R factor, Science, 157:1559-1561.
Wallroth, M., Gerats, A.G.M., Rogers, S. G., Fraley, R. T., and Horsch, R. B., 1986, Chromosomal localization of foreign genes in Petunia hybrida, Mol. Gen. Genet., 202:6-15.
Wang, K., Stachel, S. E., Timmerman, B., Van Montagu, M., and Zambryski, P. C., 1987, Site-specific nick in the T-DNA border sequence as a result of Agrobacterium vir gene expression, Science, 235:587-591.

Ribosomal DNA in Maize[1]

R.L.Phillips, M.D.McMullen[2], S.Enomoto & I.Rubenstein

Depts. of Agron. & Pl. Genet., Genet. & Cell Biol., & Pl. Molec. Genet. Inst. Univ. of MN, St. Paul, MN 55108

SUMMARY

The nucleolus organizer region (NOR) of maize chromosome 6 is now known to be the location of the 17/5.8/26S rDNA (hereafter termed 17/26S rDNA); this molecular information extends the NOR function to one of synthesizer as well as organizer. The extensive variation in ribosomal DNA (rDNA) multiplicity and the finding that much of the rDNA is in the NOR-heterochromatin and probably highly methylated raises questions as to the role of such high rDNA multiplicities in maize development. Deficiencies for parts of the NOR that include up to 98-99% of the rDNA are transmissible through ovules. The 5S rDNA is located in chromosome 2, it is apparently less variable in repeat number, and its multiplicity is not correlated with that of the 17/26S rDNA. The 17/26S and 5S sequences are highly conserved in maize, but sequence variations such as base substitutions, deletions, and methylations occur. Methylation patterns change during development. The relationship of 17/26S rDNA multiplicity with RNA concentration is not straightforward.

INTRODUCTION

The nucleolus organizer region (NOR) was defined by McClintock in 1934 as an organizer of nucleolar material. When the NOR was

---

[1]This work was supported by the Minnesota Agricultural Experiment Station and by a grant from The McKnight Foundation.

[2]Current address: USDA-ARS, Dept. of Agronomy, OARDC, Wooster, Ohio 44691

absent in microspores as the result of meiotic segregation of a heterozygous NOR-deficiency, the nucleolar material was present but dispersed and not organized into a single nucleolus. Thus, the term nucleolus organizer was applied to the "nucleolar-organizer body", the heterochromatic body in the short arm of chromosome 6 associated with the nucleolus. McClintock identified this body as the chromosomal structure involved in organizing a nucleolus by her use of an interchange with a break in the heterochromatic portion of the NOR. The interchange divided the "nucleolar-organizing body" into two parts and each possessed the capacity to organize a nucleolus, albeit ones of different sizes.

The 17/26S ribosomal RNA genes were shown via DNA/rRNA hybridization to be located at the NOR of maize (Phillips et al., 1971). This finding was not surprising because Lin (1955) had shown by cytophotometric techniques that the RNA content of the maize microsporocyte nucleolus was correlated with the number of NORs present. Since most RNA in a cell is ribosomal RNA, the localization of the 17/26S rRNA genes to the NOR was anticipated. In addition, Ritossa and Spiegelman (1965) had reported a linear relationship between rDNA amount and NOR dosage in *Drosophila*.

Localization of 17/26S and 5S rRNA Genes

The 17/26S rDNA is now precisely localized to the NOR, although the possibility exists that a few rRNA genes (orphons) may be located else where in the genome. The tandemly repeated rRNA genes are present throughout the heterochromatic portion (Phillips et al., 1971, 1974a, 1979; Ramirez and Sinclair, 1972; Givens and Phillips, 1976; Doerschug, 1976) as well as the secondary constriction portion (Phillips et al., 1979, 1983) of the NOR. Whether the rRNA genes occur in clusters within each of these NOR regions is not clear at this time, nor is it known if any other repeated sequences are interspersed.

The rRNA/DNA saturation hybridization experiments by Givens and Phillips (1976) indicated that at least 90% of the rDNA is located in the heterochromatic portion of the NOR and 10% or less is in the secondary constriction portion. Subsequently, *in situ* rRNA/DNA hybridizations with pachytene chromosomes of various NOR-interchange homozygotes indicated that about 70% of the rDNA was in the NOR-heterochromatin and 30% in the NOR-secondary constriction (Phillips et al., 1983). The former estimate of 90% rDNA in the NOR-heterochromatin is more accurate due to the possible differential hybridization efficiencies of heterochromatin versus the diffuse euchromatin of the NOR-secondary constriction. The *in situ* hybridization experiments demonstrate, however, that rDNA is present in the NOR-secondary constriction. Later, the possibility that the rDNA in the NOR-heterochromatin may represent inactive rDNA that is methylated will be discussed. *In situ* hybridization also showed that at least 20% of the rDNA is in the proximal 10% of the NOR-heterochromatin.

The 5S rDNA of plants is often not in the same chromosome as the NOR. The maize 5S rRNA genes were shown by Wimber et al. (1974) to be located either in chromosome 2 or 3. To determine the exact chromosome, Mascia et al. (1981) used interchange homozygotes with a break in the NOR and in chromosome 2 (short arm), 2 (long arm), or 3. The interchange with a break in chromosome 2S had a chromosome attached to the nucleolus with 5S rRNA hybridization near the end of the long arm. The 2L interchange had a chromosome attached to the nucleolus with 5S rRNA hybridization to its satellite region. Pachytene cells of the T3-6 interchange showed 5S rRNA hybridization to a chromosome independent of the nucleolus. These results localized the 5S rDNA to 2L, at about 88% of the distance from the centromere to the end of the long arm. The maize 17/26S rDNA and the 5S rDNA, therefore, are in different chromosomes. The evidence to date indicates that the tandemly repeated copies in each case are localized at a single site, the 17/26S rDNA at the chromosome 6 NOR and the 5S rDNA at about 2L.88.

Gene Multiplicity

Ribosomal DNA is tandemly repeated in all eukaryotic organisms, presumably due to the great requirement for rRNA at certain points in the life cycle. The maize 17/26S rDNA is likewise multiply repeated. One interesting feature is the vast amount of line to line variation in rRNA gene multiplicity, even for agronomically elite lines.

Gene number estimates are based in part on the DNA amount per nucleus. The most recent cellular DNA values for maize are about one-half earlier estimates. Although maize lines differ in their DNA content, the average value for the inbred lines reported by Laurie and Bennett (1985) was 5.35 pg per 2C nucleus, which agrees closely with the value reported by Galbraith et al. (1983). Using this DNA estimate, the rRNA gene copy number per 2C nucleus for maize is from 1650 for the *sticky* chromosome mutant to 11,500 for the University of Illinois Reverse High Protein line (Phillips, 1978). A survey of 21 inbred lines revealed variation in rRNA gene copy number from 2500 to 6000 per 2C nucleus (Phillips, 1978). Rivin et al. (1986) reports rRNA gene numbers in the same range for 10 strains of maize. This wide variation in rRNA gene multiplicity among elite maize lines may indicate that not all of them are needed.

The rRNA gene number may vary as the result of some mechanism such as unequal crossing over that would give variation in number unrelated to their importance. Most rRNA genes are present in the NOR-heterochromatin probably in a non-transcribing state. However, it is difficult to prove that additional gene copies are not needed at specific stages of the life cycle, or perhaps serve a purpose under certain stress conditions, or have some other role. Although no evidence exists at the present time proving the occurrence of unequal crossing over among rRNA genes in maize, studies on progenies of Illinois High Protein (IHP) x Illinois Low Protein (ILP) and Illinois Reverse High Protein (RHP) x IHP both backcrossed to RHP indicated that the copy number did not segregate as an indivisible block of genes (Phillips, 1977). Ribosomal RNA gene numbers outside the range of the parents as well as between the anticipated numbers were reported. Further studies are needed.

Ribosomes contain 5S rRNA as well as the rRNA products of the 17/26S genes. Variation in gene multiplicity among maize strains seems to be much less for the 5S rRNA genes, ranging from 2000 to 5000 copies per 2C nucleus (Rivin et al., 1986). One might expect that the number of 5S rRNA genes would correlate with the 17/26S rRNA gene number. Rivin et al. (1986) found no correlation between the relative number of 17/26S and 5S rRNA genes. Crosses of *Zea diploperennis* x *Zea mays* F_1 to *Zea mays* indicate that the 5S gene arrays of the two species are not inherited in a co-dominant fashion. The 5S genes of one species or the other were underrepresented in certain progeny (Zimmer and DeSalle, 1983).

Transmission of rDNA Deficiencies and Duplications

Deficiencies of all or part of the NOR might help define how much of the rDNA is necessary for normal plant development. The gametophytic screen in plants makes it unlikely that a mutation like the *Xenopus* NOR deficiency would be transmissible. However, partial NOR deficiencies can be generated via meiotic segregation in heterozygotes carrying an interchange with one break in the short arm of chromosome 6. Twenty-five interchanges are available in maize with a break either in the NOR-heterochromatin or the NOR-secondary constriction. By crossing NOR-interchange heterozygotes as the female parent with plants heterozygous for *polymitotic* (*po*), the transmission of a partial NOR deficiency is revealed by the production of *polymitotic* plants among the progeny. This system depends on the fact that *po* is located in the satellite region beyond the NOR and expresses its phenotype (male-sterility) when hemizygous. Smaller kernels can be selected in these crosses to enhance the recovery of duplicate-deficient types. In a test reported in 1983 by Phillips and coworkers, 23 of the 25 NOR-interchanges were shown to transmit the duplicate-deficient gamete deficient for a portion of the NOR. The two NOR-interchanges that initially gave a negative test have subsequently been shown to transmit the partial NOR deficiency (Phillips, unpublished results). The largest partial NOR deficiency to be transmitted through the ovule (but not the pollen) is from T6-9(067-6); this partial NOR deficiency accounts for 98-99% of the region (Phillips et al., 1984). The T5-6c with a break in the short arm near the chromosome 6 centromere does not transmit the deficiency for the entire NOR, based on the _polymitotic_ test (Phillips, unpublished results).

The information available at the present time indicates that maize embryo sacs can develop with only 1-2% of the NOR-heterochromatin, probably representing about 1-2% of the rDNA, and that resultant hemizygous endosperm tissues and sporophytes can survive when most of the rDNA is contributed only by the male parent. The survival of plants with such a dramatic reduction in rDNA indicates that the normal high level of rDNA in maize is not essential.

A true breeding, homozygous, apparent duplicate-deficient (Dp-Df) product of a NOR-interchange (Fig. 1) in which the NOR region is present in duplicate has been described in an alien addition plant of maize (Pasupuleti, personal communication). This

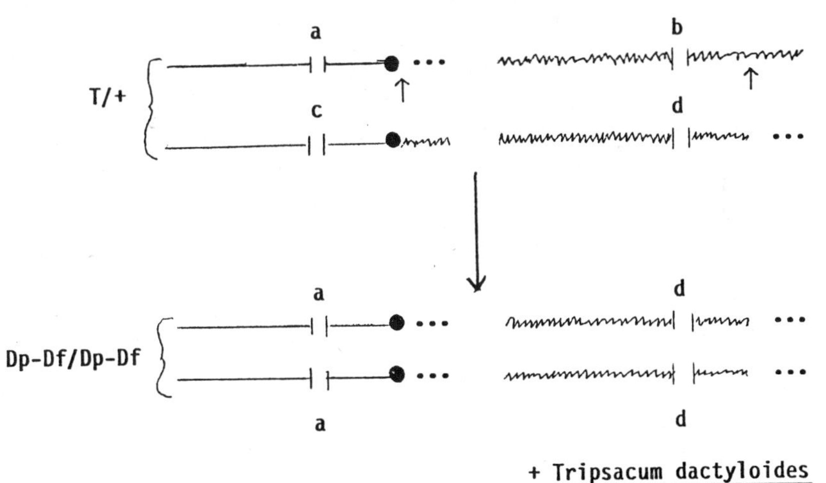

Fig. 1. Possible origin of the NOR duplication discovered by C.V. Pasupuleti. Upper diagram shows a heterozygous interchange with one break in the NOR-secondary constriction and the other in an unidentified chromosome. Lower diagram shows a homozygous duplication-deficiency (Dp-Df) derived from the heterozygous interchange. This Dp-Df plant also carried a single _Tripsacum dactyloides_ chromosome 9.

plant was derived from an intergeneric cross between maize and Tripsacum dactyloides L. and carried a normal chromosome 6 bivalent plus an apparently homozygous Dp-Df chromosome that terminated with a NOR-secondary constriction and the chromosome 6 satellite. About 75% of the pachytene and diakinesis cells had a single nucleolus with two associated bivalents and about 25% possessed two nucleoli each with an associated bivalent. Spore quartets usually possessed two nucleoli of different size in each microspore. An interesting aspect of this plant was that it carried a univalent Tripsacum chromosome 9, which has homologies with the short arm of maize chromosome 2. Possibly the presence of this alien chromosome compensated for the deficiency in the Dp-Df chromosome allowing it to be pollen transmissible. The other chromosome in this interchange involving chromosome 6 has not been identified, but might be hypothesized to be chromosome 2. This homozygous duplication for the NOR-secondary constriction may be especially useful for studying the rDNA in that region.

Ribosomal DNA Sequences

Structure: The rDNA repeat (Fig. 2) in maize is about 9.1kb in length (Benton, 1984) and nearly the entire repeat has now been sequenced (Messing et al., 1984; McMullen et al., 1986; Toloczyki and Feix, 1986; Oleson, personal communication). The various restriction enzyme digests indicate that the maize rDNA is present as a tandem "head-to-tail" repeat (Benton, 1984). The 17S rRNA has been sequenced by Messing et al. (1984) and is 1809 nucleotides in length with 50.9% GC. The 5.8S (about 165bp) and 26S rDNA (about 3360bp) sequences have been determined by Oleson (personal communication). He reports the internal transcribed spacers I and II to be 192 and 233bp, respectively.

The complete nucleotide sequence of the rDNA spacer region of approximately 3000 bases has been determined for Black Mexican Sweet (BMS) suspension culture cells (McMullen et al., 1986) and for inbred A619 (Toloczyki and Feix, 1986). Sequences from both sources

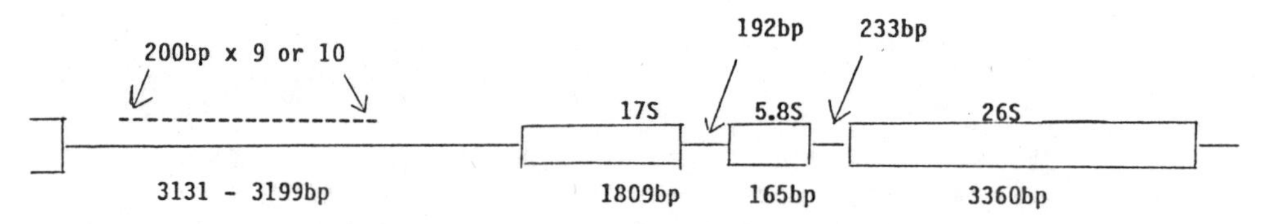

Fig. 2. The rDNA repeat indicating the sequence length of each component including the external non-transcribed and the two internal transcribed spacers, and the 17S, 5.8S, and 26S sequences.

reveal a subrepeat within the spacer region. The subrepeat is about 200 nucleotides (range between 165 and 232) in length and is present in at least 9 copies and perhaps 10, depending on the level of homology specified to define the subrepeat. These subrepeats may have a role in regulating rRNA transcription (McMullen et al., 1986, 1987; Toloczyki and Feix, 1986). About two-thirds of the spacer region is comprised of these subrepeats.

While the majority of rDNA repeat units appear uniform, heterogeneity for certain restriction sites and for length of the spacer region (Rivin et al., 1983; Hunter, 1984) is present in some lines. Hunter (1984) isolated 32 EcoRI generated clones of rDNA from Black Mexican Sweet suspension culture genomic DNA. She found that 23 were identical with the conserved EcoRI site near the 3' end of the 26S gene, 6 differed at a SmaI restriction site, 1 varied at a XhoI site, and 2 possessed deletions in the spacer region. Other experiments showed that rDNA repeats in A632, B37, W22, Illinois High Protein, Illinois Low Protein, and Illinois Reverse High Protein contained one additional EcoRI site per repeat or a HindIII site (Hunter, 1984). Variation for restriction endonuclease cleavage sites and for non-transcribed spacer lengths also were observed among inbred lines by Rivin et al. (1983). Inbred H25 is an example of several lines with an additional EcoRI site in the 26S gene near the 3' end such that 8kb and 1.1kb fragments are generated. Another EcoRI site that generates 6.5kb and 2.6kb fragments is found in *Zea diploperennis*, *Zea perennis*, and *Zea luxurians*.

The 5S rRNA repeat is 320bp in length and the repeats differ both in post-replicational modification and divergence of the nucleotide sequence (Mascia et al., 1981). Zimmer and DeSalle (1983) reported that the 5S gene repeat of inbreds B37N, H25, Tx303 and WKF, of Ohio Yellow Popcorn, of the South American maize, Confite Puneno, and of the Mexican teosinte Balsas was identical in size. The 5S repeat of *Zea luxurians*, *Zea diploperennis*, *Zea perennis*, five *Tripsacum* species, and *Sorghum* were 354bp, 300bp, 300bp, 260bp, and 280bp, respectively. Almost all of the 5S rDNA repeats have the internal C of the CCGG site methylated. The spacer region is 200bp and the coding region is 120bp; divergence in nucleotide sequence appears to occur in both regions.

Transcription: A maize rRNA precursor has been identified by hybridizing a probe specific to the spacer with rRNA from BMS suspension cells (Fig. 3). The precursor appears to be 6.5kb in length. The start of transcription is probably 823 bases upstream of the 17S region; the precursor includes these 823 bases as well as the regions ultimately processed to give the 17S, 5.8S, and 26S rRNA species. S1 protection experiments indicated two possible spacer 5' termini for the rRNA precursor, one at 519 and another at 823 bases upstream of the 5' end of the mature 17S rRNA (McMullen et al., 1986). The site 519 bases from the 17S rRNA is probably a

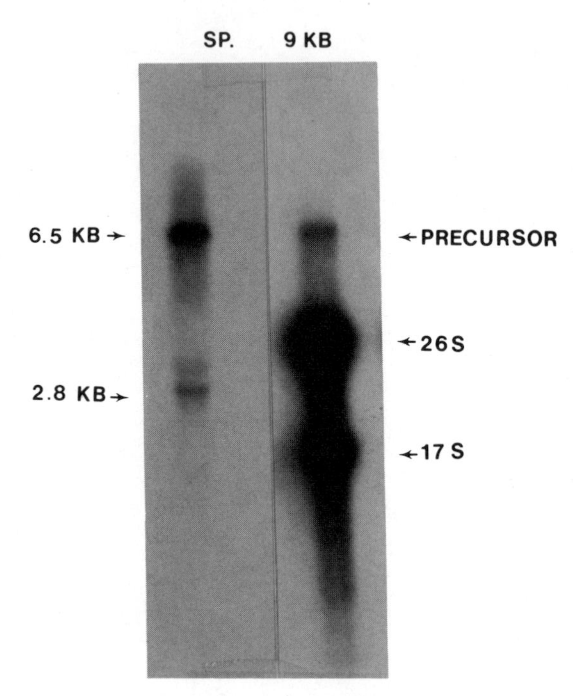

Fig. 3. Identification of a rRNA precursor by northern analysis. Total cell RNA was isolated from BMS suspension cells, denatured with glyoxal, electrophoresed in an agarose gel, transferred to nitro-cellulose and hybridized with nick translated plasmid probe of either the rDNA spacer region (SP.) or the entire 9.1kb repeat (9kb). The autoradiogram of the RNA hybridized with the entire 9.1kb repeat was overexposed to reveal the rRNA precursor in addition to the cytoplasmic 17S and 26S rRNA. The autoradiogram of the RNA hybridized with the spacer probe indicates a 6.5kb rRNA precursor as well as a 2.8kb RNA which we believe is a processing intermediate (McMullen and Rubenstein, unpublished results).

processing site. These results suggest that the maize rRNA promoter may occur in a region 800-900 bases upstream from the 5' end of the 17S rRNA coding region and about 144 bases 3' of the end of the last spacer subrepeat.

Methylation: The activity of genes may be related to the degree of methylation. We had previously noted the possible correlation of the proportion of rRNA genes in the NOR-heterochromatin and the amount of rDNA methylated--both are quite high.

The level of cytosine methylation in the rDNA region of inbred A188 plant DNA was examined using HpaII, a methylation sensitive restriction endonuclease which does not cleave the sequence CCGG when the internal C is methylated. Based on the sequence data, there are 75 CCGG sites in the 9.1kb rDNA repeat unit.

A genomic blot (Southern, 1975) was made of a HpaII complete digest of DNA isolated from the tissue of 10-day-old etiolated seedlings. The genomic blot was hybridized with a radioactive probe consisting of the spacer 3.2kb MboI-MboI fragment (Fig. 4). Most of the rDNA in the blot evidenced a molecular weight of greater than 18kb. A small fraction of the rDNA, however, generated a 9.1kb band. If a substantial fraction of the rDNA repeats were

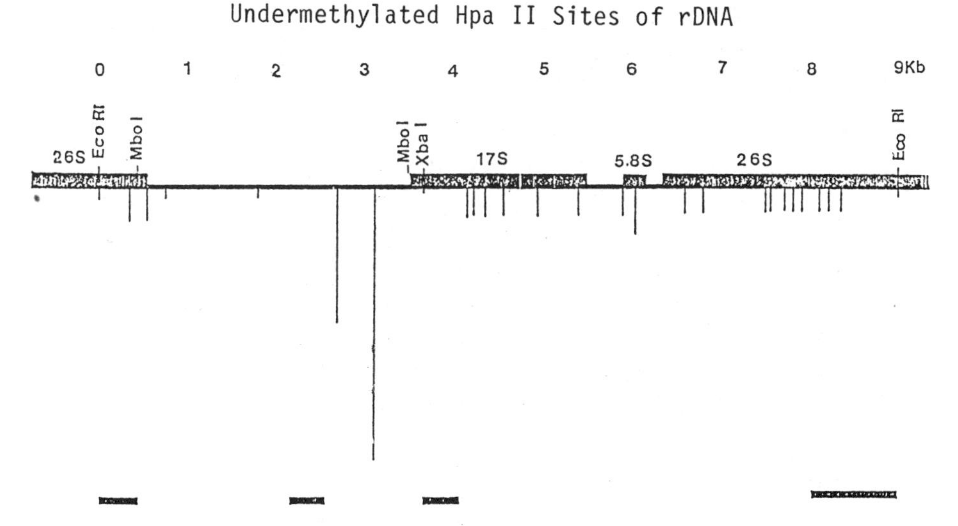

Fig. 4. The rDNA repeat showing restriction sites (EcoRI, MboI, and XbaI), day 17 endosperm undermethylated sites and their relative frequencies (vertical lines), and probes (solid horizontal bars).

unmethylated, then one would expect to detect the HpaII fragment from the region of the spacer where several CCGG sites are separated by 1kb (McMullen et al., 1986; Fig. 4). These results indicate that most of the rDNA repeats have the majority of their HpaII sites methylated, but that some of the rDNA repeats which are adjacent to one another have HpaII sites that are not methylated. Furthermore, these unmethylated HpaII sites must be located in a specific region of the 9.1kb rDNA sequence.

A similar experiment involving the double digestion of seedling maize DNA with the two restriction endonucleases EcoRI and HpaII was done to locate the unmethylated HpaII sites. Two size classes of bands were seen when the genomic blot was hybridized with a probe consisting of the entire spacer region (the MboI-MboI probe, Fig. 4). Most of the rDNA was present as a 9.1kb band. This 9.1kb band indicates that there is only one cleavable EcoRI site in each of the A188 rDNA repeats and that all HpaII sites are methylated in these particular sequences. A cluster of bands about 3kb in size was also detected. This experiment was repeated using the EcoRI-MboI probe from the 26S rRNA region of the repeat. Two bands were seen, one about 2.7kb and the other 2.9kb in size. This experiment demonstrated that the CCGG sequences where the preferential cleavage was occurring (and had been detected in the previous experiments) were in the spacer region and located close to the bases where the 5' termini of rRNA precursor(s) had been found (McMullen et al., 1986).

When the restriction endonuclease MspI is used to digest maize DNA, different results are obtained. This restriction endonuclease cleaves the CCGG sequence irrespective of whether the internal C is methylated. Genomic blots of maize DNA cleaved with MspI and probed with the 9.1kb EcoRI rDNA probe indicated a series of bands with molecular weights smaller than 9.1kb. This result indicated that CCGG sequences are present in most of the rDNA 17/26S repeats. The actual base sequence for this region confirms this result (Messing et al., 1984; McMullen et al., 1986; Oleson, personal communication).

We hypothesize that two major classes of rDNA repeats exist: a class of 17/26S rRNA genes that are methylated at most, if not all, of their HpaII sites, and a class of rDNA repeats that are preferentially undermethylated in one region, the putative promoter region. The two classes of rDNA repeats are not interspersed, but are most likely organized into clusters with similar degrees of methylation in each cluster.

Next, we asked if the methylation pattern is the same for different tissues. Developing endosperm was examined because (1) endosperm nucleoli are typically enlarged indicative of high rDNA activity, and (2) we wished to correlate any methylation changes with the DNA amplification that occurs during normal endosperm development (Kowles and Phillips, 1985).

Ears were collected every other day starting 7 days after pollination up to day 23. Endosperm was dissected and the DNA purified. Southern blots of DNA cut with HpaII were probed with the spacer probe. DNA samples of younger ears (day 7 and day 10) give the same results as the seedling DNA. This pattern seems to be conserved among different inbreds (Jupe and Zimmer, 1986). In contrast, the older endosperm DNA shows four diffuse bands of 3.9, 3.3, 2.6 and 2.1kb and less of the 9.1kb band. This shift in the pattern occurs at day 10 to 12, the same time at which the DNA increase starts (Kowles and Phillips, 1985).

To map the undermethylated CCGG sequences, day 17 endosperm DNA was double digested with either HpaII and EcoRI, or HpaII and XbaI; DNA blots were hybridized with small probes abutting EcoRI and XbaI sites (Fig. 4). Preferential undermethylation occurred in the promoter region, preserving the overall pattern seen in younger endosperm and seedling DNA. A third class of rDNA which is undermethylated at multiple sites within a repeat comprises a major fraction of the older endosperm DNA. Of the 75 CCGG sequences within the 9.1kb unit repeat, five of the CCGG sequences are clustered in the putative promoter region.

The internal C in a CCGG sequence of the rDNA is apparently methylated/demethylated in a precise order, but the underlying mechanism is unknown. We propose that the specific undermethylation pattern of the rDNA is related to the high activity of these undermethylated templates either for replication or transcription. The class of rDNA that is completely methylated may constitute the heterochromatinized rDNA.

To test this proposal we have developed a procedure for isolating nucleoli from endosperm tissue. Isolated nucleoli should be enriched for active rDNA because the NOR-heterochromatin will be sheared from the NOR-secondary constriction. Genomic blots of such DNA digested with HpaII show bands that are smaller compared to total DNA blots (Enomoto, unpublished results).

The proportion of rDNA relative to total genomic DNA also appears to change during endosperm development (Phillips et al., 1983). The reason for the rDNA proportionality changes is not known, but it has been repeatable over a four year period in field grown material. Hybridization of rRNA to whole endosperm DNA from kernels of self-pollinations of the single cross Wf9 x B37 was performed. The rDNA proportion remained constant from 4 to 10 days after pollination but then increased to a peak value around day 16. By 20 days after pollination the proportion of rDNA returned to the initial level. Whether this change in rDNA proportion represents disproportionate replication of rDNA during part of endosperm development is not clear.

Ribosomal RNA Gene Multiplicity and rRNA Levels

Exciting breeding and selection protocols could be devised if rRNA gene multiplicity were correlated with rRNA or protein levels. We evaluated nine inbred lines for rRNA concentration and rRNA gene number (Buescher et al., 1984). Only two inbreds differed significantly from inbred W23 which was used as the standard. Studies on strains with an apparent duplication of the NOR and the University of Illinois Protein lines that differed significantly in rRNA gene multiplicity also showed no obvious correlations with rRNA levels (Buescher et al., 1984). In addition, no correlation was found with 13 self-pollinated lines of the cross Illinois Low Protein x (Illinois High Protein x Illinois High Protein). The conclusion was that rRNA levels are not correlated with rRNA gene multiplicity in maize. This lack of correlation may be due to measuring the total rRNA gene number and not the number of active rRNA genes. Perhaps a correlation could be detected if the multiplicity of hypomethylated rDNA repeats were measured. Or the level of rRNA transcription might be related to the spacer subrepeats present that could serve as enhancers of transcription. The active sequences would be expected to be in the NOR-secondary constriction (McMullen et al., 1987).

A phenomenon that could be interpreted as nucleolar dominance occurs in maize heterozygous for NOR-interchanges (McMullen et al., 1987). Microspores receiving a chromosome 6 with only the proximal portion of the NOR - no matter how small - organize a normal sized nucleolus. If the distal NOR portion is present, the proximal NOR portion forms only a small nucleolus. Therefore, the distal NOR portion appears to suppress the activity of rRNA genes in the proximal heterochromatin. An understanding of this suppression of nucleolar activity might lead to methods for altering rRNA levels.

Impact of New Concepts

Ribosomal RNA genes are essential for plant development and represent fundamental components of the protein synthetic machinery. Because the various rDNA's and their RNA products are readily detectable, and even manipulatable, correlations with interesting cell, organ, or whole plant traits should lead to useful genetic and breeding protocols. Understanding the relationship of rDNA methylation and gene activity may lead to methods for determining active rRNA gene numbers. Correlations of active rRNA gene numbers with rRNA and protein levels might become evident. Important aspects of development may be related to rDNA methylation patterns. If the subrepeats of the rDNA nontranscribed spacer are enhancers, we may have ready access to an important regulatory sequence.

Further research defining the structure, regulatory controls, and genetic behavior of rDNA is clearly warranted.

LITERATURE CITED

Benton, W. D., 1984, Sequence organization of the ribosomal RNA genes in _Zea mays_, Ph.D. Thesis, St. Paul, University of Minnesota.

Buescher, P. J., Phillips, R. L., and Brambl, R., 1984, Ribosomal RNA contents of maize genotypes with different ribosomal RNA gene numbers, _Biochem. Genet._, 22:923-930.

Doerschug, E. B., 1976, Placement of the genes for ribosomal-RNA within the nucleolar organizing body of _Zea mays_, _Chromosoma_ (Berl.), 55:43-56.

Galbraith, D. W., Harkins, K. R., Maddox, J. M., Ayres, N. M., Sharma, D. P., and Firoozabady, E., 1983, Rapid flow cytometric analysis of the cell cycle in intact plant tissues, _Science_, 220:1049-1051.

Givens, J. F., and Phillips, R. L., 1976, The nucleolus organizer region of maize (_Zea mays_ L.): Ribosomal RNA gene distribution and nucleolar interactions, _Chromosoma_ (Berl.), 57:103-117.

Hunter, B. G., 1984, Examination of the heterogeneity among maize rDNA repeats and preliminary analysis of the transcription of the rDNA, Ph.D. Thesis, St. Paul, University of Minnesota.

Jupe, E. R., and Zimmer, E. A., 1986, Heterogeneity of ribosomal gene methylation in inbred lines, _Maize Genet. Coop. News Lett._, 60:22-23.

Kowles, R. V., and Phillips, R. L., 1985, DNA amplification patterns in maize endosperm nuclei during kernel development, _Proc. Nat. Acad. Sci._ USA, 82:7010-7014.

Laurie, D. A., and Bennett, M. D., 1985, Nuclear DNA content in the genera _Zea_ and _Sorghum_. Intergeneric, interspecific, and intraspecific variation, _Heredity_, 55:307-313.

Lin, M., 1955, Chromosomal control of nuclear composition in maize, _Chromosoma_ (Berl.), 7:340-370.

Mascia, P. N., Rubenstein, I., Phillips, R. L., Wang, A. S., and Xiang, L. Z., 1981, Localization of the 5S rRNA genes and evidence for diversity in the 5S rDNA region of maize, _Gene_, 15:7-20.

McClintock, B., 1934, The relation of a particular chromosomal element to the development of the nucleoli in _Zea mays_, _Z. Zellforsch. Mikrosk. Anat._, 21:294-328.

McMullen, M. D., Hunter, B., Phillips, R. L., and Rubenstein, I., 1986, The structure of the maize ribosomal DNA spacer region, _Nucl. Acids Res._, 14:4953-4968.

McMullen, M. D., Phillips, R. L., and Rubenstein, I., 1987, Molecular analysis of the nucleolus organizer region in maize, _in_: "Chromosome Engineering in Plants: Genetics, Breeding, and Evolution," T. Tsuchiya and P. K. Gupta, eds., Elsevier Science Publ., Amsterdam (in press).

Messing, J., Carlson, J., Hagen, G., Rubenstein, I., and Oleson, A., 1984, Cloning and sequencing of the ribosomal RNA genes in maize: The 17S region, DNA, 3:31-40.

Phillips, R. L., 1977, Genetic engineering for crop improvement, 32nd Annu. Corn and Sorghum Res. Conf., pp. 6-20.

Phillips, R. L., 1978, Molecular cytogenetics of the nucleolus organizer region, in: "Maize Breeding and Genetics," D. B. Walden, ed., John Wiley and Sons, Inc., New York, pp. 711-741.

Phillips, R. L., Kleese, R. A., and Wang, S. S., 1971, The nucleolus organizer region of maize (Zea mays L.): Chromosomal site of DNA complementary to ribosomal RNA, Chromosoma (Berl.), 36:79-88.

Phillips, R. L., Kowles, R. V., McMullen, M. D., Enomoto, S., and Rubenstein, I., 1985, Developmentally-timed changes in maize endosperm DNA, in: "Plant Genetics," M. Freeling, ed., Alan R. Liss, New York, pp. 739-754.

Phillips, R. L., Wang, A. S., and Bullock, W. P., 1984, Transmission of a deficiency for nearly the entire nucleolus organizer region, Maize Genet. Coop. News Lett., 58:181.

Phillips, R. L., Wang, A. S., and Kowles, R. V., 1983, Molecular and developmental cytogenetics of gene multiplicity in maize, Stadler Symp., 15:105-118.

Phillips, R. L., Wang, A. S., Rubenstein, I., and Park, W. D., 1979, Hybridization of ribosomal RNA to maize chromosomes, Maydica, 24:7-21.

Phillips, R. L., Weber, D. F., Kleese, R. A., and Wang, S. S., 1974, The nucleolus organizer region of maize (Zea mays L.): Tests for ribosomal gene compensation or magnification, Genetics, 77:285-297.

Ramirez, S. A., and Sinclair, J. H., 1975, Ribosomal gene localization and distribution (arrangement) within the nucleolar organizer region of Zea mays, Genetics, 80:505-518.

Ritossa, F. M., and Spiegelman, S., 1965, Localization of DNA complementary to ribosomal RNA in the nucleolus organizer region of Drosophila melanogaster, Proc. Nat. Acad. Sci. USA, 53:737-745.

Rivin, C. J., Cullis, C. A., and Walbot, V., 1986, Evaluating quantitative variation in the genome of Zea mays, Genetics, 113:1009-1019.

Rivin, C. J., Zimmer, E. A., Cullis, C. A., Walbot, V., Huynh, T., and Davis, R.W., 1983, Evaluation of genomic variability at the nucleic acid level, Plant Mol. Biol. Reporter, 1:9-16.

Southern, E. M., 1975, Detection of specific sequences among DNA fragments separated by gel electrophoresis, J. Mol. Biol., 98:503-517.

Toloczyki, C., and Feix, G., 1986, Occurrence of 9 homologous repeat units in the external spacer region of a nuclear maize rRNA gene unit, Nucl. Acids Res., 14:4969-4986.

Zimmer, E. A., and DeSalle, R., 1983, Structure and inheritance of 5S DNA genes of maize and teosinte, Genetics, 104:573-74.

QUANTITATIVE CLADISTIC ANALYSES OF CHROMOSOMAL BANDING DATA AMONG SPECIES IN THREE ORDERS OF MAMMALS: HOMINOID PRIMATES, FELIDS AND ARVICOLID RODENTS

William S. Modi

Program Resources, Incorporated
Frederick Cancer Research Facility
Frederick, MD

Stephen J. O'Brien

Laboratory of Viral Carcinogenesis
National Cancer Institute
Building 560, Room 21-105
Frederick, MD

INTRODUCTION

Over the past two decades a number of technical advances have been made which have improved the resolving power of differential chromosome staining procedures. For example, Q-banding (Caspersson, *et al.*, 1970), G-banding (Seabright, 1971), and R-banding (Dutrillaux and Lejeune, 1971) all produce longitudinal chromosomal differentiation, enabling the identification of homologous elements both within and between species. C-banding provides determination of the amount and location of constitutive heterochromatin (Sumner, 1972), whereas a number of fluorochromes, when used in conjunction with the appropriate counterstain, produce regional banding patterns or highlight specific heterochromatic regions (Schweizer, 1981). Staining with silver nitrate has been shown to identify the chromosomal locations of transcriptionally active 18S + 28S ribosomal genes (rDNA) (Bloom and Goodpasture, 1976). Primate chromosomes have been shown to stain differentially against a rodent background in somatic cell hybrids using the alkaline Giemsa (G-11) staining procedure (Bobrow and Cross, 1974).

Incorporation of tritiated thymidine or bromodeoxyuridine (BrdU) followed by the appropriate staining allows for the visualization of sister chromatic exchanges (Perry and Wolff, 1974).

These technical developments in chromosome banding methodology, when coupled with progress in cell culture, heterologous cell fusion techniques and molecular biology, have led to rapid advances in gene mapping (Ruddle, et al., 1971; Harper and Saunders, 1981; Shows et al., 1982; O'Brien et al., 1985b), mammalian dosage compensation (Lock and Martin, 1986), chromosome structure (Adolph, 1986), and cell cycle dynamics (Prescott, 1987). Cytogenetic morphology has also been studied in a comparative sense among related biological species (Bush et al., 1977; Baker et al., 1987). The use of cytogenetic characters in three divergent mammalian taxa and the evaluation of cytogenetic methodology in taxonomic inference form the subject of this paper.

Sokal (1985) has suggested that taxonomy has three principle goals: 1) to establish an information storage and retrieval system, 2) to generate hypotheses about evolutionary relationships, and 3) to serve as a model against which to test general evolutionary hypotheses such as those dealing with rates, modes of speciation and character modification. Conceptual differences in the treatment of comparative biological data have led to the emergence of three distinct schools of taxonomic thought. The phenetic school (Sneath and Sokal, 1973) bases classification upon the overall similarity of the organisms being considered. The cladistic school (Eldredge and Cracraft, 1980; Wiley, 1981) establishes relationships based upon genealogy by using shared, derived character states to define monophyletic groupings. The evolutionary systematists (Simpson, 1961; Mayr, 1969) base classifications upon observed similarities and differences among organisms in light of their inferred evolutionary histories. One assumption perhaps shared by all three schools is that a natural taxonomy should be a hierarchial, nonoverlapping arrangement of organisms (Sokal, 1985).

The present study focuses upon cladistic or phylogenetic approaches to systematically assessing chromosomal banding patterns among species belonging to three mammalian orders. Paramount to the study is the quantitative representation of G-bands using discrete numerical codes, and the subsequent usage of computer based algorithms to construct cladistic trees using a parsimony criterion.

MATERIALS AND METHODS

Phylogenetic analyses of G-banded karyotypes were carried out for species belonging to three different groups of mammals. Each of these three data sets has certain unique characteristics and will be described separately.

Carnivora: Felidae

Cytogenetic preparations (G-bands) were obtained from a total of 18 species (Modi *et al.*, 1987a). All methods used in cell line initiation, chromosome harvest and differential staining are discussed elsewhere (Modi *et al.*, 1987b). Briefly, metaphase chromosomes were obtained from fibroblast cell lines that were initiated from skin explants. Fibroblasts were transformed with feline sarcoma virus and propagated in MEM alpha medium supplemented with 10% fetal bovine serum. Metaphase chromosomes were harvested and fixed following standard procedures.

G-banded karyotypes for the 18 felid species were compared with those summarized by Wurster-Hill and Centerwall (1982) who examined G-banded karyotypes from 33 of the 37 species of cats. Eight felid autosomes exhibit interspecific variation. The combination of polymorphic autosomes was used to assign each of the 33 species to one of 13 groups. Because of their interrelationships, the presence or absence of each of eight chromosomes could be expressed by a binary code (0 or 1) for each of six chromosomes (B4, D2, E4, E5, F1, and C3). These binary codes (0 or 1) for each species appear on the right-hand side of Table 1.

The chromosome characters for each species group were treated in a cladistic analysis to construct minimum-length evolutionary trees using a parsimony criterion through the PAUP computer program (Swofford, 1984). The PAUP program also has a branch swapping option that calculates multiple tree topologies of equal length, if they exist, for any given data set. A consensus tree that represents a summary of the information contained within the family of equally parsimonious rival trees was also generated.

Primates: Pongidae and Hominidae

Chromosomal banding data including late prophase karyotypes and ideograms for the orangutan, gorilla, chimpanzee and man have been described by Yunis and Prakash (1982). They found that 22 pairs of autosomes and the X and Y chromosomes were homologous among the four species and that interspecific differences are attributed to rearrangements involving these homologous elements.

Table 1. Summary of cytogenetic data for 33 species of Felidae.

Species	Common Name	Species Code	2n	Felid Chromosome[a]				
				B4	D2	E4	F1	C3
Felis catus	Domestic cat	FCA	38	0	0	0	1	0
Felis libyca	African wild cat	FLI	38	0	0	0	1	0
Felis sylvestris	European wild cat	FSI	38	0	0	0	1	0
Felis margarita	Sand cat	FMA	38	0	0	0	1	0
Felis chaus	Jungle cat	FCH	38	0	0	0	1	0
Felis nigripes	Black-footed cat	FNI	38	0	0	0	1	0
Prionailurus bengalensis	Leopard cat	PBE	38	0	1	1	0	0
Prionailurus viverrinus	Fishing cat	PVI	38	0	1	1	0	0
Puma concolor	Puma, cougar, mountain lion	PCO	38	0	1	1	0	0
Mayailurus iriomotensis	Iriomote cat	MIR	38	0	1	1	0	0
Herpailurus yagouaroundi[b]	Jaguarundi	HYA	38	0	1	1	0	0
Panthera leo	Lion	PLE	38	1	1	0	0	0
Panthera tigris	Tiger	PTI	38	1	1	0	0	0
Panthera pardus	Leopard	PPA	38	1	1	0	0	0
Panthera onca	Jaguar	PON	38	1	1	0	0	0
Panthera uncia	Snow leopard	PUN	38	1	1	0	0	0
Lynx lynx	Lynx	LLY	38	1	1	0	0	0
Lynx rufus	Bobcat	LRU	38	1	1	0	0	0
Pardofelis marmorata	Marbled cat	PMA	38	1	1	0	0	0
Leopardus pardalus	Ocelot	LPA	36	0	2[c]	0	0	1
Leopardus weidi	Margay	LWE	36	0	2[c]	0	0	1

Leopardus tigrina	Tiger cat	LTI	36	0	1	0	0	1
Leopardus geoffroyi	Geoffroy's cat	LGE	36	0	1	0	0	1
Lynchailurus colocolo	Pampas cat	LCO	36	0	0	0	0	1
Ictailurus planiceps	Flat-headed cat	IPL	38	1	1	1	0	0
Caracal caracal	Caracal	CCA	38	0	1	0	1	0
Otocolobus manul	Pallas's cat	OMA	38	0	0	1	0	0
Prionailurus rubiginosa[b]	Rusty spotted cat	PRU	38	0	0	1	0	0
Leptailurus serval	Serval	LSE	38	0	1	0	0	0
Profelis temmincki[b]	Asian golden cat	PTE	38	0	1	0	0	0
Neofelis nebulosa[b]	Clouded leopard	NNE	38	1	0	0	0	0
Acinonyx jubatus	Cheetah	AJU	38	1	0	1	0	0
Profelis aurata	African golden cat	PAU	38	1	1	0	1	0

[a]See Fig. 2 for 0,1,2 designations.
[b]Each of these species possesses unique character states for one or more chromosomes that may or may not be listed in the above five chromosomes that are binarily coded. Since these polymorphisms are not shared with other taxa, they are excluded from this table and from the quantitative analyses.
[c]A unique character state for D2 received a code of 2 in these species.

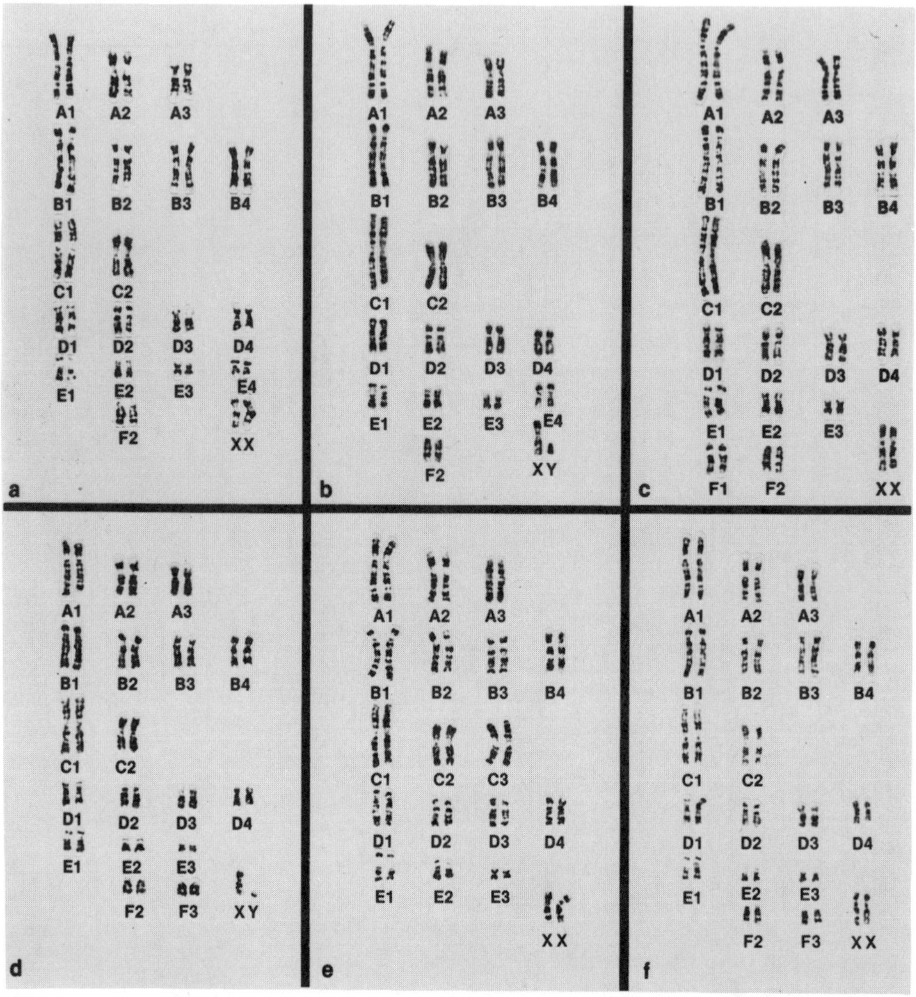

A1 A2 A3
B1 B2 B3 B4
C1 C2
D1 D2 D3 D4
E1 E2 E3 E4
F2 XX
a

A1 A2 A3
B1 B2 B3 B4
C1 C2
D1 D2 D3 D4
E1 E2 E3 E4
F2 XY
b

A1 A2 A3
B1 B2 B3 B4
C1 C2
D1 D2 D3 D4
E1 E2 E3
F1 F2 XX
c

A1 A2 A3
B1 B2 B3 B4
C1 C2
D1 D2 D3 D4
E1 E2 E3
F2 F3 XY
d

A1 A2 A3
B1 B2 B3 B4
C1 C2 C3
D1 D2 D3 D4
E1 E2 E3
XX
e

A1 A2 A3
B1 B2 B3 B4
C1 C2
D1 D2 D3 D4
E1 E2 E3
F2 F3 XX
f

Fig. 1. G-banded karyotypes of a) leopard cat (Prionailurus bengalensis), b) cheetah (Acinonyx jubatus), c) domestic cat (Felis catus), d) tiger (Panthera tigris), e) tiger cat (Leopardus tigrina), f) lion (Panthera leo).

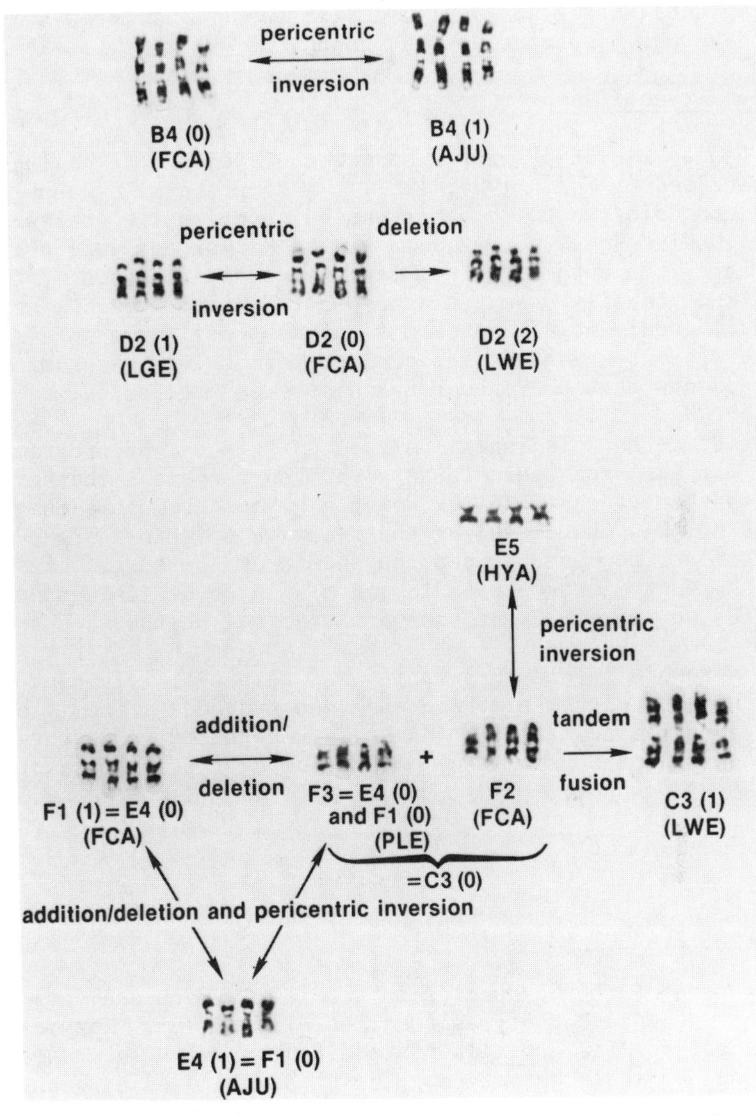

Fig. 2. The polymorphic autosomes found among the different species of cats that were used in assigning each species to one of the species groups listed in Table 1. The interrelationships among the chromosomes are indicated by arrows and accompanying rearrangements. Chromosomal character states and species codes appear in parentheses.

Each chromosomal variant within each of these homologous pairs was assigned a multistate character code (ranging from 1-4) and the resulting data matrix was analyzed with the PAUP program using an exhaustive search subroutine that found the shortest possible tree.

Rodentia: Arvicolidae

G-banded karyotypes for 22 species of voles and lemmings have been described by Modi (1986; 1987). A comprehensive description of the methodology used in an extensive quantitative analysis is given by Modi (1987). Briefly, G-banded karyotypes were prepared by arranging the autosomes from large to small for each of the 22 species individually. Next, interspecific karyotypic comparisons were carried out and all autosomes and autosomal arms were scored as being present or absent in each species. Also, euchromatic rearrangements that differed between species were noted.

Six different cladistic analyses using the PAUP program were carried out in which the weights given rearrangement characters were less than or equal to the weights given chromosome characters (defined below), thus enabling the relative weights assigned to rearrangements to vary between one and zero. Further, differences in branch lengths between FARRIS and MINF hypothetical taxonomic unit (HTU) optimization options were examined on the same tree.

As a result of branch swapping and retaining multiple solutions, each of the six PAUP analyses produced a family of equally parsimonious trees. Adams-2 consensus trees were subsequently constructed using the CONTREE computer program (Swofford, 1984).

RESULTS

Carnivora: Felidae

Eighteen of the 37 species in the family were analyzed by preparing G-banded karyotypes (Modi *et al.*, 1987b). Karyotypes for six of these species are illustrated in Figure 1. Interspecific comparisons of these karyotypes plus those prepared by Wurster-Hill and colleagues (Wurster-Hill and Gray, 1975; Wurster-Hill and Centerwall, 1982; Wurster-Hill *et al.*, 1987) (a total of 33 species) indicate that 13 chromosomal species groups may be established based upon eight polymorphic autosomes.

Variability among these eight felid autosomes is tabulated in Table 1 and illustrated in Figure 2. Chromosomes B4 and D2 occur in all species and exist in two and three different forms,

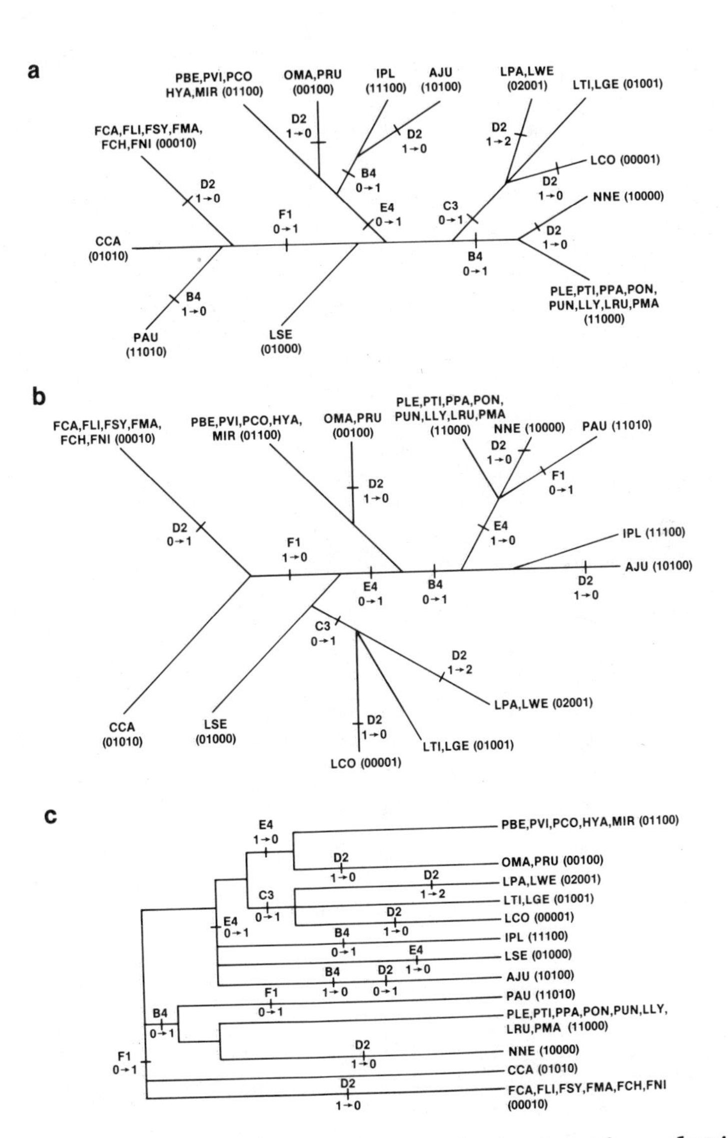

Fig. 3. Unrooted evolutionary trees depicting the relationships among the 13 chromosomal species groups of cats based upon the five binarily coded polymorphic autosomes appearing in Table 1. (a-b) rival trees of equal length, yet differing in topology. (c) Adams-2 consensus tree.

respectively. Chromosomes C3, E4, E5, F1, F2, and F3 are present in different combinations in different species. F2 and F3 have combined through tandem fusion to produce C3. In one species, the jaguarundi, F2 is modified to become E5. Similarly, E4, F1, and F3 appear to be related to one another via one or two rearrangement events (Fig. 2). The presence or absence of C3, E4, F1, F2, and F3 in the karyotype of any species may then be represented by coding for C3, E4, and F1; if these three chromosomes are missing, then F2 and F3 are automatically present. Thus, the binary codes for these three chromosomes, for B4 and for D2 are given in Table 1.

Rooting of evolutionary trees by the traditional cladistic method of defining a most primitive character state for each character was not possible because several of the polymorphic versions of each chromosome are found in non-felid families (Modi et al., 1987a). For this reason, unrooted evolutionary trees depicting the relationships among the 13 species groups based upon minimum overall tree length were constructed using the global branch swapping procedure available in the PAUP program. Over 50 trees of length 12, yet varying in branching pattern, were obtained. Two equally parsimonious trees and an associated Adams-2 consensus tree are portrayed in Fig. 3. The consistency index, the total number of character states minus the number of characters divided by the total tree length, is 0.50. This indicates that one-half of the 12 steps on the tree involve convergent changes. The most convergent chromosomal character is D2, which changes up to six times on some trees. C3 is wholly consistent, showing only one change, while the other characters change two or three times per tree.

Since a large number of different tree topologies was obtained, it is possible to view the relationships among the chromosomal species groups in slightly different ways. Several groups of species differ from one another only with respect to which form of B4 is present: the group of eight species containing the lion (PLE) versus the group containing the serval (LSE), the cheetah (AJU) versus the group containing Pallas' cat (OMA) and the group containing the leopard cat (PBE) versus the flat-heated cat (IPL). The flat-headed cat (IPL) is also only one step away from the cheetah (AJU) or from the group containing the lion (PLE). The caracal (CCA) is only one step away from the domestic cat (FCA) group on the one hand, and from the serval (LSE) group on the other. The five South American species, ocelot (LPA), margay (LWE), tiger cat (LTI), Geoffroy's cat (LGE) and the pampas cat (LCO) which are contained in three species groups, can be allied based upon the common presence of chromosome C3.

Table 2. Multistate character codes for 17 phylogenetically informative polymorphic chromosomes in each of four higher primate species (human, chimpanzee, gorilla and orangutan) and in the outgroup Old World Monkeys, baboon and rhesus monkey (data from Yunis and Prakash, 1982).

	Human chromosome homologue[a]																
Species	1	2p[b]	2q[b]	3	5	7[b]	8	9[b]	10	11	12	14	15	16	18	20	Y
Human	1	1	1	1	1	1	1	2	1	1	1	1	1	1	1	1	1
Chimpanzee	2	2	2	1	2	1	1	1	1	1	2	1	2	2	2	1	1
Gorilla	3	3	2	1	3	2	2	3	2	1	2	2	1	3	2	1	1
Orang-utan	2	3	3	2	1	3	3	3	3	2	1	1	1	1	2	2	2
Baboon, rhesus monkey	2	3	2	2	1	3	1	3	3	1	1	1	1	1	2	1	2

[a]Chromosomes 4 and 17 are unique to each of the four species, while chromosomes 6, 13, 19, 21, 22, and the X are invariant among the four species. None of these is subsequently useful in the identification of monophyletic groups.

[b]Chromosomes 2p, 2q, 7 and 9 are ordered, the others are unordered (see text and Fig. 4a).

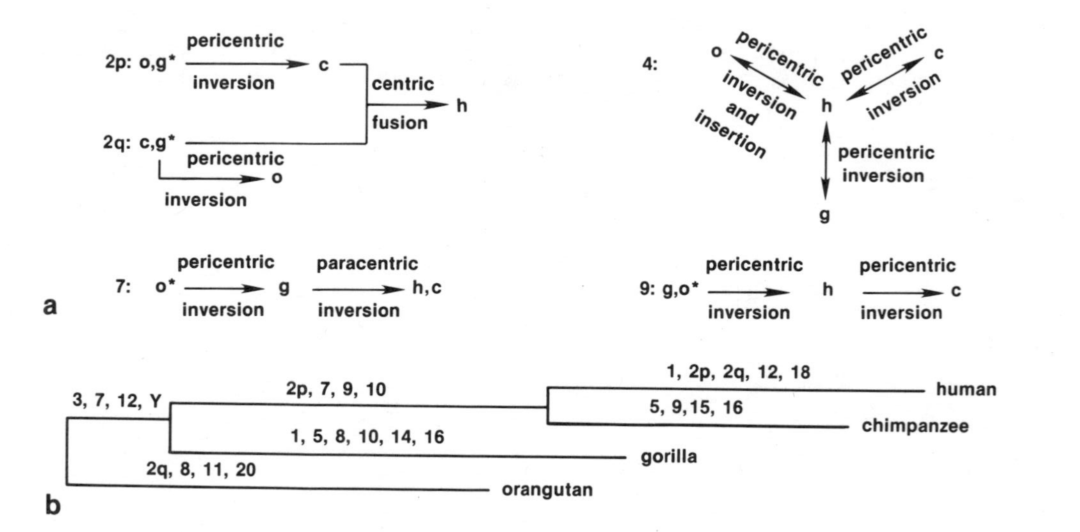

Fig. 4. a) Transition series representation of interspecific karyotypic variability for five primate chromosomes that are treated as ordered characters. Human, chimpanzee, gorilla and orangutan are abbreviated as h, c, g and o, respectively. The primitive condition, where known, is indicated by the asterisk. Each arrow represents one character step. Chromosome 4 is included here but is excluded from the accompanying analysis because each species possesses a unique variant. b) Phylogenetic tree produced by PAUP analysis of the primate data set appearing in Table 2. Changes in chromosomes from the primitive condition are indicated by their numbers along the branches.

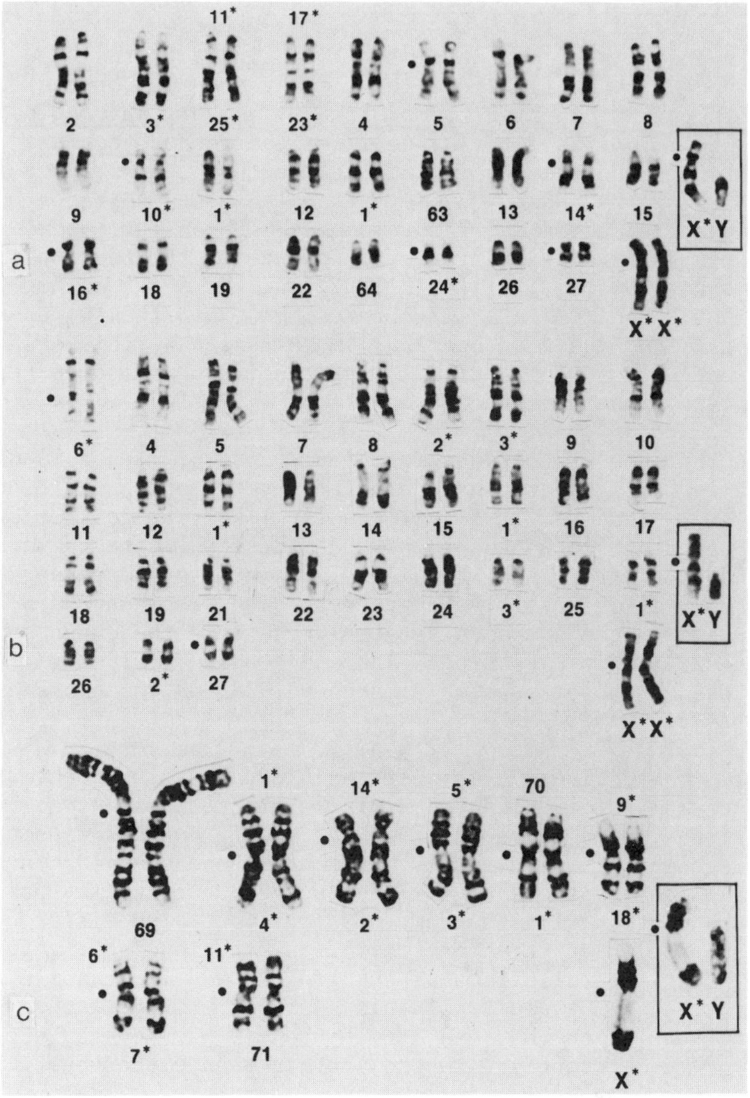

Fig. 5. G-banded karyotypes of three species of arvicolid rodents. a) Complete karyotype of a female and sex chromosomes of a male Microtus ochrogaster (2n = 54); b) complete karyotype of a female and sex chromosomes of a male M. pinetorum (2n = 62); c) complete karyotype of a female and sex chromosomes of a male M. oregoni (2n = 17 - 18). Numbers correspond to the chromosome characters appearing in Table 4, solid dots indicate centromere positions, and asterisks denote chromosomes that have undergone rearrangement from the primitive condition.

Primates: Pongidae and Hominidae

An analysis of the chromosomal banding patterns reported by Yunis and Prakash (1982) enables the chromosomes for the four hominoid primate species (Homo sapiens, human; Pan troglodytes, chimpanzee; Gorilla gorilla, gorilla; and Pongo pygmaeus, orangutan) to be coded using multistate characters. Each species has 23 (22 in the human due to a translocation) pairs of homologous autosomes plus the X and Y chromosomes. Human chromosomes 4 and 17 were seen to have unique variants in each of the four species, while chromosomes 6, 13, 19, 21, 22, and the X chromosome were invariant. The remaining 17 chromosomes exhibited two or three variants among the four species. Each variant within each of these 17 potentially phylogenetically informative chromosomes received a multistate, integer code between 1 and 4 (Table 2).

Among these 17 chromosomes, four (2p, 2q, 7, and 9) occur in such a way that their character states represent transition series (Fig. 4a). That is, it is not equally likely to go from one character state to another, and thus these characters must be treated as ordered. The remaining 12 characters are unordered. The ancestral or primitive character state for each character was determined by the character states present in the outgroup baboon and rhesus monkey karyotypes (Yunis and Prakash, 1982). These primitive chromosome states present in Old World monkeys are listed in Table 2.

The exhaustive search that the PAUP analysis executed using the data in Table 2 produced a single, most parsimonious evolutionary tree topology (Fig. 4b). Only one chromosome, number 12, was homoplastic; this is, this chromosome changed convergently along the chimpanzee and gorilla lineages. An equally parsimonious solution with the same topology, but with different branch lengths, would depict chromosome 12 as experiencing a reversal rather than a parallelism but changing at two alternative branches in the tree. The results of this analysis provide strong support for the conclusion that man and chimpanzee have shared a common ancestor more recently with one another than either has with the gorilla (see Fig. 4b).

Rodentia: Arvicolidae

Twenty-two species of arvicolid rodents were compared for banding homology (Modi, 1986; 1987). Diploid numbers among all 22 species examined range from 18 to 64. Karyotypes of three species are presented in Figure 5. As a result of karyotypic comparisons between species, 38 autosomes or autosomal arms were found to be homologous in two or more species. These homologous elements were

Table 3. Binary data matrix of 214 karyotypic characters (columns) and 22 species of arvicolid rodents (rows). Codes are 1 = present, 0 = absent, ? = missing datum. Species abbreviations: Cga - *Clethrionomys gapperi*, Cca - *C. californicus*, Dto - *Dicrostonyx torquatus*, Lsi - *Lemmus sibiricus*, Sco - *Synaptomys cooperi*, Pin - *Phenacomys intermedius*, Lcu - *Lagurus curtatus*, Nal - *Neofiber alleni*, Mlo - *Microtus longicaudus*, Mcl - *M. californicus*, Mto - *M. townsendii*, Mpe - *M. pennsylvanicus*, Mbe - *M. breweri*, Mmx - *M. mexicanus*, Mri - *M. richardsoni*, Mag - *M. agrestis*, Moe - *M. oeconomus*, Moh - *M. ochrogaster*, Mmn - *M. montanus*, Mcn - *M. canicaudus*, Mpm - *M. pinetorum*, Mor - *M. oregoni*, Per - *Peromyscus* (outgroup).

```
    1abcdefghijklmnop2abcdefghijklm3abcdefghi4abcdefghi5abc6abcdef7abcd8abcd
Cga 111000000000000001000000000000010000000001000000000100010000001000010000
Cca 111000000000000001000000000000010000000001000000000100010000001000010000
Dto 100000000000000001100000000000010100000001100000000110011000001110000????
Lsi 100000000000000001000000000000010000000001000000000100010000001000010000
Sco 100000000000000001000000000000010000000001000000000100010000001000011000
Pin 100110000000000001010000000000010000000001010000000100010000001000000????
Lcu 100000000000000001000000000000010000000001000000000100010000001000010000
Nal 100000000000000001001000000000011000000001000000000100010000001000010000
Mlo 100001000000000001000100000000011010000001001000000101010100001001010100
Mcl 100000000100000001000000010000011000000001001000000100010000001000010011
Mto 100010100000000001000000010000011000000001000110000100010000001000010000
Mpe 100000100000000001000000011000011001000001000100000100010000001100010000
Mbr 100000100000000001000000011000011001000001000100000100010010001100010000
Mmx 100000000010000001000000000010011000100001000000100100010001001000010000
Mri 100010010000000001000000000000011000000001000000000100010000001000010000
Mag 100000000000100001000000000000011000000001000000010100010000001000010000
Moe 100000000000110001000000000001010000000101000000000100110000101000110000
Moh 100110000000000001000000000000011000000001000000000101010000001000010000
Mmn 100000100000001001000000010100011000001001000111000100010000011000010000
Mcn 100000100000001001000000011000011000001001000111000100010000011000010000
Mpm 100110001000000001000000000000011100000101000000001100010000001000010000
Mor 100010100000000111000111000000011000000011000000000100010000001100000????
Per 100000000000000001000000000000010000000001000000000100010000001000010000

            1    1        1  1  1         1 1   1    1  12   2   2  2 2   2
    efg9abcd0abcd1abcdefgh2ab3ab4abcdefghi5a6abc7abcd8ab90abc1abc2ab3a4abc5
Cga 00010000100001000000001001001000000000101000100001001100010001001010001
Cca 00010000100001000000001001001000000000101000100001001100010001001010001
Dto ???10000110001100000001000??1100000000110???1000011010???0???1000?10001
Lsi 00010000110001010000001001001000000000101000100001001100010001101 0???1
Sco 00010000110001000000001001001000000000101000100001001100010001001010000
Pin ???10000110001000000001001001010000000101000100001001 0???10001001010001
Lcu 0001000010000100000000100100100000000010100010000100110000???1001010001
Nal 0001000000????100000000100100100000000010100010000101110001000100101000 1
Mlo 00011000101001000000001011001001000000101100101001001111000???1010?11001
Mcl 00010000100001000000001001001000000000101000110001000100010000??1010101
Mto 0001000010010100000000110100100000000000?1010100000??00???11000??0?10001
Mpe 0001000010100100100000110100100010000 00?1010100001000 0???11000??0?10001
Mbr 0001000010100100100000110100100010000 00?1010100001000 0???11000??0?10001
Mmx 0001000000????10000000010010010000000000101000100001001 0???10000??0?10001
Mri 00010000100001000000001001001000010000101000100001001100010001000?10101
Mag 00010000100001000000001001001000000000101001100001001100110001001110001
Moe 10010100100001000100001001011000000000101000 0????10010???10100??1010010
Moh 00010000100011000000101001001000001000101000100101001 0???10100??1010101
Mmn 01010010101001000011001101101000000110100???0????1000 0???10010??0?0???1
Mcn 01110010101001000010001101101000000110100???0????1000 0???10010??0?0???1
Mpm 00010000100001000000001001001000000000101000100011001 0???10000??1010001
Mor ???100010????1000000010??0??10000000010?0???0????1000 0???0???0??0?0???0
Per 00010000100001000000001001001000000000101000 0????1001 0???10000??0?10001

             222333333333344444444445555555555666666666677
    abc6abcde78901234567890123456789012345678901Xabcdefghijklmn Ya
Cga 00010000010000000000000000000000000000000000000000000010000000000000010
Cca 00010000010000000000000000000000000000000000000000000010000000000000010
Dto 10011000010000000000001111000000000000000000000000000011000000000000011
Lsi 00010100000000000000000000000000000000000000000000000010000000000000010
Sco ???10000010000000000000001000000000000000000000000000010010000000000010
Pin 00010000010000000000000000111000000000000000000000000010100000000000010
Lcu 00010000010000000000000000000000000000000000000000000010100000000000010
Nal 00010000010000000000000000000000000000000000000000000010000000000000010
Mlo 00010010010000000000000000000111111000000000000000000010010000000000010
Mcl 00010000010000000000000000000000000010000000000000000010001000000000010
Mto 01010000001111100000000000000000000010000000000000000010000100000000010
Mpe 01010000001111111110000000000000000000000000000000000010000010000000010
Mbr 01010000001111111110000000000000000000000000000000000010000010000000010
Mmx 00010000010000000000000000000000000000111000000000000010000001000000010
Mri 00010000010000000000000000000000000000000100000000000010001000000000010
Mag 00010000010000000000000000000000000000000000000000000010000000000000010
Moe ???10000000000000000000000000000000000000011100000000010000000011000010
Moh 00010000010000000000000000000000000000000000011000000010000000000100010
Mmn 00010001000000000000110000000000000000000000000110000010000000000000101 0
Mcn 0010?????000000000001100000000000000000000000000000110010000000000001101 0
Mpm 00010000110000000000000000000000000000000000000000000010001000000000010
Mor ???0?????00000000000000000000000000000000000000000011110000000000001110
Per 00010000010000000000000000000000000000000000000000000010000000000000010
```

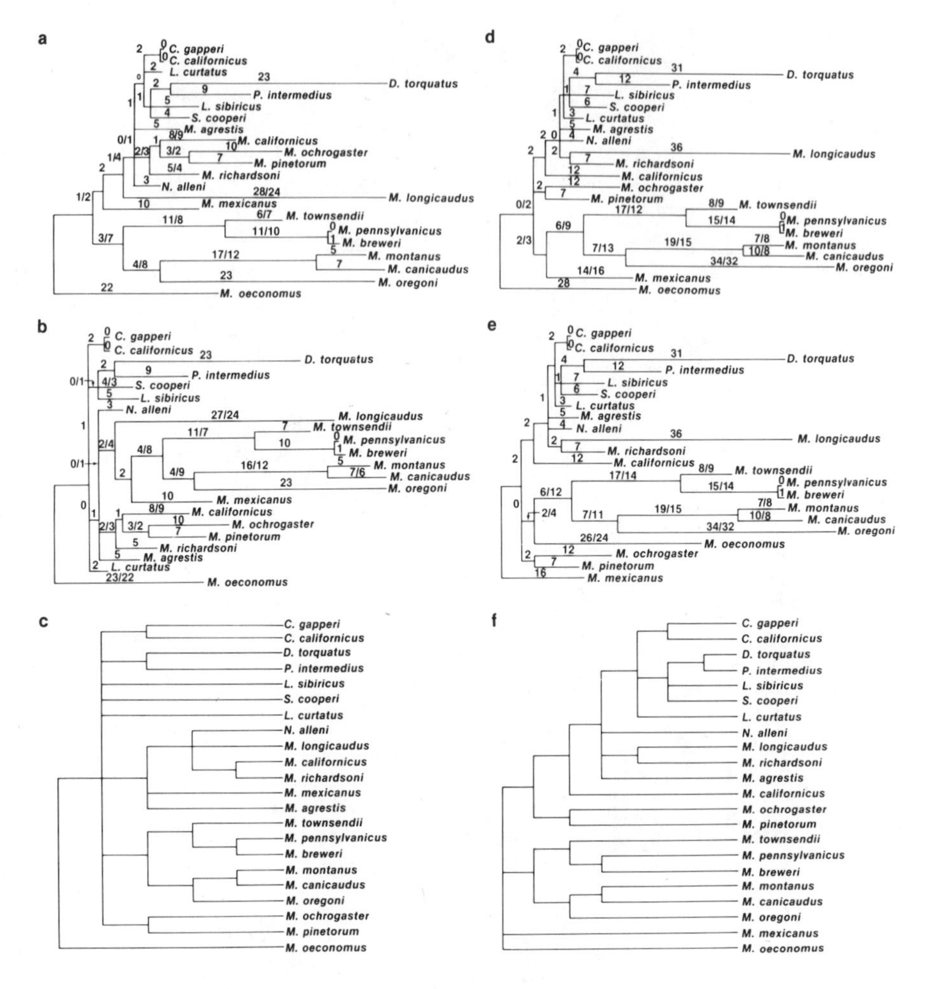

Fig. 6. Phylogenetic trees of 22 species of arvicolid rodents produced by parsimony analyses of the data matrix appearing in Table 4 using the PAUP program. Shown for each of two weighting schemes are two rival trees of equal length but differing in topology and an associated Adams-2 consensus tree. Branch lengths (the amount of karyotypic divergence along a lineage from a proposed primitive condition) are included for all rival trees under two different hypothetical taxonomic unit (HTU) optimization options, FARRIS and MINF (see Modi, 1987; and Swofford, 1984 for a discussion). In cases where the two options yielded different lengths, MINF values precede and FARRIS values follow the slash. In cases where the two options yielded similar lengths only a single value is given. a–c) 1:1 weighting scheme, d–f) 1:2 weighing scheme.

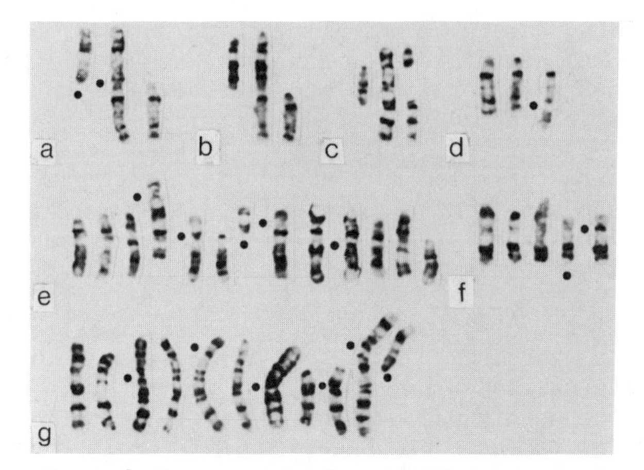

Fig. 7. A series of chromosomes from different species of arvicolid rodents arranged to illustrate different autosomal rearrangements. Solid dots indicate centromere locations of biarmed or inverted acrocentric elements. a) Centric fusion involving pairs 7 and 6 from M. californicus and M. canicaudus. b) Tandem fusion involving pairs 8 and 6 from L. curtatus and M. mexicanus. c) Tandem fission of large acrocentric pair 1 of L. curtatus to three small acrocentrics in M. pinetorum. d) Geographic variation in pair 6 of M. pinetorum due to centric transposition and addition; elements from left to right are from Peromyscus gratus (outgroup), M. pinetorum (population from New York population) and M. pinetorum (Virginia population). e) Chromosome 2, examples from left to right are from P. gratus, M. longicaudus (illustrating addition type of rearrangement), M. californicus (paracentric inversion), M. oregoni (addition, deletion and tandem fusion with 14), M. mexicanus (pericentric inversion), M. pinetorum, two chromosomes (tandem fission), M. oeconomus (centric transposition), N. alleni (paracentric inversion), M. canicaudus (deletion and centric transposition), M. townsendii (deletion), P. intermedius, two chromosomes (non-reciprocal translocation with 18). f) Chromosome 14 from P. gratus, L. sibiricus (unrearranged), P. intermedius (addition), M. pennsylvanicus (centric transposition), M. ochrogaster (pericentric inversion). g) Chromosome 1 from P. gratus, C. gapperi (reciprocal translocation with 9), M. mexicanus (centric transposition), L. curtatus (unrearranged), M. californicus (centric transposition), M. agrestis (deletion), M. canicaudus (tandem fission and centric fusion with pair 5), M. pennsylvanicus (tandem fission and deletion), M. longicaudus (tandem fission and deletion), M. oeconomus (paracentric inversion and centric fusion with pair 4, example from C. gapperi).

given identical numbers ranging from 1 to 38 (Fig. 5). Further, 141 chromosomal rearrangements involving these 38 autosomes and the two sex chromosomes were found and were given alphabetic designations. Thirty-three additional elements, found in only one of the 22 species, were given unique numbers between 39 and 71. The total number of characters used in the numerical analyses is the sum of these four subsets, namely, 38 + 141 + 2 + 33 = 214. This total of 214 characters is divisible into chromosome (n = 38 + 2 + 33 = 73) and rearrangement (n = 141) characters (Tables 3-4).

Two different types of karyotypic characters were defined: a) chromosome characters, which consist of entire chromosomes or chromosomal arms, and b) rearrangement characters. A species by character data matrix was then established by assigning a value of 1 or 0 for the presence or absence, respectively, of each of both types of characters in the karyotype of each species. Among rearrangement characters, if the associated chromosome character was absent in a particular species, then that taxon received a value of question mark (?) for that rearrangement character, which symbolizes a missing datum. Character states for all of the characters were assessed as being primitive or derived by using the karyotype of the murid rodent *Peromyscus* as an outgroup (Committee, 1977).

Six different parsimony analyses were carried out on the data matrix in Table 3 where *a priori* character weights assigned to chromosomes and rearrangements varied. In these analyses the weights given the two types of characters were (rearrangement characters: chromosome characters), 1:1, 1:2, 1:4, 1:10, 0:1, and 1:1 with the addition of a dummy variable that forced certain species to cluster together. Differences among these analyses are described elsewhere (Modi, 1987).

Each of the six PAUP analyses yielded a minimum of 50 equally parsimonious trees. Two trees of equal length yet varying in topology and an associated Adams-2 consensus tree are shown for two of these analyses (Fig. 6). Differences among topologies and branch lengths were observed both within and between weighting regimes (Fig. 6a-b). Further, the six consensus trees yielded slightly different results from one another. An extensive discussion of the complete analysis appears in Modi (1987). The cytosystematic results may be summarized as follows. Thirteen species (*Clethrionomys gapperi*, *C. californicus*, *Lemmus sibiricus*, *Synaptomys cooperi*, *Phenacomys intermedius*, *Lagurus curtatus*, *Neofiber alleni*, *Microtus californicus*, *Microtus mexicanus*, *M. richardsoni*, *M. agrestis*, *M. ochrogaster* and *M. pinetorum*) possess between 48 and 62 largely acrocentric chromosomes many of which are homologous with chromosomes contained in the outgroup. All, except *L.*

Table 4. Frequency distribution of ten different euchromatic chromosomal rearrangements involving autosomes and sex chromosomes among 22 species of arvicolid rodents. Percentages of the total number of rearrangements represented by each of the ten types are given in parentheses.

	Autosomes	Sex chromosomes	Autosomes and sex chromosomes
Addition	15(11.9)	0(0.0)	15(10.6)
Deletion	7(5.5)	4(26.7)	11(7.8)
Pericentric inversion	11(8.7)	7(46.7)	18(12.8)
Paracentric inversion	11(8.7)	1(6.7)	12(8.5)
Centric fusion	36(28.6)	2(13.3)	38(27.0)
Tandem fusion	20(15.9)	0(0.0)	20(14.2)
Nonreciprocal translocation	1(0.8)	0(0.0)	1(0.7)
Reciprocal translocation	1(0.8)	0(0.0)	1(0.7)
Centric transposition[a]	16(12.7)	1(6.7)	17(12.1)
Tandem fission[b]	8(6.3)	0(0.0)	8(5.7)
Total	126	15	141

[a]Observed in species where biarmed chromosomes show identical G-banding patterns to the primitive acrocentric condition and no evidence of pericentric inversion.

[b]Observed in species where two or three small-sized acrocentrics are homologous to the primitive condition, a single, large-sized acrocentric.

curtatus, contain several derived characters that may be shared with one or more other species, but which are largely unique. *M. ochrogaster* and *M. pinetorum*, or *Dicrostonyx torquatus* and *P. intermedius* appear as sister taxa based upon four or two homoplastic autosomal rearrangements, respectively. *D. torquatus*, *M. oeconomus* and *M. longicaudus* have karyotypes more highly derived than those cited above. Six species of *Microtus* cluster together and are further subdivided into two groups containing *M. pennsylvanicus*, *M. breweri* and *M. townsendii* on the one hand, *M. montanus*, *M. canicaudus* and the more distantly related *M. oregoni* on the other (Fig. 6).

Fig. 7 presents individual chromosomes from the karyotypes of several species that exemplify at least one of each type of autosomal rearrangement listed in Table 4. Chromosomes 3, 6 and 14 are portrayed in order to exhibit the nature of variability found among several of the most polymorphic autosomes. The most commonly occurring autosomal rearrangements were centric fusion, tandem fusion, and centric transposition (Table 4). Pericentric inversion was the most frequent sex chromosome rearrangement.

DISCUSSION

Phylogenetic or cladistic assessments of systematic relationships are founded upon several basic tenets. Character variability between taxa is expressed in the form of character states, and these states are subsequently identified as being primitive or derived. The outgroup method is generally employed for the determination of character state polarity (see Watrous and Wheeler, 1981). Genealogical relationships are thus based upon the presence of shared, derived character states in monophyletic sister taxa, whereas shared, primitive states are non-utilitarian in estimating relationships. Most cladists accept the idea that the basis for genealogical inference rests upon the parsimony principle, that is, that the construction of evolutionary trees should minimize "ad hoc hypotheses" of homoplasy or require a minimum number of character state changes. Parsimony methods were first introduced by Camin and Sokal (1965), while useful modifications were developed shortly thereafter (Kluge and Farris, 1969; Farris, 1970). Perhaps the principle theoretical criticism levied against the cladistic approach stems from the problem of homoplasy; that is, how may shared derived character states, purportedly indicative of recency of common ancestry, be distinguished from convergent or reversal events in character state evolution? Computationally, the parsimony problem or the problem of finding the tree of global minimum length becomes increasingly difficult as the number of taxa increases. It appears to be unlikely that an efficient algorithm for computational problems such as this one, termed NP - complete, will ever be designed (Day, 1983).

A number of studies have utilized chromosomal G-bands for systematic assessments of mammals (reviewed by Baker, et al., 1987). None of these reports, however, utilized numerical computerized approaches in tree construction. The present study illustrates that, in general, interspecific variation in mammalian chromosomal banding patterns can meaningfully be treated as discrete character states, which can subsequently be coded and utilized in tree building computer programs. As a result, chromosomal data can be quantitatively analyzed the same way in which morphological and molecular data have been treated. Cladistic analyses were executed here under the parsimony hypothesis; however, these same data could have been subjected to phenetic analyses by using the unweighted pair group method of analysis (UPGMA) clustering on the Manhattan distance matrices. Some of the more important ramifications of the present study are highlighted.

Critical to all three analyses was accurate coding. The situations in the cat and primate data sets were similar to one another in that the same chromosomes are present, in one form or another, in all species in each respective group. On the other hand, in the rodent data the presence or absence of different chromosomes and chromosomal rearrangements in different species prompted the usage of the missing datum code for rearrangements in addition to the binary representation of their presence or absence. This situation results from the fact that the rodents exhibit greater interspecific chromosomal differences than do the cat or primate species.

In the cats, most of the chromosomes showed no interspecific variability and thus were not of systematic interest. Among the eight that did vary, it turns out that six of these are mutually exclusive (i.e., C3, E4, E5, F1, F2, and F3 are not all found together in one karyotype) derivatives of one another. This means that the presence or absence of these six potentially informative chromosomes can be represented by fewer than six (only three were needed) binary characters.

In the primates, all four species possessed the same homologous chromosomes. Interspecific differences were due to rearrangements within these homologous elements. The 17 phylogenetically informative chromosomes could then be coded as integer, multistate characters. Also, four of these 17 chromosomes exist in the form of a transition series. This prompted the usage of both ordered and unordered multistate characters in the same data set.

In the cat and rodent data sets, since several characters were not consistent on the final trees, multiple equally parsimonious solutions were obtained after branch swapping. This led to consensus tree construction in order to summarize the information content of the rival trees. Since the number of primate taxa was low, an exhaustive search could be carried out, and since only one character was inconsistent, multiple solutions do not exist. This enables a fairly unambiguous interpretation of the primate data.

In the rodent data set, character weighting was executed. The pros and cons of using character weighting in taxonomy have been reviewed elsewhere (Modi, 1987). It is acknowledged that the assignment of character weights may be quite subjective; however, an argument might be made that chromosomal banding variation of the nature seen in the Arvicolidae may be a situation where the application of variable *a priori* weights is valid.

Finally, the idea of taxonomic concordance warrants mention. In all three data sets, it can be seen that reasonable good congruence exists between the systematic affinities evidenced by the present study compared with the relationships derived from other types of data. These results confirm earlier reports that the cat family is one of the most karyotypically conservative groups yet surveyed. The presence of several polymorphic pairs of autosomes allows for the definition of 13 species groups; however, the fact that some of these chromosomes evolve convergently on the trees makes it difficult to interpret the relationships among some of the groups. The primary systematic conclusions of the study identify four groups of related species: a) the domestic cat group, which contains six species having identical karyotypes; b) the *Prionailurus* group of four species having identical karyotypes (the leopard cat, the fishing cat, the puma and the Iriomote cat); c) the *Panthera* group, which contains the lion and seven other species all having identical karyotypes; d) the ocelot lineage, which is comprised of five South American species (ocelot, margay, tiger cat, Geoffroy's cat, and the pampas cat), all of which have the C3 chromosome, yet differ with respect to chromosome D2.

The species contained within the domestic cat species group agree with the taxonomic classification of these species based upon other kinds of data. For example, morphological comparisons (Hemmer, 1978; Herrington, 1983), DNA-DNA hybridization of endogenous retroviral sequences (Benveniste, 1985), allozyme genetic distance (O'Brien *et al*., 1987) and albumin immunological distance studies (Collier and O'Brien, 1985) all suggest that these six species are closely related to one another, although Collier and O'Brien (1985) also include Pallas' cat (OMA) in this group. According to Kurten (1968) these species are thought to be derived from a common ancestor that inhabited the Mediterranean basin.

The second group of species contains the leopard cat, the fishing cat, the puma, and the Iriomote cat. Herrington (1983) united these first three species along with the rusty spotted cat and the flat-headed cat. These latter two species differ from the first three by two or one chromosomal rearrangements, respectively. The immunological distance study (Collier and O'Brien, 1985) also united the fishing cat, the leopard cat and the flat-headed cat.

The *Panthera* group contains eight large-sized species that also ally based upon an allozyme genetic distance and immunological distance (Collier and O'Brien, 1985; O'Brien *et al*, 1987). Five of these species (excluding the lynx, bobcat and marbled cat) were also grouped by Herrington (1983). Herrington (1983) placed the lynx and bobcat together and allied the marbled cat with the clouded leopard. Ewer (1973) notes that a uniquely ossified hyoid unites the lion, tiger, leopard, and jaguar.

As far as the South American species are concerned, Glass and Martin (1978) united the margay, ocelot and tiger cat while Herrington (1983) grouped the five chromosomally similar species along with the Andean mountain cat (*Oreailurus jacobita*) and the kodkod (*Oncifelis guigna*). Further, the immunological distance study of Collier and O'Brien (1985) also indicated relationships among six South American cats (ocelot, margay, tiger cat, Geoffroy's cat, kodkod, and pampas cat). Affinities among the remaining species groups are not well resolved. The presence of multiple, equally parsimonious trees limits the utility of G-bands in estimating the systematic relationships of these species.

Regarding the primates, evidence derived from DNA reassociation (Benveniste and Todaro, 1976; Sibley and Ahlquist, 1984; O'Brien *et al*., 1985b), isozymes (O'Brien *et al*., 1985b), restriction enzyme induced chromosome banding (Bianchi et al., 1985) mitochondrial DNA (Ferris *et al*., 1981), and two-dimensional protein electrophoresis (Goldman *et al*., 1987) all support the conclusion that man, chimpanzee and gorilla share a common ancestor more recently with one another than with the orangutan. The phylogenetic tricotomy among human-chimpanzee-gorilla has been vigorously debated. The divergence of the gorilla line from a common ancestor before man and chimpanzee split, which we conclude, is consistent with molecular topologies derived from DNA hybridization (Sibley and Ahlquist, 1984) and from genetic distance using proteins separated by two-dimensional gel electrophoresis (Goldman *et al*., 1987). On the other hand, the mitochondrial DNA (Ferris *et al*., 1981) and the restriction enzyme chromosome banding data (Bianchi *et al*., 1985) support the view that the human lineage branched off before the chimpanzee and gorilla lineages diverged. Bianchi *et al*. (1985) attributed this discrepancy to the fact that mitochondrial DNA and the restriction enzyme banding patterns are measuring DNA sequences that are evolving more rapidly than the average rate of change for single-copy nuclear genes.

The systematics of the present subset of arvicolid species (the family contains approximately 125 species) was reviewed by Modi (1987). Some of the main systematic points may be highlighted as follows. *Dicrostonyx* has a karyotype consisting primarily of uniquely derived chromosomes. This taxon is thought to represent a highly differentiated and isolated lineage based upon fossil and comparative anatomical criteria (Hooper and Hart, 1962; Gromov and Poliakov, 1977; Carleton, 1981). *Clethrionomys*, *Synaptomys*, *Lemmus*, *Neofiber*, and *Lagurus* all have karyotypes containing largely primitive elements and a few uniquely derived ones, thus making it difficult to infer their relationships. Other studies suggest that *Lemmus* and *Synaptomys* are related (Hooper and Hart, 1962; Hinton, 1926; Gromov and Poliakov, 1977; Carleton, 1981). *Neofiber* is thought to be only distantly related to any of these other genera (Hooper and Hart, 1962; Carleton, 1981). *Clethrionomys* is regarded as most closely related to several Old World genera not examined in the present study, and perhaps distantly to *Phenacomys* (Hooper and Hart, 1962).

Among the 14 species of *Microtus* analyzed, *M. pennsylvanicus*, *M. breweri*, *M. townsendii*, *M. montanus*, *M. canicaudus*, and *M. oregoni* were seen to form a loosely defined group. The karyotypes of several other species, including *M. richardsoni*, *M. mexicanus*, *M. agrestis*, and *M. californicus* are fairly primitive. Traditional arrangements place *M. pennsylvanicus*, *M. breweri*, *M. townsendii*, *M. montanus*, *M. canicaudus*, *M. californicus*, *M. mexicanus*, *M. longicaudus*, *M. oeconomus*, and *M. agrestis* in the subgenus *Microtus* (Hall and Cockrum, 1953). The remaining four species are generally assigned to separate subgenera, *M.* (*Pedomys*) *ochrogaster*, *M.* (*Aulacomys*) *richardsoni*, *M.* (*Pitymys*) *pinetorum*, and *M.* (*Chilotus*) *oregoni* (Hall and Cockrum, 1953; Chaline, 1974). However, certain disagreements do exist (discussed in Modi, 1987).

REFERENCES

Adolph, K. W., 1986, Organization of mitotic chromosomes, *in*: "Chromosome Structure and Function," M. S. Risley, ed., Van

Nostrand Reinhold Co., New York, pp. 92-125,
Baker, R. J., Qumsiyeh, M. B., and Hood, C. S., 1987, Role of chromosome banding patterns in understanding mammalian evolution, in: "Current Mammalogy", Vol. I, H. H Genoways, ed., Plenum Publishing Co., New York, pp. 67-96.
Benveniste, R. E., 1985, The contributions of retroviruses to the study of mammalian evolution, in: "Molecular Evolutionary Genetics," R. J. MacIntyre, ed., Plenum Publishing Co., New York, pp. 359-417.
Benveniste, R. E., and Todaro, G. J., 1976, Evolution of type C viral genes: Evidence for an Asian origin of man, Nature, 261:101-109.
Bianchi, N. O., Bianchi, M. S., Cleaver, J. E., and Wolff, S., 1985, The pattern of restriction enzyme-induced banding in the chromosomes of chimpanzee, gorilla, and orangutan and its evolutionary significance, J. Mol. Evol., 22:323-333.
Bloom, S. E., and Goodpasture, C., 1976, An improved technique for selective silver staining of nucleolar organizer regions in human chromosomes, Hum. Genet., 34:199-206.
Bobrow, M., and Cross, J., 1974, Differential staining of human and mouse chromosomes in interspecific cell hybrids, Nature, 251:77-79.
Bush, G. L., Case, S. M., Wilson, A. C., and Patton, J. L., 1977, Rapid speciation and chromosomal evolution in mammals. Proc. Natl. Acad. Sci. U.S.A., 74:3942-3946.
Carleton, M. D., 1981, A survey of gross stomach morphology in Microtinae (Rodentia: Muroidea), Z. Säugertierk., 46:93-108.
Caspersson, T., Zech, L., and Johansson, C., 1970, Differential banding of alkylating fluorochromes in human chromosomes, Exp. Cell Res., 60:315-319.
Camin, J. H., and Sokal, R. R., 1965, A method for deducing branching sequences in phylogeny, Evolution, 13:311-326.
Chaline, J., 1974, Esquisse de l'évolution morphologique, biométrique et chromosomique du genre Microtus (Arvicolidae, Rodentia) dans le Pléistocené de l' hémisphère nord, Bull. Soc. geol. Fr., 16:440-450.
Collier, G. E., and O'Brien, S. J., 1985, A molecular phylogeny of the Felidae: immunological distance, Evolution, 39:473-487.
Committee for standardization of chromosomes of Peromyscus, 1977, Standardized karyotype of deer mice, Peromyscus (Rodentia), Cytogenet. Cell Genet., 19:38-43.
Day, W. H. E., 1983, Computationally difficult parsimony problems in phylogenetic systematics, J. Theor. Biol., 103:429-438.
Dutrillaux, B., and Lejeune, J., 1971, Sur une nouvelle technique d'analyse du caryotype humaine, C.R. Acad. Sci. Paris Ser. D., 272:2638-2640.

Eldredge, N., and Cracraft, J., 1980, "Phylogenetic Patterns and the Evolutionary Process", Columbia University Press, New York.
Ewer, R. F., 1973, "The Carnivores," Cornell University Press, Ithaca.
Farris, J. S., 1970, Methods for computing Wagner trees, Syst. Zool., 19:83-92.
Ferris, S. D., Wilson, A. C., and Brown, W. M., 1981, Evolutionary tree for apes and human based on cleavage maps of mitochondrial DNA, Proc. Natl. Acad. Sci. U.S.A., 78:2432-2436.
Glass, G. E, and Martin, L. D., 1978, A multivariate comparison of some extant and fossil Felidae, Carnivore, 1:80-87.
Goldman, D., Giri, G., and O'Brien, S. J., 1987, A molecular phylogeny of the hominoid primates as indicated by two-dimensional protein electrophoresis, Proc. Natl. Acad. Sci. U.S.A., 84:3307-3311.
Gromov, I. M., and Poliakov, I. A., 1977, "Polevki (Voles) (Microtinae): Fauna SSSR (Mammals)", Nauka, Moscow.
Hall, E. R., and Cockrum, E. L., 1953, A synopsis of the North American microtine rodents, Univ. Kansas Publ. Mus. Nat. Hist., 5:373-498.
Harper, M. E., and Saunders, G., 1981, Localization of single copy DNA sequences on G-banded human chromosomes by in situ hybridization, Chromosoma (Berl.), 83:431-439.
Hemmer, H., 1978, The evolutionary systematics of living Felidae: present status and current problems, Carnivore, 1:71-79.
Herrington, S. J., 1983, Systematics of the Felidae: a quantitative analysis, Unpublished Master's Thesis, Univ. of Oklahoma, Norman.
Hinton, M., 1926, Monograph of the voles and lemmings (Microtinae) living and extinct, Vol. I., Brit. Mus. Nat. Hist., London.
Hooper, E. T., and Hart, B., 1962, A synopsis of recent North American microtine rodents, Misc. Publ. Mus. Zool. Univ., Michigan, 120:1-68.
Kluge, A. G., and Farris, J. S., 1969, Quantitative phyletics and the origin of anurans, Syst. Zool. 18:1-32.
Kurten, B., 1968, "Pleistocene Mammals of Europe," Aldine Press, Chicago.
Lock, L. F., and Martin, G. R., 1986, Dosage compensation in mammals: X chromosome inactivation, in: "Chromosome Structure and Function," M. S. Risley, ed., Van Nostrand Reinhold Co., New York, pp. 187-220.
Mayr, E., 1969, "Principles of Systematic Zoology," McGraw-Hill, New York.
Modi, W. S., 1986, Karyotypic differentiation among two sibling species pairs of New World microtine rodents, J. Mammal,. 67:159-165.

Modi, W. S., 1987, Phylogenetic analyses of chromosomal banding patterns among the Nearctic Arvicolidae (Mammalia: Rodentia), Syst. Zool. (in press).

Modi, W. S., Nash, W. G., Miyake, Y.-I., and O'Brien, S. J., 1987b, Comparative cytogenetic analyses of the Felidae (Carnivora), (submitted for publication).

Modi, W. S., Nash, W. G., Ferrari, A. C., and O'Brien, S. J., 1987a, Cytogenetic methodologies for gene mapping and comparative analyses in mammalian cell culture systems, Gene Anal. Tech., 4:75-85.

O'Brien, S. J., Collier, G. E., Benveniste, R. E., Nash, W. G., Newman, A. K., Simonson, J. M., Eichelberger, M. A., Seal, U. S., Bush, M., and Wildt, D. E., 1987, Setting the molecular clock in Felidae: The great cats, Panthera, in: "Tigers of the World: The Biology, Biopolitics, Management and Conservation of an Endangered Species," R. L. Tilson, ed., Noyes Publications, Park Ridge (in press).

O'Brien, S. J., Nash, W. G., Wildt, D. E., Bush, M. E., and Benveniste, R. E., 1985a, A molecular solution to the riddle of the giant panda's phylogeny, Nature, 317:140-144.

O'Brien, S. J., Seuanez, H. N, and Womack, J. E., 1985b, On the evolution of genome organization in mammals, in: "Molecular Evolutionary Genetics (Monographs in Evolutionary Biology Series)," R. J. MacIntyre, ed., Plenum Press, New York, pp. 519-589.

Perry, P., and Wolff, S., 1974, New Giemsa method for the differential staining of sister chromatids, Nature, 251:156-158.

Prescott, D. M., 1987, Cell reproduction, Int. Rev. Cytol., 100: 93-128.

Ruddle, F. H., Chapman, V. M., Riccuiti, F., Murnane, M., Klebe, R., and Meera-Khan, P., 1971, Linkage relationships of seventeen human gene loci as determined by man-mouse somatic cell hybrids, Nature, 232:69-73.

Schweizer, D., 1981, Counter-stain enhanced chromosome banding. Hum. Genet., 57:1-14.

Seabright, M., 1971, A rapid banding technique for human chromosomes, Lancet, ii 971-972.

Shows, T. B., Sakayuchi, A. Y., and Naylor, S. L., 1982, Mapping the human genome, cloned genes, DNA polymorphisms and inherited disease, in: "Advances in Human Genetics," H. Harris, ed., Plenum Press, New York, Vol. 12, pp. 341-452.

Sibley, C. G., and Ahlquist, J. E., 1984, The phylogeny of the hominoid primates as evidenced by DNA-DNA hybridization, J. Mol. Evol., 20:2-15.

Simpson, G. G., 1961, "Principles of Animal Taxonomy, Columbia University Press, New York.

Sneath, P. H. A., and Sokal, R. R., 1973, "Numerical Taxonomy," W.H. Freeman Co., San Francisco.

Sokal, R. R., 1985, The continuing search for order, Am. Nat., 126:729-749.

Sumner, A. T., 1972, A simple technique for demonstrating centromeric heterochromatin, Exp. Cell Res., 75:304-306.

Swofford, D. L., 1984, PAUP: phylogenetic analysis using parsimony. Users manual, version 2.3. Illinois Nat. Hist. Survey, Champaign.

Watrous, L. E., and Wheeler, Q. D., 1981, The outgroup comparison method of character analysis, Syst. Zool., 30:1-11.

Wiley, E. O., 1981, "Phylogenetics: The Theory and Practice of Phylogenetic Systematics," J. H. Wiley and Sons, New York.

Wurster-Hill, D. H., and Centerwall, W. R., 1982, The interrelationships of chromosome banding patterns in canids, mustelids, hyena, and felids, Cytogenet. Cell Genet., 34:178-192.

Wurster-Hill, D. H., and Gray, C. W., 1975, The interrelationships of chromosome banding patterns in procyonids, viverrids, and felids, Cytogenet. Cell Genet., 15:306-331.

Wurster-Hill, D. H., Doi, J., Izawa, M., and Ono, Y., 1987, A banded chromosome study of the Iriomote cat, Felis iriomotensis, J. Hered., 78:105-107.

Yunis, J. J., and Prakash, O. M., 1982, The origin of man: a chromosomal pictorial legacy, Science, 215:1505-1530.

REGULATION OF DROSOPHILA CHORION GENE AMPLIFICATION

Terry Orr-Weaver[1] and Allan Spradling

Department of Embryology
Carnegie Institution of Washington
115 W. University Pkwy
Baltimore, MD 21210

INTRODUCTION

The ability of animal and plant chromosomes to faithfully reduplicate remains poorly understood. In contrast, bacterial and viral chromosome replication has been elucidated in detail. DNA replication in these systems initiates at specific origin sequences following the binding of proteins that facilitate assembly of an active complex of replication enzymes. RNA transcripts produced at or near the origin sometimes also participate in controlling replicon initiation. Today, the question of whether similar control mechanisms govern the replication of chromosomes in higher organisms represents one of the most important issues in eukaryotic chromosome research.

A typical eukaryotic chromosome initiates DNA synthsis at hundreds of sites along its length during each S phase. If each initiation site is controlled in a manner similar to model viral replicons, eukaryotic chromosomes should be interspersed with replication origins and cis-regulatory sequences recognized by proteins controlling initiation. Furthermore, mechanisms must exist which coordinate the replicons along each chromosome, guaranteeing that each functions once and only once

---

[1]Current address:
Whitehead Institute for Biomedical Research
Department of Biology
Massachusetts Institute of Technology
Cambridge, MA 02142

per cell cycle. Since patterns of DNA replication change during development, and differ between tissues, cis and trans regulators of initiation must undergo developmental regulation as well. However recent experiments have challanged this view. Specific DNA sequences are not required for regulated replication of DNAs injected into Xenopus embryos (Harland and Laskey, 1980; Mecheli and Kearsey, 1984), demonstrating that this conceptual framework may not apply, at least during rapid embryonic cleavage divisions where chromosomes must replicate in just a few minutes.

Studies of other systems have supported the idea that chromosomal replicons are specific regulated units. Identifying individual replicons is a prerequisite to testing whether initiation is controlled at specific origin sequences. Unfortunately, specific units of replication in eukaryotic chromosomes have for the most part remained elusive. In this review we summarize studies on an exceptional process of chromosome replication during Drosophila development where specific replicons can be defined and studied. During oogenesis, the intense requirements for chorion (eggshell) protein sythesis in the ovarian follice cells are satisfied by the tissue-specific amplification of two chromosome regions containing the clustered chorion genes (reviewed in Spradling and Orr-Weaver, 1987). Our studies have revealed that amplification of the chorion domains is controlled by specific cis-regulatory sequences called amplification control elements (ACEs). The amplification control elements defined by these experiments show a number of similarities to sequences regulating animal viral chromosome replication. Of particular interest, sequences controlling chorion gene transcription may also function to developmentally regulate replicon activation, as in the case of certain viruses.

CHORION GENE AMPLIFICATION

The follicle cells surrounding each developing Drosophila oocyte synthsize and secrete the eggshell during the last 16 hours of oogenesis (see Mahowald and Kambysellis, 1980, Margaritis, 1985). The production of the two outer eggshell layers, the protein-rich endochorion and exochorion, has been studied in particular detail (reviewed in Kalfayan et al., 1985; Kafatos et al., 1985; Spradling and Orr-Weaver, 1987). There are two clusters of chorion genes in the genome. The six major chorion proteins and four minor proteins are encoded by single copy genes in tandem arrangement within the two clusters. All the genes in both clusters amplify prior to becoming active at specific times during the process of eggshell production. Amplification is required for normal levels of chorion protein production; mutations disrupting amplification lead to partial or complete female sterility associated with the production of thin eggshells (Spradling and Mahowald, 1981; Orr et al., 1984).

The usefulness of chorion gene amplification in studies of chromosome replication is a consequence of the amplification mechanism. Prior to the onset of amplification, follicle cells cease mitotic divisions and begin to polytenize. Genomic replication continues, although measurements of total nuclear DNA content (Mahowald et al., 1979; Hammond and Laird, 1985) and of specific DNAs (Hammond and Laird, 1985) suggests that underreplication of some sequences takes place during this time, as in the case of the polytene salivary gland (see Spradling and Orr-Weaver, 1987). Genomic replication ends at about the time amplification is first detected in stage 9. Amplification of the two chorion gene clusters appears to result from the continued activation in the ovarian follicle cells of two specific chromosomal replicons after all other chromosome replication has ceased.

Figure 1 summarizes evidence that repeated initation of bidirectional replication forks at the site of the gene clusters is responsible for amplification. A gradient of decreasing amplifiction is observed in the chromosomal regions flanking the gene clusters. Elongation of the forks along the chromosome on either side of the gene cluster following initiation would be expected to produce such gradients. Four rounds of initation would be required to produce the 16-fold amplification of the X-linked genes, while the sixty-fold amplification on chromosome 3 is consistent with six rounds. Strong independent support for the reinitiation model of amplification shown in Fig. 1 has been obtained by directly visualizing amplifying chorion genes associated with replication forks in the electron microscope (Osheim and Miller, 1983). Nested repliction forks were also observed with the approximate dimensions expected from the measurements shown in Fig. 1.

The amplified regions resemble chromosomal replicons used for general replication in several respects. The total size of the amplified region is 80-100 kb. This is a typical replicon size in the polytene chromosomes of third instar *D. hydei* salivary glands (Steinemann, 1982). The average rate of fork elongation during amplifiction can be calculated based on the assumption that forks initiate within the gene clusters. DNA sequences increasing distances along the chromosome from the gene clusters will undergo amplification with a time lag that depends on the average rate of fork movement. As expected,

Fig. 1. Chromosomal domains of chorion gene amplification. A. A schematic representation of the reinitiation model for amplifiction. Replication forks initiate in the region of the "amplification control element" (ACE) and proceed in both directions. The figure is drawn approximately to scale to illustrate the origin of the amplification gradients directly below, however the number of rounds of initiation has been reduced for clarity. Measurements of the amplification levels of DNA sequences surrounding chorion gene cluster on the X chromosome (B.) or on chromosome three (C.) are plotted as a function of the distance (in kilobase pairs) from the cluster. The insets show the cytological distance on salivary gland polytene chromosomes spanned by each amplified region. (Data from Spradling, 1981; reproduced from Spradling, 1987).

A

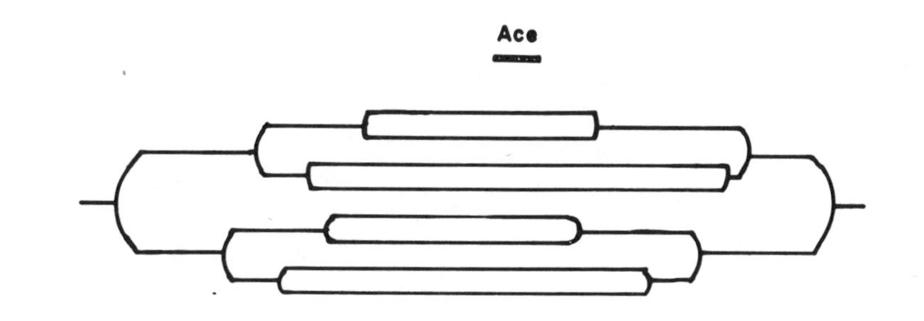
Ace

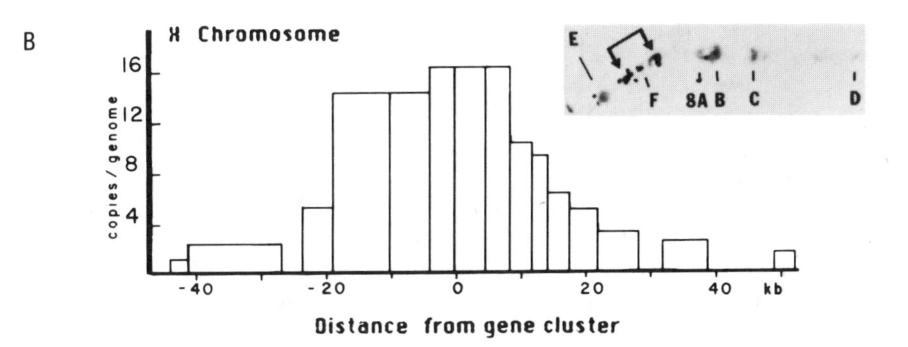
B
X Chromosome
16
12
8
4
copies / genome
-40
-20
0
20
40
kb
Distance from gene cluster
E
F
8A B
C
D

C

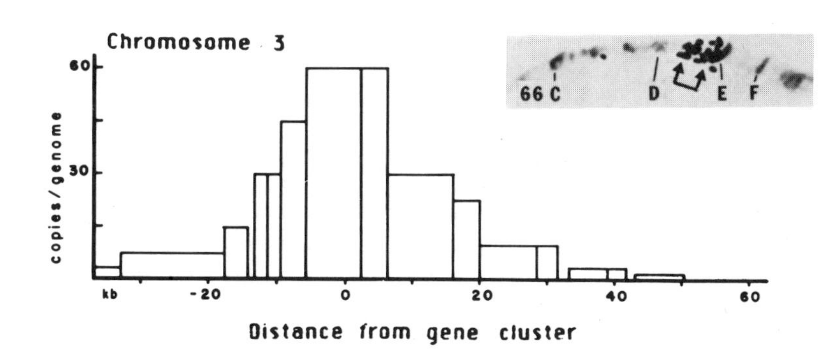
Chromosome 3
60
30
copies/genome
kb
-20
0
20
40
60
Distance from gene cluster
66 C
D
E
F

Fig. 2. Overlapping regions control both amplification and chorion gene transcription. The thick lines represent physical maps of the two chorion gene clusters. The position of ten chorion gene mRNAs (A, B, C, E, s36, s38, s18, s15, s19, and s16) are shown, as well as the location of a adjacent gene proposed to represent ovary tumorous (otu). The wavy line shows the position of the In(1)ocelliless breakpoint (In(1)oc). Above each map, the limits of uncertainity regarding the location of a replication origin as determined by electron microscopy (EM) or by "peak" mapping using Southern blots is shown. The large box labelled ACE1 or ACE3 summarizes genetic studies on amplification control. The DNA sequences represented by the box (in the case of ACE1 only one of the two hatched regions need be present) induce amplification in transformants. The small black section of each box represents the sequences essential for amplification. Below each map, the region containing the essential sequences is shown on an expanded scale. Below the expanded map is shown the essential amplification sequences (black line), and also transcriptional control sequences (filled boxes). The different hatching within the transcriptional control sequences illustrates region which differ in function as described in the text.

X

Limits of Origin (EM)
Limits of Origin (blots)
ACE 1
1 kb
In(1)oc
C
A
B
E
s36
s38
otu
-700
-500
-300
-100
5'
amplification
s38 expression

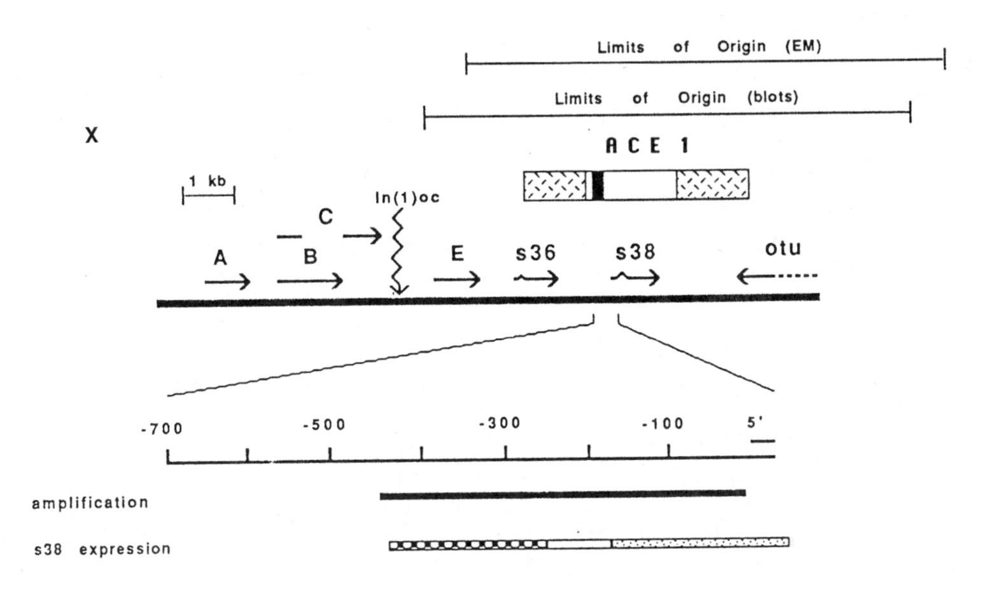

3

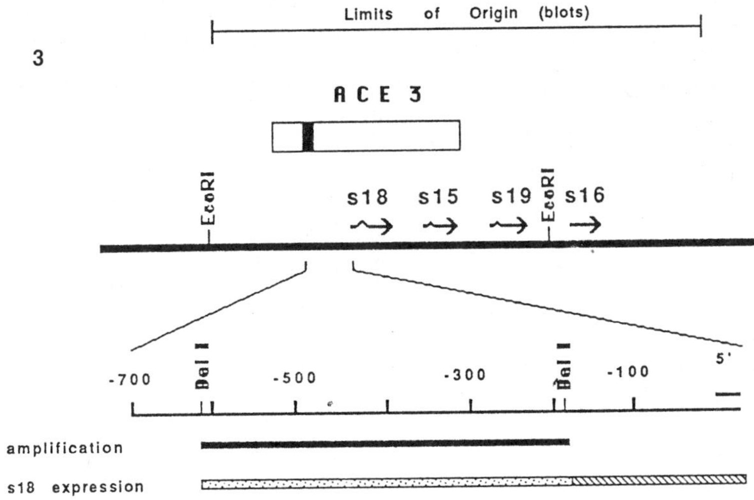

Limits of Origin (blots)
ACE 3
EcoRI
s18
s15
s19
EcoRI
s16
-700
Bal I
-500
-300
Bal I
-100
5'
amplification
s18 expression

sequences surrounding the third chromosome gene cluster initated amplification at times that depended on their distance from the gene cluster (G. Leys and A. Spradling, unpublished). These experiments yielded an average rate of fork movement of 80-110 bp/min in flies cultured at room temperature. This is similar to the rates of fork movement measured previously in polytene salivary gland cells (Corodeiro and Menghenini, 1975; Steinemann, 1982). However the chorion regions differ dramatically from normal replicons in containing multiple nested replication forks. In a normal replicon an unknown mechanism prevents the initiation of additional rounds of replication until the cell cycle is completed and a new S phase begins.

CIS-REGULATORY SEQUENCES CONTROLLING REPLICATION

One hallmark of prokaryotic and viral replicons is the use of specific replication origins. Several lines of evidence argue that chorion gene amplification utilizes specific origins near or within the two gene clusters. Osheim and Miller observed that in their electron micrographs of stage 11 follicle cell chromatin, replication forks downstream of the s38 gene were always moving away from the gene cluster, while forks upstream of the s36 transcription unit were always moving in the opposite direction (Fig. 2). The simplest interpretation is that initiation occurs specifically at a site or sites lying between these divergently oriented forks, a distance of approximately 12 kb. The peaks of the amplification gradients must occur at the site of a replication origin according to the reinitiation model. Careful quantitation of amplification levels in DNA from stage 13-14 egg chambers localized these peaks within about a 10 kb region encompassing part of each gene cluster. The site of origin sequences defined by electron microscopy and by "peak" measurements on the X chromosome are consistent (Fig. 2).

When viral chromosomes are studied genetically, replication origins are frequently cis-essential, as expected if they serve a specific structural function required for replication initiation. Genetic studies of the chorion gene regions have revealed cis-essential sequences for amplification. A specific chromosome inversion, In(1)ocelliless, disrupts the X chromosome amplified region, relocating the s36 and s38 gene region and causing developmentally regulated amplification at the new chromosomal location (Spradling and Mahowald, 1981). The location of the cis-regulatory sequences defined by the inversion breakpoint is consistent with the origin measurements (Fig. 2A). P element-mediated transformation allows detailed genetic characterization of amplification control sequences. DNA sequences from the chorion regions are cloned within P element transposons marked with the rosy$^+$ gene and introduced into rosy mutant embryos. Independent insertions that confer wild type eye color are selected for further study. The inserted DNA in each line is located at a single site that lies essentially at random somewhere on the euchromatic chromosome arms. The ability of the integrated transposon to amplify is studied by analyzing the copy number of transposon sequences in egg chambers from the transformants during oogenesis when the endogenous chorion genes undergo amplification.

The genetic requirements for amplification of both the third chromosome and X chromosome gene clusters have been extensively studied (deCicco and Spradling, 1984; Orr-Weaver and Spradling, 1986; Spradling et al., 1987; Deladakis and Kafatos, 1987). The results are summarized in Fig. 2. DNA from most regions of the gene clusters does not induce amplification when integrated at new chromosomal sites on P element transposons. However both gene clusters contain a single region (ACE1 or ACE3) which causes a developmentally normal program of follicle cell specific amplification at many sites of integration. Replication forks which initate under the control of these sequences elongate into adjacent DNA, as in the case of normal amplification, although a detailed reconstruction of the amplification gradients induced by a chorion transposon has not yet been carried out. The level of amplification is frequently lower than normal at ectopic sites; at certain sites amplification is completely inhibited. Mobilization of an inactive insertion to new sites can restore amplification, demonstrating that the suppression is the result of a position effect (Spradling et al., 1987).

The boxes shown in Fig. 2 define the minimum control elements that have been shown to induce amplification. When additional chorion DNA that normal flanks the ACE elements is included in transposons, the fraction of insertions which amplify and the average amplification levels are increased. These results suggest that multiple elements with additive quantitative effects are located at multiple sites throughout the gene cluster. Several such elements have been roughly mapped to lie upstream of several chorion genes (Orr-Weaver and Spradling, 1986; Deladakis and Kafatos, 1987). However even large transposons encompassing most of the gene cluster are still subject to position effects. At rare insertion sites even such transposons are completely unable to amplify.

REGULATION OF REPLICATION IN MODEL SYSTEMS

A more detailed examination of model viral replicons provides insight into possible mechanisms for the developmental control of the chorion gene replicons during amplification. In comparing such studies to analysis of chorion gene amplification two factors should be considered. In model systems, only when very detailed mutagenesis is carried out has it been possible to distinguish between sequences required in cis for replication, and the biochemically determined site where replication initates, or origin proper. Second, concomitant biochemical studies have been critically important in determining the role played by specific chromosome regions in replication.

In several animal viruses, detailed genetic studies have defined sequences required in cis for efficient replication that are separable from the replication origin itself. In several systems, a linkage between the control of replication and the control of transcription has been demonstrated. The linkage between replication and transcription is manifested by cis-acting control regions that affect both processes. These cis-acting elements are transcriptional enhancers whose regulatory effects are orientation and distance independent. Although the mechanism by which enhancers control both replication and transcription is presently not understood, it is likely that regulation is exerted by protein binding. Interaction of proteins bound at the enhancer with origin binding proteins might facilitate the initiation of DNA replication. However, the cis-acting control element might act indirectly, for example, by directing a specific chromatin segment containing a replication origin to a permissive region of the nucleus.

Polyoma Polyoma replication requires the virally encoded T antigen and two cis-acting components, the origin of replication and the enhancer. Two regions of the origin are important, a 32 base pair inverted repeat and an A-T rich tract. These are separated by 60 base pairs of nonessential sequences. The enhancer is required in cis for polyoma replication, even when T antigen is provided in trans in COP-5 cells (Tyndall et al., 1981; de Villiers et al., 1984). The SV40 enhancer or the mouse immunoglobulin heavy chain enhancer can be substituted for the polyoma enhancer (de Villiers et al., 1984). The polyoma enhancer is composed of functionally repeated elements, although the sequences differ. The A component (26 base pairs) is homologous to the adenovirus ElA enhancer; multiple copies of this sequence function as the enhancer both for transcription and replication. Element B has homology to the SV40 enhancer, and element C may be a part of element B (Veldman et al., 1985).

Experiments with polyoma indicate that the tissue specificity of replication may be controlled by an enhancer.

When the immunoglobin heavy chain enhancer was substituted for the polyoma enhancer the recombinant virus was capable of replication in a myeloma cell line (de Villiers et al., 1984). Naturally occurring viral variants with altered host specificity have mutations in the A and B enhancer elements. Most of these are duplications; although one is a deficiency and duplication (de Simone et al., 1985). However, these experiments must be interpreted cautiously, since T antigen was not provided in trans. Consequently the alteration in tissue specificity may be the result of the transcriptional requirement for T antigen, rather than a direct effect on origin usage. Following injection into mouse embryos polyoma will replicate. Constructs with the A and B enhancers and the origin of replication or with the B enhancer alone and the origin of replication will replicate in embryos when T antigen is provided in trans. In contrast the A enhancer and origin will not replicate in 2 day old embryos, even though this construct will replicate in differentiated cells (Wirak et al., 1985). In addition to the above examples of an enhancer exerting a positive influence to induce replication in a novel tissue, elements within the polyoma enhancer can exert negative effects on replication. A deletion of the B and C elements of the polyoma enhancer (the PvuII D deletion) resulted in an expanded host range, suggesting these elements repress replication in some tissues (Campbell and Villarreal, 1986).

While the tissue specificity of polyoma replication is regulated by the enhancer, the species specificity is regulated, at least in part, by the α polymerase and primase. Substitution of the SV40 enhancer for the polyoma enhancer failed to permit replication of polyoma in simian cells; the converse construct also did not allow SV40 to replicate in mouse cells. However polyoma replicated after the addition of purified mouse replicase to a monkey CV-1 cell extract (Murakami et al., 1986 a, b).

The large T antigen is required both to activate replication and transcription, through its binding to the early gene side of the origin core. In addition to this viral product, a number of trans acting cellular factors are expected to participate in the initiation of replication and transcription. These are currently beginning to be identified by the initial criterion that they bind to the enhancer region. Two such factors have been described, both of which footprint the B enhancer; their role in transcription and/or replication awaits further characterization (Fujimura, 1986; Piette and Yaniv, 1986).

Bovine papilloma virus BPV undergoes two types of regulation of replication. Early in infection the genome amplifies to 100 copies per cell. Replication subsequently occurs in step with the host cell cycle, with the virus being stably maintained extrachromosomally. Two types of cis-acting elements are essential for BPV replication, a plasmid maintenance sequence (PMS) and an enhancer (Lusky and Botchan, 1986a). BPV contains 2 PMS sequences; these may be functionally redundant, although the *in vivo* initiation site has been mapped within PMS1 by electron microscopy (Waldeck et al., 1984). Lusky and Botchan (1986a) generated a series of linker insertion mutants within the PMS-1 region and assayed their phenotype in a transient replication assay. They identified two essential domains for replication, one being PMS1 and the other being a transcriptional enhancer. This enhancer lies close to PMS-1 at the 5' end of the early transcription unit. The replication requirement for this enhancer appears to be met by replacement of it with the β-globin or retroviral LTR enhancers (Chen et al., 1985; DiMaio et al., 1982). Another enhancer is present 3' to the polyadenylation site of the early mRNA. However this enhancer is needed for maintenance or transformation by the viral genome only if the BPV DNA is linked to prokaryotic pBR322 DNA (Howley et al., 1985). In addition to the role of an enhancer in regulating BPV replication, PMS-1 contains both the origin of DNA replication and a promoter. While this promoter is required for BPV replication, its effect is most likely mediated through protein-protein interaction rather than transcriptional activation. Some mutations inactivate the promoter without effecting replication (Stenlund et al., 1987).

Three, and possibly five, viral products regulate BPV replication. The E2 product affects the amplification of the virus after infection, possibly indirectly as a consequence of its interaction with PMS1 (Lusky and Botchan, 1986a). E2 binds to PMS1 (Androphy et al., 1987). The 5' end of the E1 gene (the M gene) represses early amplification (Berg et al., 1986). A hybrid plasmid with both the SV40 origin and the BPV genome replicates at low copy number in COS cells. This repression of the normal replicative properties of the SV40 origin requires the M gene *in trans*, and two cis acting regions linked to PMS1 and PMS2 of BPV (Roberts and Weintrab, 1986). In contrast to the repressive effects on the amplification phase of replication, the M gene is required for stable replication (Lusky and Botchan, 1986a) as are the 3' end of E1 (the R gene) and the E6/7 gene products (Lusky and Botchan, 1986b).

SV40 The cis-acting control region of SV40 is composed of the origin of replication which contains a T antigen binding site, and two types of promotor elements, the G-C rich 21 base pair repeats, and the 72 base pair repeats. Unlike polyoma, the 72 base pair enhancers are not essential for replication.

However, if both the 72bp or 21bp repeats are deleted replication levels are decreased 100 fold in vivo (Hertz and Mertz, 1986; Li et al., 1986). Therefore either the 72bp or 21bp repeat regions can supply the replication function, but one must be present. In contrast to BPV and EBV, the enhancer effect on SV40 replication is not position independent and the repeated regions must be adjacent to the origin core to function efficiently.

The identification of trans-acting factors and the elucidation of their roles in transcription and/or replication may prove to be most tractable with SV40, because an efficient in vitro replication system has been developed (Li and Kelley, 1984). The sole viral-encoded factor required for transcription and replication is large T antigen. T Ag binds to the replication origin and also to the 21 base pair repeats. An intensive effort is currently ongoing in several laboratories to identify cellular factors that bind to the enhancer. The in vitro replication system for SV40 will ultimately allow testing of the effect of these enhancer binding factors on replication. The factor SP1 binds to the 21 base pair repeats to stimulate transcription (for review Dynan and Tjian, 1985). SP1 will also bind to and facilitate transcription of murine cellular sequences isolated based on their homology to the replication origin and 21 base pair repeats (Dynan et al., 1985). The mouse dhfr promoter is also activated by SP1 (Dynan et al., 1986). These two observations raise the possibility that the association of origins with enhancers may be common, and SP1 could activate both processes. Its effects on replication remain to be examined. Another transcriptional binding factor, CTF, has been shown to be the same protein as NF1 (Jones et al., 1987). NF1 is required for the initiation of adenovirus replication (for review, Campbell, 1986), and may facilitate cellular replication as well. AP1 is another cellular factor that binds the SV40 enhancer; many others will undoubtedly soon be identified (Lee et al., 1987).

EBV The origin of replication for Epstein-Barr virus has been isolated by its ability to allow extrachromosomal maintenance of plamids bearing it in cells expressing the viral nuclear antigen EBNA-1. It contains two components, an enhancer that has 20 imperfect repeats of a 30 base pair sequence and a 65bp region of dyad symmetry that is the presumed origin of DNA replication. These two elements can be separated by 1.3Kb without affecting replication and the two elements are orientation independent (Reisman et al., 1985; Lupton and Levine, 1985). EBNA-1 binds both to the enhancer and the presumed origin of replication, suggesting that binding of this protein may effect both replication and transcription (Reisman and Sugden, 1986).

REGULATION OF AMPLIFICATION

Genetic dissection of amplification control elements have revealed that transcription control elements may control amplification in a manner analogous to the enhancer-like elements described above. The potential role of chorion gene regulatory sequences was discovered in studies which mapped elements essential for amplification (Orr-Weaver and Spradling, 1986). A series of small deletions was constructed in the large EcoRI fragment containing ACE3 (Fig. 2). The constructs also contained insertions of the E. coli lacZ gene to mark the s18 or s15 genes on the transposon. Transformed lines containing each deletion were tested for their ability to amplify and to express s18 or s15. The results showed that a within the entire ACE3 region, only a 427 bp BalI deletion (see Fig. 2, filled box) eliminated amplification. The effect of this mutations on chorion gene transcription was suggestive. Transcription of s18-lacZ could not be detected in strains carrying the BalI deletion transposon. The loss of s18 transcription and amplification is consistent with a role for an upstream s18 transcriptional control element in amplification. However, more detailed mutational analysis required to prove this point. Alternatively, sequences controlling amplification and s18 might lie close together on the chromosome, but be functionally separable.

Similar experiments were also carried out on the X chromosome (Spradling et al., 1987). Although the entire ACE1 region was not tested, a single region containing sequences essential for amplification was located between position -9 and -488 upstream from s38 (Fig. 2). This region contains several elements that regulate s38 transcription (Wakimoto et al., 1986; Spradling, A., Kalfayan, L., and Wakimoto, B., manuscript in preparation). An element required for normal levels of transcription is located between -476 and -266 (hatched box). In addition, elements necessary for correct timing of transcription (open box: -266 to -109) and further promoter elements (stippled box: -109 to +74) are present. Determining whether these elements play a role in amplfication as well as transcription will require the construction of transposons in which they have been specifically mutated or deleted, but which retain all other sequences required for amplification.

Enhancer-like transcription control elements that stimulate replication from a nearby replication origin during amplification would provide explanations for several aspects of chorion gene amplification. Amplifiction is observed in follicle cells beginning in about stage 9, but in no other tissue or time of development. A requirement for an active transcriptional element could explain the developmental regulation of amplification. For example, an enhancer binding protein might be produced only in the follicle cells at the time of amplification initiation. The fact that the s18 or s38 genes do not actually produce transcripts until after amplification is underway does not detract from this proposal. The production of transcription factors other than the putative enhancer binding protein may be limiting.

A second aspect of chorion gene regulation that could be explained by this model concerns the additive stimulatory effects on amplification observed with DNA segments flanking the ACE elements themselves. The effects of the transcriptional enhancers may be additive. If each chorion gene contains such an element upstream, then DNA flanking the ACE elements will contain such sequences, thus explaining their stimulatory effect. A requirement for multiple copies of the enhancer-like element might suggest a functional significance for the clustered arrangement of the genes in both amplified regions.

CONCLUSION

We have summarized the current status of our knowledge of the regulation of chorion gene amplification. Comparisons with the regulation of several animal virus replicons suggested a role for chorion gene transcriptional control elements in stimulating amplification from specific replication origins located within the gene clusters. These studies strongly support the general utility of model replicons in understanding the regulation of eukaryotic chromosome replication. However the proposed role for transcriptional control elements in amplification remains unproven; we now know that such elements are located within areas that are genetically important for amplification. Determining whether their association is functional or fortutious will require further study.

REFERENCES

Androphy, E., Lowy, D. and Schiller, J., 1987, Bovine papillomavirus E2 trans-activating gene product binds to specific sites in papillomavirus DNA, Nature, 325:70-73.

Berg, L., Lusky, M., Stenlund, A. and Botchan, M., 1986, Repression of bovine papilloma virus replication is mediated by a virally encoded trans-acting factor, Cell, 46:753-62.

Campbell, J., 1986, Eukaryotic DNA replication, Ann. Rev. Biochem., 55:733-71.
Campbell, B., Villarreal, L., 1986, Lymphiod and other tissue-specific phenotypes of polyomavirus enhancer recombinants: Positive and negative combinational effects on enhancer specificity and activity, Mol. Cell Biol., 6:2068-79.
Chen, E., Howley, P., Levinson, A., and Seeburg, P., 1982, The primary structure and genetic organization of the bovine papilloma virus type 1 genome, Nature, 299:529-34.
Cordeiro, M., Menghenini, R., 1975, The rate of DNA replication in the polytene chromosomes of Rhynchosciara angelae, J. Mol. Biol., 78:261-74.
de Cicco, D., Spradling, A., 1984, Localization of a cis-acting element responsible for the developmentally regulated amplification of Drosophila chorion genes, Cell, 38:45-54.
de Simone, V., La Mantia, G., Lania, L., and Amati, P., 1985, Polyomavirus mutation that confers a cell-specific cis advantage for viral DNA replication. Mol. Cell Biol., 5:2142-46.
de Villiers, J., Schaffner, W., Tyndall, D., Lupton, S., Kamen, R., 1984, Polyomavirus DNA replication requires an enhancer, Nature, 312:242-46.
Delidakis, C., Kafatos, F.C., 1987, Amplification of a chorion gene cluster in Drosophila is subject to multiple cis-regulatory elements and to long range position effects, J. Mol. Biol., in press.
DiMaio, D., Treisman, R., and Maniatis, T., 1982, Bovine papilloma virus vector that propagates as a plasmid in both mouse and bacterial cells, Proc. Nat. Acad. Sci. USA, 79: 4030-34.
Dyan, W. and Tjian, R., 1985, Control of eukaryotic messenger RNA synthesis by sequence-specific DNA-binding proteins, Nature, 316:774-78.
Dyan, W., Saffer, J., Lee, W. and Tjian, R., 1985, Transcription factor Spl recognizes promoter sequences from the monkey genome that are similar to the simian virus 40 promoter, Poc. Natl. Acad. Sci. USA, 82:4915-19.
Dyan, W., Sazer, S., Tjian, R. and Schimke, R., 1986, Transcription factor Spl recognizes a DNA sequence in the mouse dihydrofolate reductsae promoter, Nature, 319:246-48.
Fujimura, F., 1986, Nuclear activity from F9 embryonal carcinoma cells binding specifically to the enhancers of wild-type polyoma virus and PyEC mutant DNAs, Nucl. Acids Res., 14:2845-61.
Hammond, M.P., Laird, C.D. 1985, Chromosome structure and DNA replication in nurse and follicle cells of Drosophila melanogaster, Chromosoma, 91:279-86.
Harland, R., and Laskey, R.A., 1980, Regulated replication of DNA microinjected into eggs of X. laevis, Cell, 21: 761- 71.

Hertz, G., and Metz, J., 1986, Bidirectional promoter elements of simian virus 40 are required for efficient replication of the viral DNA, Mol. Cell Biol., 6:3513-22.
Howley, P.M., Schenborn, E.T., Lund, E., Byrne, J.C. and Dahlberg, J.E., 1985, The bovine papillomavirus distal enhancer is not cis essential for transformation or for plasmid maintenance. Mol. Cell. Biol., 5:3310-15.
Jones, K., Kadonaga, J., Rosenfeld, P., Kelly, T. and Tjian, R., 1987, A cellular DNA-binding protein that activates eukaryotic transcription and DNA replication, Cell, 48:79-89.
Kafatos, F.C., Mitsialis, S.A., Spoerel, N., Mariani, B., Lingappa, J.R., Delidakis, C., 1985, Studies on the developmentally regulated expression and amplification of insect chorion genes, Cold Spring Harbor Symp. Quant. Biol., 50:537-48.
Kalfayan, L., Levine, J., Orr-Weaver, T., Parks, S., Wakimoto, B., de Cicco, D., and Spradling, A., 1986, Localization of sequences regulating Drosophila chorion gene amplification and expression, Cold Spring Harbor Symp. Quant. Biol., 50: 527-535.
Lee, W., Haslinger, A., Karin, M. and Tjian, R., 1987, Activation of transcription by two factors that bind promoter and enhancer sequences of the human metallothionein gene and SV40, Nature, 325:368-372.
Li, J. and Kelly, T., 1984, Simian virus 40 replication in vitro, Proc. Natl. Acad. Sci. USA, 81:6973-77.
Li, J.J., Peden, K.W.C., Dixon, R.A.F. and Kelly, T. 1986, Functional organization of the simian virus 40 origin of DNA replication, Mol. Cell. Biol., 6:1117-28.
Lupton, S., and Levine, A., 1985, Mapping genetic elements of Epstein-Barr virus that facilitate extrachromosomal persistence of Epstein-Barr virus-derived plasmids in human cells. Mol. Cell Biol., 5:2533-42.
Lusky, M., Botchan, M., 1981, Inhibition of SV40 DNA replication in simian cells by specific pBR322 DNA sequences, Nature, 293:79-81.
Lusky, M, Botchan, M., 1986a, Transient replication of bovine papilloma virus type 1 plasmids: cis and trans requirements, Proc. Nat. Acad. Sci. USA, 83:3609-16.
Lusky, M, Botchan, M., 1986b, A bovine papilloma virus type 1-encoded modulator function is dispensable for transient viral replication but required for the establishment of the stable plasmid state, J. Virol., 60:729-742.
Mahowald, A.P. and Kambysellis, M.P., 1980, Oogenesis, in: "The Genetics and Biology of Drosophila," M. Ashburner and T.R.F. Wright, eds., Academic Press, New York, pp. 141-224.
Mahowald, A.P., Caulton, J.H., Edwards, M.K., Floyd, A.D., 1979, Loss of centrioles and polyploidization in follicle

cells of Drosophila melanogaster, Exp. Cell Res., 118:404-10.

Margaritis, L.H., 1985, Structure and physiology of the eggshell, in: "Comphrensive insect physiology, biochemistry and pharmacology," vol. 1., G.A. Kerkut and L.I. Gilbert, eds., Pergamon Press: Elmsford, pp. 113-151.

Mecheli, M. and Kearsey, S., 1984, Lack of specific sequence requirements for DNA re;lication in Xenopus eggs compared with high sequence specificity in yeast, Cell. 38:55-64.

Murakami, Y., Wobbe, C.R., Weissbach, L., Dean, F. and Hurwitz, J., 1986, Role of DNA polymerase a and DNA primase in simian virus 40 DNA replication in vitro, Proc. Narl. Acad. Sci. USA. 83:2869-73.

Murakami, Y., Eki, T., Yamada, M., Prives, C. and Hurwitz, J., 1986, Species-specific in vitro syynthesis of DNA containing the polyoma virus origin of replication, Proc. Natl. Acad. Sci. USA, 83:6347-51.

Orr, W., Komitopoulou, K., Kafatos, F., 1984, Mutants supressing in trans chorion gene amplification in Drosophila, Proc. Nat. Acad. Sci. USA, 81:3773-77.

Orr-Weaver, T., Spradling, A.C., 1986, Drosophila chorion gene amplicifation requires an upstream region regulating s18 transcription, Mol. Cell. Biol., 6:4624-33.

Osheim, Y.N., Miller, O.L., 1983, Novel amplification and transcriptional activity of chorion genes in Drosophila melanogaster follicle cells, Cell, 33:543-53.

Piette, J. and Yaniv, M., 1986, Molecular analysis of the interaction between an enhancer binding factor and its DNA target, Nucl. Acids Res., 14:9595-11.

Reisman, D. and Sugden, B., 1986, Trans-activation of an Epstein-Barr viral transcriptional enhancer by the Epstein-Barr viral nuclear antigen 1, Mol. Cell. Biol. 6:3838-46.

Reisman, D., Yates, J., Sugden, B., 1985, A putative origin of replication of plasmids derived from Epstein-Barr virus is composed of two cis-acting components, Mol. Cell. Biol., 5:1822-32.

Roberts, J. and Weintraub, H., 1986, Negative control of DNA replication in composite SV40-bovine papilloma virus plasmids, Cell, 46:741-52.

Spradling, A., Mahowald, A.P., 1981, A chromosome inversion alters the pattern of specific DNA replication in Drosophila follicle cells, Cell, 27:203-09.

Spradling, A.C., 1981, The organization and amplification of two clusters of Drosophila chorion genes, Cell, 27:193-202.

Spradling, A.C., deCicco, D.V., Wakimoto, B.T., Levine, J.F., Kalfayan, L.J., and Cooley, L., 1987, Amplification of the X-linked chorion gene cluster requires a region upstream from the s38 chorion gene, EMBO J, 6:1045-1053.

Spradling, A. and Orr-Weaver, T., 1987, Regulation of DNA

replication during Drosophila development, Ann. Rev. Genetics, 21:in press.
Steinemann, M., 1981, Chromosomal replication in Drosophila virilis III. Organization of active origins in the highly polytene salivary gland cells, Chromosoma, 82:289-307.
Stenlund, A., Bream, G. and Botchan, M., 1987, A promoter with an internal regulatory domain is part of the origin of replication in BPV-1, Science, 236:1666-71.
Tyndall, C., LaMantia, G., Thacker, C.M. Favalora, J. and Kamen, R., 1981, A region of the polyomavirus genome between the replication origin and late protein coding sequences is required in cis for both early gene expression and viral DNA replication, Nucl. Acids. Res., 9:6231-50.
Veldman, G., Lupton, S. and Kamen, R., 1985, Polyomavirus enhancer contains multiple redundant sequence elements that activate both DNA replication and gene expression, Mol. Cell. Biol., 5:649-58.
Wakimoto, B., Kalfayan, L., and Spradling, A., 1986, Developmentally regulated expression of Drosophila chorion genes introduced at diverse chromosomal locations, J. Mol. Biol., 187:33-45.
Wakimoto, B., Kalfayan, L., Levine, J., and Spradling, A., 1986, Localization of region controlling Drosophila chorion gene expression, in: "44th Symposium of the Society for Developmental Biology," J. Gall, ed., A.R. Liss, Inc., New York, pp 43-54.
Waldeck, W., Rosl, F. and Zentgraf, H., 1984, Origin of replication in episomal bovine papilloma virus type 1 DNA isolated from transformed cells, EMBO J., 3:2173-78.
Wirak, D., Chalifour, L., Wassarman, P., Muller, W., Hassell, J. and De Pamphilis, M., 1985, Sequence-dependent DNA replication in mouse embryos, Mol. Cell Biol., 5:2924-35.

MOLECULAR CLONING OF THE MAIZE *R-nj* ALLELE BY TRANSPOSON TAGGING WITH *Ac*

Stephen L. Dellaporta*, Irwin Greenblatt+, Jerry L. Kermicle++, James B. Hicks# and Susan R. Wessler[δ]

*Yale University
Department of Biology
New Haven, CT 06511

+Molecular and Cellular Biology
University of Connecticut
Storrs, CT 06268

++Laboratory of Genetics
University of Wisconsin
Madison, WI 53706

#Scripps Clinic and Research Foundation
Department of Molecular Biology
La Jolla, CA 92037

[δ]Department of Botany
University of Georgia
Athens, GA 30602

ABSTRACT

The *R* locus of maize is one of several genes that condition the red and purple anthocyanin pigments throughout the body of the plant and seed. Many alleles of the *R* locus have been described, each determining a different pattern of pigment distribution in the plant. Genetic studies have lead to the proposal that *R* is a complex locus containing several similar but functionally distinct modules that regulate anthocyanin expression in specific organs or tissues. We report on the isolation of a segment of DNA from the *R* locus using the controlling element *Activator* (*Ac*)as a transposon tag. Molecular evidence is presented confirming the complex organization of this regulatory locus. The identification of cross-hybridizing restriction fragments corresponding to the tissue-specific *R* components suggests that these elements, while functionally divergent in tissue-specificity, have retained significant DNA homology and may have evolved from a common ancesteral component at *R*.

INTRODUCTION

The *R* gene of maize impinges on anthocyanin pigmentation in at least two ways. First, it governs the expression of unlinked genes encoding enzymes in the flavonoid biosynthetic pathway. An active *R* gene is required for normal activity of the first and a terminal step of flavonoid biosynthesis (products of the *C2* and *Bronze* genes) and, presumably for other steps in the pathway (coded by *A*, *A2*, *C*, and *Bronze-2*) (Dooner, 1983). Second, various alleles of *R* confer diverse patterns of pigment distribution among seed and plant parts. Thus, the *R* gene may regulate pigment deposition by controlling the expression of the flavonoid biosynthetic gene pathway in a tissue-specific manner. Alternatively, *R* may confer tissue-specific pigment distribution patterns by producing a product(s) in specific cells that catalyzes an early enzymatic step in flavonoid biosynthesis.

R-allelic diversity is attributable to variation in the number and kind of genetic units which L.J. Stadler (1951) termed "*R*-genic elements". Such elements occur at the *R* locus of different maize races either singly or as complexes of two or more members. Genetic analysis of one complex, designated as *R-r:standard*, reveals one element which confers anthocyanin formation to the aleurone layer of the kernel endosperm and another which confers coloration on various seedling and mature plant parts (Dooner and Kermicle, 1971). These elements, called plant or (P), and seed or (S), components, behave genetically as separate genes within the *R* complex. A schematic of this genetic model is shown in Fig. 1A. The *R-navajo* (*R-nj*) allele, utilized in the present study for molecular cloning, has not given evidence of complexity in similar tests although it pigments the crown portion of the aleurone in the kernel and various seedling parts as well as the anthers and silks of mature plants (Kermicle 1970, unpublished data).

Information concerning the function of the *R* gene product is not available. Therefore, we have begun a molecular study of this locus by cloning a portion of the *R* complex. Because the *R* gene product has not been characterized, an indirect approach to identifying an *R* molecular clone was necessary. Following the molecular approach first described by Fedoroff et al (1984) to clone the *bronze* locus, we made use of a cloned mobile sequence (Fedoroff et al; 1983), *Activator* (*Ac*), as a "transposon tag" for the *R* DNA. *Ac* transposes preferentially to nearby sites in the chromosome (Van Schaik and Brink, 1979; Greenblatt and Brink, 1962; Greenblatt, 1984), serving as a localized mutagen. A reciprocal translocation was used to bring an *Ac* element located at the *P* locus on chromosome-1 and the target allele *R-nj*, in chromosome-10 together in *cis* arrangement. An unstable allele of *R-nj*, designated *r-nj:m1*, was located from this stock and shown to contain an active *Ac* at *R-nj* by genetic and molecular analysis using an *Ac* sequence as a hybridization probe. The *Ac* containing fragment was cloned and an adjacent unique genomic fragment was used as a hybridization probe on Southern blots to correlate physical structure with the organization of the *R-r:standard* allele. Three restriction fragments with sequence homology to the *R-nj* DNA probe were identified two of which can be correlated with (P) and (S) function. The third homologous fragment indicates the presence of a cryptic component of the *R-r* allele.

MATERIALS AND METHODS

R-alleles

R-navajo (R-nj): Isolated from the Cudu strain, this stable allele is distinguished most obviously by a solid patch of aleurone pigment in the crown region of the kernel. Other regions of the aleurone are pigmented irregularly. The embryos, coeloptile and roots of seedlings, as well as silks and anthers of mature plants pigment with variable intensities.

R-r:standard and its derivatives: *R-r:standard* confers strong uniform pigmentation to the aleurone layer of kernels, to various seedling parts and to anthers of mature plants. Plant and seed effects are determined by separate genetic elements, (P) and (S), which are organized as members of a direct duplication. Unequal crossing over results in loss of the duplication, yielding products carrying only (P), designated as *r-r*, or (S), designated *R-g*. The particular such derivatives used here were *r-r:n46* and *R-g:1*.

Mutation of (S) to (s) in *R-r:standard* yielded a (P) (s) product designated as *r-r:n35*. Such noncrossover *r-r* mutants retain the duplication of *R-r:standard* as evidenced by the fact that they retain two doses of *Inhibitor of striate* (*Isr*), a second gene encompassed within the duplicated segment (Kermicle and Axtell, 1981), and recombine with (S) derivatives to reconstitute the *R-r* phenotype (Dooner and Kermicle, 1974).

r-g: This test allele pigments neither the seed nor plant parts.

Isolating *Ac*-induced *R* mutations

Detection of putative insertions of *Ac* into *R-nj* were made on the basis of the presence of variegated aleurone color in the region of the *R-nj* expression - only on

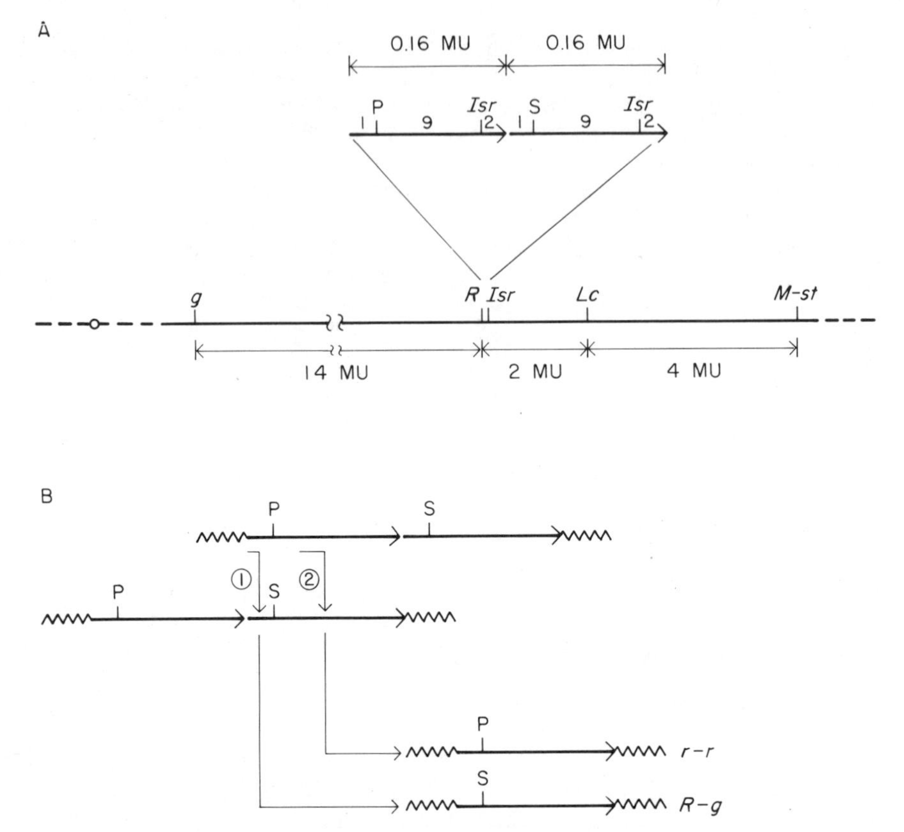

Fig. 1. Genetic Organization of the *R* Locus.

A. Genetic map of the long arm of chromosome-10 of maize showing the genetic map of the *R* locus and flanking markers: *g: golden; Isr: inhibitor of strlate; Lc: leaf color component; M-st: modifier of stippled.* The region that includes *R* is expanded to show the genetic fine structure relationship of the (P) and (S) components of the *R-r:standard* allele. Each component is carried on a chromosome duplicated segment (arrow) and includes the linked marker *Isr* as part of the duplication. The position of the *R* component indicated above the arrow is based on the relative numbers of recombinants observed in experiments of *R-r* fractionation and resynthesis (see Dooner and Kermicle, 1974, 1976). The position of *Isr* within the duplicated segment was determined relative to a distinguishable allele carried in *R-stippled* (Kermicle and Axtell, 1981, and Kermicle, unpublished).

B. Genetic model of unequal crossing over in *R-r* homozygotes. Recombination in region 1 or 2 (arrows) will yield *R-g* (S) or *r-r* (P) derivatives, respectively.

the crown portion of the kernel (Fig. 1). Kernels were planted and tested if they had a colorless aleurone with colored spots occurring and also if they had colored aleurone with colorless spots occurring. All such kernels were planted and pollinated with a *r-g* stock. It was from this population that the *R-nj* mutable analyzed here (*r-nj:m1)* was recovered. The *r-nj:m1* mutation was propagated for several generations by backcrossing into a *r-g* W22 genetic background.

DNA preparation, restriction and genomic blot analysis

Genomic DNA was isolated from mature leaf tissue by a previously described method (Shure et al., 1983). Genomic DNA was digested with a three-fold excess units of restriction enzymes according to the manufacturer's recommendation (New England Biolabs or Bethesda Research Laboratories). DNA samples were electrophoresed through 0.8% agarose (Sigma) gels, and were transferred to nitrocellulose (Schleicher and Schuell) according to the method of Southern (1975). Plasmid DNA was isolated by the method of Holmes and Quigley (1981), the plasmid insert was purified by electroelution (Dretzen et al., 1981) and nick translated as described by Rigby et al., (1977). DNA polymerase and DNAase I for nick translation was obtained from International Biotechnology Inc., and ^{32}P-α-NTP at a specific activity of 800 Ci/mM was supplied by New England Nuclear.

Filters were prehybridized for 2 hrs at 65°C in 10% dextran sulfate (Pharmacia), 6X SCP (1X SCP = 100 mM NaC1, 30 mM Na_2HPO_4, and 1 mM EDTA, pH 6.5), 2% sarcosine and 500 ug/ml heparin (Type II Sigma). Hybridization was performed for 12 hrs at 65°C in the above buffer containing denatured salmon sperm DNA (final concentration 100 ug/ml) and denatured, nick translated probe DNA (finale probe concentration = 10 ng/ml). The filters were washed three times in 2X SCP, 1% SDS at 65°C for 30 minutes and then in 0.1X SCP, 0.1% SDS at 65°C for 15 minutes.

Genomic Library Construction and Screening

Maize DNA libraries were constructed with 2 ug of Bam HI-digested genomic DNA and 5 ug of λ EMBL4 arms and co-ligated at 4°C for 24 hr in a 15 ul reaction volume containing 100 units of T_4 DNA ligase (New England Biolabs). Three ul of the ligation mixture was packaged *in vitro* as described by Hohn (1979). Phage were plated on *E. coli* LE392 (Berman et al., 1982) and transferred to nitrocellulose according to Maniatis et al., (1982). The filters were hybridized and washed using the conditions described above, except dextran sulfate was omitted from the buffer solutions. The *r-nj:m1* library was screened with nick-translated pAcH1.6 which contains the internal 1.6 kb Hind III fragment of *Ac9* (Fedoroff et al., 1983) and positive recombinant phage were further purified by rescreening. Small scale phage DNA samples prepared according to Berman et al., (1982). Phage DNA samples (approximately 1ug) were digested with 3 units of the appropriate restriction enzyme (purchased from New England Biolabs). DNA electrophoresis and blot hybridization analysis conditions were identical to those described above.

Restriction mapping and plasmid subcloning

General restriction sites in λ*Ac*-28 were mapped by digesting phage DNA to completion with the cloned *Ac9* element for comparisons. The position of the *Ac* element in the 15 kb Bam HI insert of λAc-28 was determined by the location of restriction sites predicted from published *Ac* DNA sequence data (Pohlman et al., 1984; Muller-Neumann et al, 1985). The 0.7 kb Bgl II-Hinc II DNA fragment flanking the *Ac* element was purified from agarose gels by electroelution (Dretzen et al., 1981) and subcloned into the Bam HI-Hinc II polylinker sites of the pUC119 vector (J. Viera, personal communication). This plasmid was named pR-nj:1.

RESULTS

Insertion of *Ac* at the *R* Locus

The first step in the process of transposon tagging is to identify the insertion of a particular transposon into the gene of interest. Often this involves creating a situation where both the frequency of insertion and the mode of detection are enhanced. Previous work on the mechanism of *Ac* transposition in maize (Van Schaik and Brink, 1959; Greenblatt and Brink, 1962; Greenblatt, 1984) has shown that when *Ac* transposes from the *P-vv* allele on chromosome I, two-thirds of the insertions occur at sites on the same chromosome arm and most of those occur distal to the *P* locus. The existence of such a preferred target area suggests an enrichment scheme by which the gene of interest is moved distal and as close as possible to the launch site, in this case, the *P* locus. An *Ac*-induced mutation is detected as an unstable allele of the target locus.

In order to tag the *R* locus with *Ac* the allele *R-nj* (*R-navajo*), conferring anthocyanin pigment to the crown of aleurone (Fig. 2), was placed distal to the *P-vv* allele using the chromosome translocation T1-10g. *P-vv* and *R-nj* have easily identifiable kernel phenotypes affecting pigmentation in the pericarp and aleurone, respectively, while the translocation, when heterozygous, confers a phenotype of semi-sterile seed set. The desired *P-vv* T1-10g *R-nj*] chromosome was carried heterozygous with a *P-wr* or *P-ww* allele on chromosome-1 to avoid an inhibitory effect on transposition associated with increased *Ac* dosage (McClintock, 1954); the normal chromosome-10 carried r-g (colorless seed, green plant). These plants exhibit semi-sterile seed set due to the translocation, and the medium variegated pericarp phenotype characteristic of a single active *Ac* element at the *P* locus. Such plants were then pollinated by a *r-g* stock so that the resultant progeny kernels were

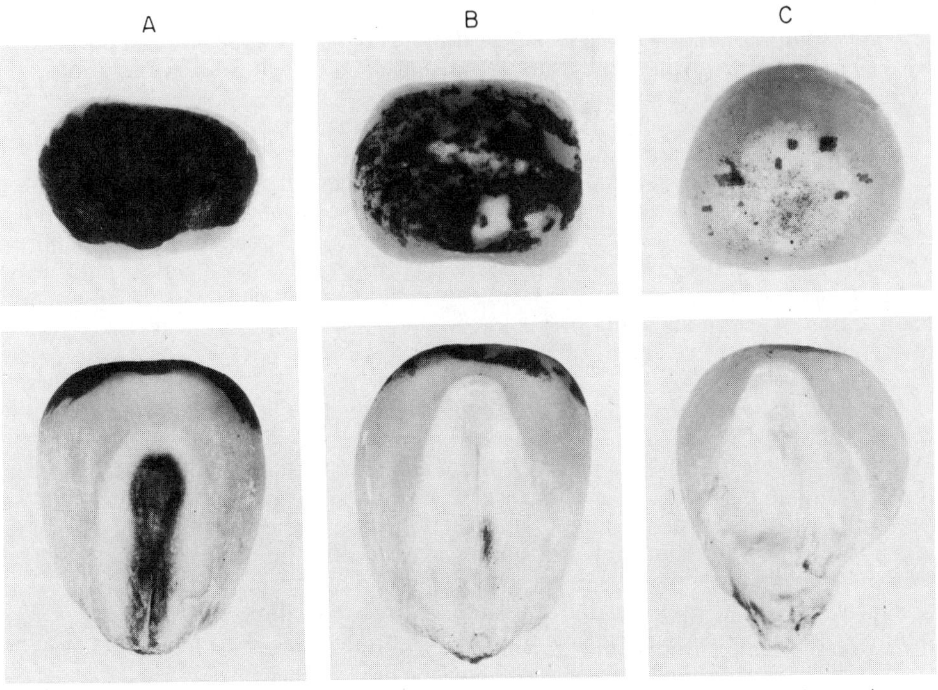

Fig. 2. Kernel phenotypes of *R-nj* and *r-nj:m1* alleles.

A. Kernel A is homozygous for the *R-nj:cudu* allele which results in aleurone pigmentation at the crown portion of the endosperm and in the embryo.

B. Kernel B contains the triploid endosperm genotype: *r-g/r-g/r-nj:m1*. The phenotype shows large, frequent somatic sectors of aleurone color in the crown portion of the endosperm and a colored sector in the embryo.

C. Kernel C contains the triploid endosperm genotype: *r-nj:m1/r-nj:m1/r-nj:m1*. Somatic mutations from colorless to colored aleurone are developmentally delayed resulting in small colored aleurone sectors. No colored embryo sectors are visible.

50 percent *R-nj* and 50 percent colorless (all other major pigment conditioning genes, *A*, *C*, *C2* for example, were homozygous in these stocks). Since the rate of *Ac* transposition from *P* also is dependent on the residual genetic background, a hybrid between two inbred lines, 4Co63 and W23, was chosen as the *r-g* parent because *Ac* moves at a higher rate in these backgrounds than any other yet measured (Greenblatt, unpublished results). This vigorous hybrid provided large ears with a high number of seed - approximately 500 seeds per ear despite the semi-sterile seed set conditioned by T1-10g heterozygosity - and, thus, more kernels to screen for new mutations.

Detection of putative insertions of *Ac* into *R-nj* were made on the basis of the presence of variegated aleurone color in the region of *R-nj* expression. Aleurone color in such cases is presumed to be the result of secondary *Ac* transpositions of the *R* locus that show up as revertant somatic sectors or spots. Approximately 78,000 kernels were screened and forty-six kernels with aleurone variegation were identified, planted, and pollinated with an *r-g* pollen source. Four resulting plants produced ears with a *R-nj* mutable phenotype and were confirmed as mutable alleles in subsequent generations. The mutable *r-nj:m1* allele, the subject of this report, was one of the four confirmed mutable alleles of *R* obtained in this study.

Ac activity varies inversely with *Ac* copy number (McClintock, 1954). According to the proposed mechanism of *Ac* transposition described by Greenblatt (1984), a chromosome carrying an *Ac*-induced *R-nj* mutant could either retain or lose *Ac* from the *P-rr*, the donor locis. In the former case the two doses of *Ac* should reduce pericarp striping resulting in a light rather than medium variegation phenotype. In the latter case, *Ac* is lost from *P-vv* resulting in an ear with solid red pericarp. In fact, all four mutants identified in the screen yielded ears showing semi-sterile seed set (characteristic of the translocation) and among these four, three produced ears with light variegated pericarp and the fourth mutation produced a red pericarp ear. The *r-nj:m1* mutant selection resulted in a semi-sterile, light-variegated ear.

Identification of *Ac* at *R-nj*

In order to place the newly generated *r-nj:m1* allele in a homozygous genetic background and to remove it from the T1-10g translocation, the original mutant lines were backcrossed at least eight times to the inbred genetic strain W22. W22 contained the *r-g* allele and no additional active *Ac* elements. It is therefore possible to determine unambiguously the condition of the *R* locus in each kernel. The *rnj:ml* allele was previously confirmed as an *Ac*-induced mutation by three genetic criteria: 1) the *r-nj:m1* allele behaves as an autonomous mutable system; 2) it serves to *trans*activate *Ds* elements; and 3) somatic instability of *r-nj:m1* shows the characteristic dosage effect of *Ac*-induced mutations (Brink and Williams, 1973). We confirmed this analysis by varying *r-nj:m1* dosage in the triploid endosperm tissue (Fig. 2) and by mating *r-nj:m1/r-g Wx/wx* plants to a *r-g/r-g wx-ml/wx-ml* tester stock. The wx-ml allele is a *Ds*-induced mutation that shows somatic and germinal instability only when *Ac* is present (McClintock, 1948). The instability of *Ds* at *wx-ml* co-segregated with the *r-nj:m1* allele in these tests (data not shown).

The genetic characteristics provide strong evidence that *r-nj:m1* was caused by insertion of *Ac* at or near the *R-nj* locus.

After backcrossing *r-nj:m1/r-g* plants to the inbred W22 *r-g* tester, DNA from 8 plants heterozygous for the *r-nj:m1* allele and 7 homozygous *r-g/r-g* siblings was subjected to genomic blot analysis in order to determine whether a particular *Ac* sequence could be correlated with genetic linkage to *r-nj:m1*. The DNA samples were digested with the restriction enzyme Sst I, chosen because it does not cut within the known active *Ac* elements so far characterized (Pohlman et al., 1984; Muller-Neumann et al., 1985). After blotting to a nitrocellulose filter, the DNAs were probed with pAcH1.6, a plasmid containing the internal 1.6 kb Hind III fragment isolated from the *Ac9* element (Fedoroff et al., 1983). An example of these blots are shown in Fig. 3. At least 8 Sst I fragments homologous to *Ac* were common to all the progeny examined. In addition, a band of approximately 11 kb was observed only in the *r-nj:m1* carrying plants, indicating genetic linkage of the *Ac* element to the *R* locus.

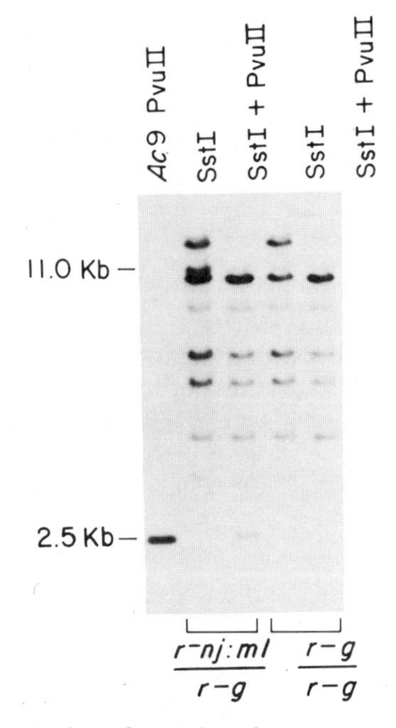

Fig. 3. Molecular Analysis of *r-nj:m1*.

Genomic analysis of DNA extracted from sibling progeny from the genetic cross *r-nj:m1/r-g x r-g/r-g*. Variegated (*r-nj:m1/r-g*) and colorless (*r-g/r-g*) kernels were planted and DNA extracted, digested with the enzymes(s) indicated, separated by gel electrophoresis, blotted to nitrocellulose filters and hybridized with pAcH1.6. Lane 1 contains a genomic copy reconstruction of pAc9 (Fedoroff et al., 1983) digested with Pvu II.

A second test established a direct relation between this 11 kb band and the *r-nj:m1* fragment. The endonuclease activity of Pvu II is sensitive to DNA methylation (McClelland, 1982). Nearly all *Ac*-homologous sequences not part of an active *Ac* element were found to be present in the heavily methylated fraction of maize DNA that is resistant to digestion with Pvu II. However, active *Ac* elements at a number of independent loci can be digested with Pvu II to generate the internal 2.5 kb Pvu II fragment of *Ac* in genomic blots (Chomet *et al.*, 1987). We found that the 11 kb Sst I fragment in *r-nj:m1* was digested with Pvu II and was replaced by a 2.5 kb Pvu II fragment that comigrates with the internal Pvu II fragment of *Ac* (Fig. 3). The other 8 Sst I fragments that hybridize to the *Ac* probe appear insensitive to Pvu II digestion except for a 15 kb band. However, since this band is detected in both *r-nj:m1* and *r-g* sibling DNA, it does not represent the Sst I fragment containing the *Ac* insertion at *R-nj*.

The 2.5 kb Pvu II fragment thus provides a unique indicator for the active Ac element. In an additional linkage study, we followed the segregation of this Pvu II fragment and the *r-nj:m1* allele in the genetic backcross progeny of *r-nj:m1/r-g* progeny described above (data not shown). Once again the physical marker showed 100% linkage to the *r-nj:m1* allele in all 15 segregants examined. A more detailed study of the relationship between *Ac* activity and methylation as detected by restriction enzyme digestion will be published elsewhere.

Genomic Cloning of *r-nj:m1*

Genomic blot analysis described above indicated that the *Ac* element at *R-nj* contained the 2.5 kb internal Pvu II fragment characteristic of active *Ac* elements. In order to isolate this particular *Ac* element a library of Bam HI fragments from *r-nj:m1* DNA inserted into the bacteriophage cloning vector EMBL4 was prepared. Approximately 4 X 10^5 hybrid phage were screened by hybridization with the pAcH1.6 probe. Twenty-eight positive clones were picked and further screened for the presence of the expected Pvu II fragment by restriction enzyme mapping. The results of this secondary screen are shown in Fig. 4.

One such phage (λ*Ac*-28) was found to contain a 2.5 kb Pvu II fragment that hybridized to pAcH1.6 and by further restriction mapping was shown to contain the expected partial *Ac* fragment plus an additional 10.5 kb of genomic DNA (Fig. 5). A 0.7 kb Bgl II-Hinc II restriction fragment from λ*Ac*28 that represents sequences flanking the *Ac* element was then subcloned. This plasmid was designated pR-nj:1.

Confirmation of *pR-nj:1* as an *R*-specific probe

According to the genetic and molecular characterization of the *R* locus and the *r-nj:m1* allele described above, a *bonafide R* locus probe should exhibit several characteristics. First, it should hybridize to the 11 kb Sst I fragment associated with

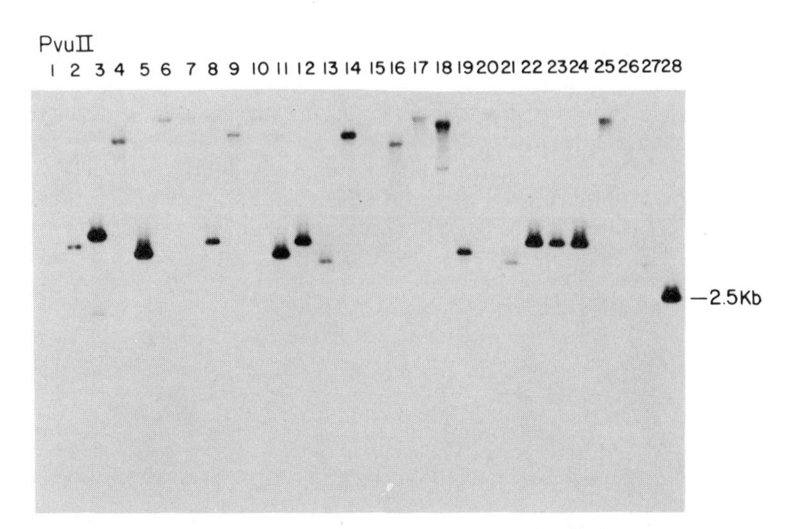

Fig. 4. Pvu II Screening of *Ac* Recombinant Clones
Southern blot of Pvu II-digested DNA purified from λ recombinant clones hybridizing to pAcH1.6 probe. Phage DNA was digested with Pvu II separated by gel electrophoresis, blotted to nitrocellulose and probed with pAcH1.6. Lane 29 contains 0.5 ug of the cloned *Ac9* element (Fedoroff et al., 1983) digested with Pvu II.

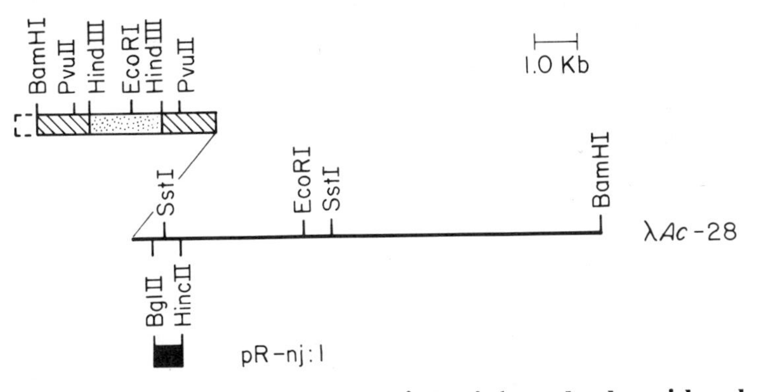

Fig. 5. Restriction maps of λ*Ac-28*, λR-nj:1 and plasmid subclone.
λ*Ac-28* contains a 15 kb Bamm HI fragment with DNA homologous to *Ac* shown as the rectangle. The pAcH1.6 probe contains the internal 1.6 kb Hind III fragment of *Ac* that corresponds to the stippled region of the rectangle. A flanking 0.7 kb DNA segment from the Bgl II to Hinc II sites (black box) was subcloned into pUC119 and designated pR-nj:1.

R. Second, since *r-nj:m1* should be the result of a 4.5 kb insertion of *Ac* into the progenitor *R-nj* allele, the *pR-nj:1* probe should hybridize to a unique 6.5 kb Sst I fragment in DNA from homozygous *R-nj* plants. Third, the 11 kb Sst I fragment in the *r-nj:m1* allele should be altered when *r-nj:m1* reverts to *R-nj*. This revertant allele was isolated from *r-nj:m1* x *r-g* backcross progeny.

Fig. 6 shows a blot of Sst I digested DNA from plants homozygous for *R-nj* (lane 1), *r-nj:m1* (lane 2), and for a *R-nj* revertant allele derived from *r-nj:m1* (lane 3). pR-nj:1 DNA hybridizes to an 11 kb Sst I fragment of *r-nj:m1* and a 6.5 kb Sst I

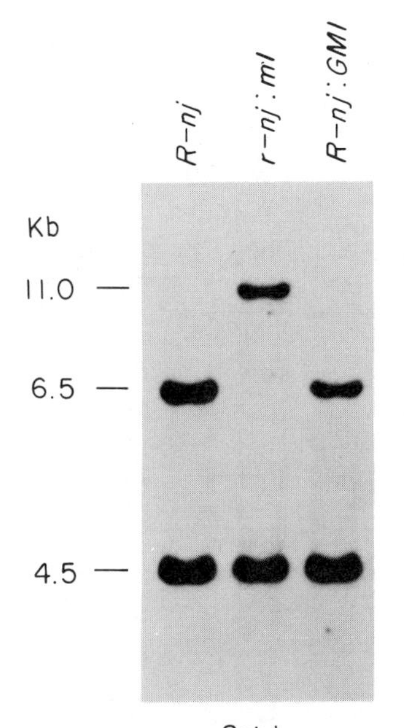

Fig. 6. Southern blot analysis of *R-nj:cudu*, *r-nj:m1*, an *R-nj* derivative allele from *r-nj:m1*, and the *r-g allele*.

Three micrograms of maize genomic DNA were digested with Sst I, separated by electrophoresis on a 0.8% agarose gel and transferred to nitrocellulose. The blot was hybridized with the pR-nj:1 probe. The 6.5 kb Sst I fragment of the *R-nj* allele is replaced by an 11.0 kb fragment of *r-nj:m1* *R-nj:GM1* is a germinal mutation of *r-nj:m1* to *R-nj*. DNA from a plant homozygous for *R-nj:GM1* shows the restoration of the 6.5 kb Sst I fragment found in the progenitor *R-nj* allele.

fragment of *R-nj* and the *R-nj* revertant allele. This is the expected result from the insertion and excision of *Ac* from a 6.5 kb genomic fragment in *R-nj*. An additional 4.5 kb Sst I band was also detected in each DNA. Because the pR-nj:1 insert contains a central Sst I site (Fig. 5), this second fragment, as expected, remains unaltered by integration and excision of *Ac*.

The most compelling evidence that pR-nj:1 is a part of the *R* locus, however, comes from physical examination of alleles derived from *R-r:standard* by meiotic

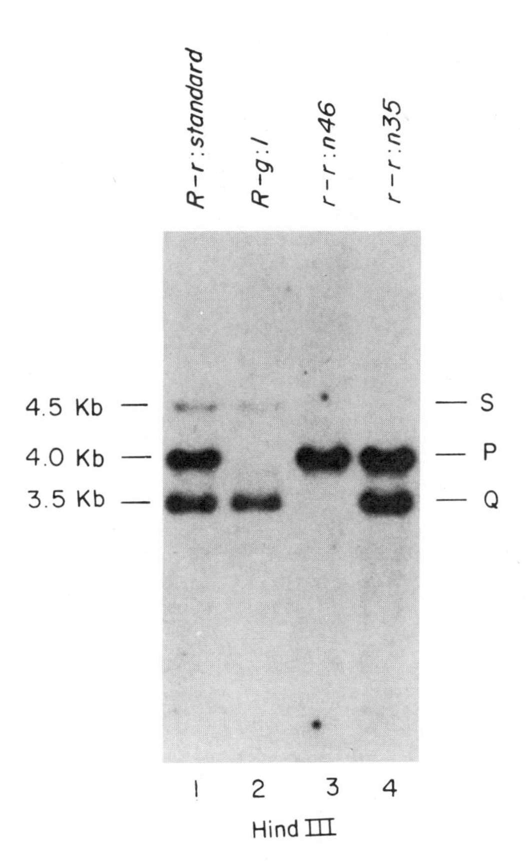

Fig. 7. Southern blot analysis of *R-r:standard* and derivative alleles.

DNA was prepared from plants homozygous for the *R-r:standard* or derivative allele indicated, digested with Hind III, separated by electrophoresis on a 0.8% agarose gel, blotted to nitrocellulose, and hybridized with the pR-nj:1 probe.

recombination. The genetic model for the structure of the *R* locus described in the Introduction and in Fig. 1A is based on the notion that each tissue-specific component of the *R* locus resided in a tandem duplication of a segment of DNA. The physical extent of the duplicated DNA is unknown but under the terms of the model it must be sufficient to allow significant levels of recombination by displaced synapsis and crossing over.

DNA from a series of plants homozygous for the *R0r:standard* or a derivative allele was subjected to genomic blotting analysis to test these predictions and the results are shown in Fig. 7. The *R-r:standard* allele containing both (P) and (S) components exhibits three Hind III fragments, of 3.5, 4.0 kb, and a weakly hybridizing 4.5 kb, homologous to pR-nj:1 (lane 1). The *R-g:1* and *r-r:n46* alleles were derived from *R-r:standard* and are believed to represent intralocus recombination events such as depicted in Fig. 1B, leading to the loss of (P) function in *R-g:1* and of (S) function in *r-r:n46*. Lanes 2 and 3 show Hind III digested *r-r:n46* and *R-g:1* DNAs probed with pR-nj:1. Only the 4.0 kb band is retained in *r-r:n46* whereas the 3.5 and 4.5 kb bands are retained and the 4.0 kb band is missing in R-g:1. Based on this analysis, we can tentatively associate (P) function to the 4.0 kb fragment. These results predict that the 3.5 or 4.5 kb bands, or both, are associated with (S) function.

The fourth allele analyzed, *r-r:n35*, was derived from *R-r:standard*, but was not associated with exchange of flanking markers. This allele was genetically characterized as a mutation of (S) to (s) without loss of the duplication based on the fact that it retains two doses of another genetic marker (*Inhibitor of striate* = *Isr*) located on the duplicated chromosomal segment. However, when *r-r:n35* DNA is digested with Hind III and probed with pR-nj:1, both the 3.5 kb band and (P) band are detected but the 4.5 kb band is lost (lane 4). This result can be explained by assigning the 4.5 kb fragment to (S) and evoking the existence of a third, phenotypically null, *R* component with the proposed name "(Q)".

DISCUSSION

In maize, several genes including the *bronze* (Fedoroff et al., 1984) *a1* (O'Reilly et al., 1985), *a2* (Martin et al., 1986), *c1* (Paz-Ares et al.., 1986; K. Cohn and B. Burr, personal communication), *c2* (Weinand et al., 1986), *bz2* (Theres et al., 1986), and *P* (Lechelt et al., 1986), have been cloned using the transposon tagging strategy. Reported here is the isolation of an *R*-specific DNA probes using the maize controlling element *Ac* with two modifications to existing cloning strategies. First, the target locus (*R*) was physically coupled to the donor element (*Ac* at *P*) using a reciprocal chromosomal translocation. The receptor site for *Ac* transposition from the *P* locus favors intrachromosomal events, usually located distal (Van Schaik and Brink, 1959; Greenblatt and Brink, 1962, 1963; Greenblatt, 1984). Therefore, placing the target gene distal to *P* should increase the probability of *Ac* insertion into *R*. Secondly, by genomic blot analysis, a restriction fragment containing the active *Ac* element at *R* was identified by using a criteria of differential methylation states of the *Ac* DNA versus homologous but genetically inactive *Ac*-like DNA sequences.

Transposition of *Ac* to *R-nj*

Little is known regarding the specificity and probability of transposition of *Ac* into a specific gene sequence. Because the pattern of *Ac* transposition from the *P* locus on chromosome-1 favors intrachromosomal events with the receptor site usually located distally to the *P* locus (Greenblatt, 1984), interchromosomal transposition of *Ac* from the *P* locus is less probable than an intrachromosomal event. In a separate study, Kermicle (1980) screened for unstable *R* mutations among kernel progeny from plants carrying an active *Ac* element on chromosome-1 as *P-vv* and the *R* allele at its normal location on chromosome-10. Of the four mutable alleles recovered in this study, all turned out to be *Ds* insertions; none were insertions of *Ac*. In the present study, a translocation to place *R-nj* distal to *P-vv* on the same chromosome was used, and four candidate mutable *R-nj* alleles were obtained. Although only one of the four was characterized further, it proved to be an authentic insertion of *Ac* at *R-nj*. Depending on how many of the other mutations prove to be *Ac* insertions, the frequency of obtaining such specific mutants may exceed one in 10^5 - an easily manageable number of kernels to screen.

According to the current model of *Ac* transposition from the *P* locus, the total number of insertions may be much higher than detected in the present study because one-half of the *Ac* transpositions should result in duplication of the *Ac* element (one at *P-vv* and one at *R-nj*). These kernels will exhibit a depressed rate of mutability due to the increase in *Ac* copy number (four copies in the triploid endosperm) and be difficult to distinguish from the *r-g* allele segregating in one-half of the F1 progeny. Since the pattern of mutability varies from mutant to mutant, only those insertion events resulting in a relatively high intrinsic level of mutability were detected in the light variegated class of progeny. Two alternative methods would circumvent this problem. One would be to cross the parent homozygous for [*P-vv T1-10g R-nj*] to a *r-g* tester. All colorless or mutable kernels could be subsequently tested for insertion of *Ac* at *R*. However, the limitation of this approach is the reduced rate of transposition events due to two copies of *Ac* at *P*. A second possibility would be to cross one parent heterozygous for [*P-vv T1-10g R-nj*] and a normal chromosome-10 carrying the dominant *R-nj* allele to a *r-g* tester. Colorless kernel progeny would be presumably mutations of the *R-nj* allele.

Based on this experience, and assuming *Ac* lacks stringent sequence specificity for insertion sites, the translocation method may be generally applicable for obtaining *Ac*-induced mutations in other genes that can be placed distal to *P*-vv using reciprocal chromosomal translocations. Moreover, *Ac* elements are found at other loci in different chromosome regions of maize. The present collection of reciprocal translocation in maize (over 1000) should allow mutagenesis of a large number of genes using this strategy.

Genetic analysis of *r-nj:m1*

The ear phenotype of the original *r-nj:m1* plant was semi-sterile seed set and light variegated pericarp, in addition to the colorless to colored spotting pattern in the aleurone. This phenotype is expected when a second active *Ac* element, here resident at *r-nj:m1*, is added to the genome of *P*-vv and T1-10g. It is one phenotype expected by a single transposition of *Ac* from *P* to *R*. Analysis of transpositions of *Ac* from the *P-rr-Ac* complex (Greenblatt, 1984) suggests that an *Ac* remains at the *P* locus in the chromatid that receives the transposed *Ac* element [light variegated or *P-rr-Ac* + transposed *Ac* (*tr-Ac*)]. The twinned chromatid (potentially a red sector co-twin) loses its *Ac* at *P* and (68-75% of the time) carries a tr-*Ac* at the receptor site (same receptor site as found in the light variegated co-twin) (see Greenblatt and Brink, 1962; Greenblatt, 1968, 1974, 1984 for details of the mechanism of *Ac* transposition from *P*).

Confirmation that *r-nj:m1* did indeed represent at *Ac* insertion in the *R-nj* allele was based on three criteria. First, in genetic crosses involving *r-nj:m1,* this allele was inherited as an autonomous unstable mutation. Secondly, the somatic instability pattern of *r-nj:m1* showed a characteristic dosage effect of *Ac*-induced mutations. Third, the presence of *r-nj:m1* caused instability of an unlinked *Ds* element as the *waxy* locus. The present results on *r-nj:m1* agree with a previous report on this allele by Brink and Williams (1973).

Genomic cloning of *r-nj:m1*

The evidence that the *R* locus has indeed been cloned relies on a comparison of various *R* alleles. A comparison of siblings with and without *r-nj:m1* (Fig. 3) indicated a single difference in hybridization patterns - a 11 kb Sst I fragment that hybridizes to the internal 1.6 kb Hind III fragment of *Ac9*. A second confirmation that this Sst I fragment represents the *Ac* at *r-nj:m1* was its sensitivity to Pvu II digestion. Active *Ac* elements at several loci, including *waxy*, C2, *P* and *bronze* are susceptible to Pvu II cleavage at two internal sites while Pvu II sites in homologous *Ac*-like sequences are absent or protected form Pvu II digestion presumably by DNA methylation (Chomet et al., 1986).

Of the 28 recombinant clones identified with an internal *Ac* probe, one was shown to contain the expected *Ac* homologous fragments. A 700 bp region adjacent to the *Ac* element was subcloned to obtain the *pR-nj:1* probe. This probe detects a 11 kb Sst I fragment in *r-nj:m1* and a 6.5 kb Sst I fragment in both *R-nj* and a *R-nj* revertant allele derived from *r-nj:m1*. This is the expectation (if the probe is *R*-specific) since *Ac* integration and excision into *R-nj* should result in gain and loss, respectively, of 4.5 kb of *Ac* DNA. This probe also detects a polyadenylated mRNA in tissue expressing anthocyanin pigmentation (S. Wessler, unpublished results). The final evidence presented indicating that the probe is *R* specific comes from the molecular analysis of the *R-r* allele described below.

Organization of the *R-r* allele

R-r:standard represents a compound allele of the *R* locus that conditions anthocyanin pigmentation in the aleurone of the endosperm, certain seedling tissues and the anthers in mature plants. The seed-pigmenting component (S) and the plant-pigmenting component (P) are separable by meiotic recombination. The distribution of flanking markers among (P) loss and (S) products fractionated from *R-r*/*R-g* heterozygotes indicates that (P) is carried in the proximal member of a duplication and (S) in the distal member (Dooner and Kermicle, 1971, 1974). In *R-r* homozygotes, *R-g* or (S) and *r-r* or (P) derivatives are thought to arise by displaced synapsis and recombination within the duplication members containing the (P) and (S) components since these events also are associated with recombination of heterozygous flanking markers (see Fig. 1B). The relative positions of (P) and (S) within the duplication and the genetic length of the duplication segment has been estimated (Fig. 1A) by the frequency and type of recombinants recovered. If enough synaptic homology remains between the duplication harboring (P) and (S) components, one might expect that an *R*-specific probe would cross-hybridize among various *R* components. Although pR-nj:1 represents a probe derived from the *R-nj* allele, it detects three Hind III fragments in genomic DNA from *R-r* homozygotes at 4.5, 4.0 and 3.5 kb (Fig. 7). The intensity of hybridization of each fragment appears to vary, with the 4.0 kb fragment showing the highest signal of hybridization, followed by the 3.5 and 4.5 kb bands. If each fragment represents an *R* component, these differences may reflect the various degrees of homology between the (Nj) probe and each *R-r* component although the significance of this observation remains unclear at present.

DNA from an *R-g* derivative characterized genetically as an allele with only the (S) component remaining contains the 4.5 and 3.5 kb bands. When the (S) component is lost by meiotic recombination between (P) and (S) in a *R-r* homozygote, as in the case of the *r-r:n46* allele, both the 4.5 and 3.5 kb bands are lost and only the 4.0 kb fragment remains. However, the 3.5 kb as well as the 4.0 kb band remains detectable in the *r-r:n35* allele which was derived from *R-r* as loss of (S) function without flanking marker recombination. Hence, the 4.5 kb band is correlated one to one with (S) function and the 4.0 kb band with (P) function in the *R-r* derivatives characterized.

How can the 3.5 kb *R*-homologous fragment be explained. We offer the following explanation. This homology could represent an additional *R* component capable of displaced synapsis between (P) or (S). It is possible that this component represents only a partial duplication of *R* sequences. We propose to name this component "(Q)" because its location would be between (P) and (S) for the following reasons. If the order of these components was (P) (S) (Q) and each component was capable of synapsis, then reciprocal recombination could yield a (Q) only derivative and would be classified as a *r-g* allele. Such *r-g* offspring have never been observed from *R-r:standard* homozygotes (Dooner and Kermicle, 1971). The lack of a *r-g* class is consistent with the (P) (Q) (S) order. According to this model *R-g:1* would then be classified as a (S) (Q) allele and *r-r:n46* would be classified as a (P) allele from the data shown in Fig. 7.

Can this model explain *r-r:n35*, which appears to be a (P) (Q) allele based on the results shown in Fig. 7? If such derivatives result from non-reciprocal exchange one can postulate that gene conversion or intrachromosomal recombination could result in loss of (S). Alternatively, a double recombination event could generate such alleles without flanking marker exchange.

This model predicts two types of *r-r* and *R-g* derivatives derived by reciprocal recombination in *R-r* homozygotes - those with (Q) and those lacking (Q). We are presently testing this hypothesis by examining several independently derived *R-g* and *r-r* alleles. Interestingly, a second recombination or conversion event in (S) (Q) or (P) (Q) alleles could generate (Q) alleles. This may explain the requirement of two-steps for generating *r-g* derivatives from *R-r*.

ACKNOWLEDGEMENTS

We thank Paul Chomet, Jychian Chen, Christine Mitchell, Mark Alfenito, and Amar Klar for their helpful discussions and comments on the manuscript. We are grateful to Bettina Harris and Laurie Lowman for expert preparation of the manuscript and the Cold Spring Harbor Art Department for the illustrations and photographs. This work was supported by grants from the National Institute of Health to S.L.D. and S.W. and by a research contract with the Department of Energy to S.L.D.

REFERENCES

Behrens, U., N. Fedoroff, A. Laird, M. Muller-Neumann, P. Starlinger, and J. Yoder, 1984, Cloning of the Zea mays controlling element *Ac* from the *wx-m7* allele, Mol. Gen. Genet. 194:346-347.

Berman, M.L., L.W. Enquist,, and T.J. Silhavy, 1982, Advanced Bacterial Genetics, Cold Spring Harbor, N.Y. City.

Brink, R.A. and E. Williams, 1973, Mutable *R-navajo* alleles of cyclic origin in maize, Genetics 73:273-296.

Chomet, P.S., S. Wessler, and S.L. Dellaporta, 1986, Inactivation of the maize transposable element *Activator* is associated with its DNA modification. EMBO J.

Dooner, H.K., 1983, Coordinate genetic regulation of flavonoid biosynthetic enzymes in maize, Mol. Gen. Genet. 189:136-141.

Dononer, H.K., and J.L. Kermicle, 1976, Displaced and tandem duplications in the long arm of chromosome 10 in maize, Genetics 82:309-322.

Donner, H.K., and J.L. Kermicle, 1974, Reconstitution of the *R-r* compound allele in maize, Genetics 78:691-701.

Dooner, H.K., and J.L. Kermicle, 1971, Structure of the *R-r* tandem duplication in maize, Genetics 67:427-436.

Dretzen, G., M. Bellard, P. Sassone-Corsi, and P. Chambon, 1981, A reliable method for the recovery of DNA fragments from agarose and acrylamide gels, Anal. Biochem. 112:295.

Fedoroff, N.V., D.B. Furtek,, and O.E. Nelson, Jr., 1984, Cloning of the *bronze* locus in maize by a simple and generalizable procedure using the transposable controlling element *Ac*, Proc. Natl. Acad. Sci., U.S.A. 81:3825-3829.

Fedoroff, N., S. Wessler, and M. Shure, 1983, Isolation of the transposable maize controlling elements *Ac* and *Ds*, Cell 35:235-242.

Greenblatt, I.M., 1984, A chromosomal replication pattern deduced from pericarp phenotypes resulting from movements of the transposable element, *Modulator*, in maize, Genetics 108:471-485.

Greenblatt, I.M., and R.A.Brink, 1963 Transpositions of *Modulator* in maize into divided and undivided chromosome segments, Nature 197:412-413.

Greenblatt, I.M., and R.A. Brink, 1962, Twin mutations in medium variegated pericarp maize, Genetics 47:489-501.

Hohn, B., 1979 *In vitro* packaging of λ and cosmid DNA, Methods Enzymol. 68:299.

Holmes, D.S., and M. Quigley, 1981, A rapid boiling method for the preparation of bacterial plasmids, Anal. Biochem. 119:193.

Kermicle, J.L., 1985, Alternative tests of allelism. in: "Plant Genetic," UCLA Symp. on Molecular and Cellular Biology, vol. 35, M. Freeling ed, Alan R. Liss, Inc., City, pp. 491-507.

Kermicle, J.L., 1980, Probing the component structure of a mize gene with transposable elements, Science 108:1457-1459.

Kermicle, J.L., 1970, Somatic and meiotic instability of *R-strippled,* an aleurone spotting factor in maize, Genetics 64:247-258.

Kermicle, J.L., and J.D. Axtell, 1981, Modification of chlorophyll striping by the *R* region, Maydica 26:185-197.

Lechelt, C., A. Laird, and P. Starlinger, 1986, Cloning DNA from the *P* locus, Maize Genet. Coop. Newsletter 60:40.

Maniatis, T., E.F. Fritsch,, and J. Sambrook, 1982, "Molecular Cloning: A Laboratory Manual," Cold Spring Harbor Laboratory, City.

Martin, B., A. Gierl, P. Peterson, and H. Saedler, 1986 Molecular cloning of the *A2* locus, Maize Genet. Coop. Newsletter 60:41-42.

McClelland, M., 1982 The effect of sequence specific DNA methylation on restriction endonuclease cleavage, Nucleic Acid Res. 9:5859-5866.

McClintock, B., 1954, Mutations in maize and chromosomal aberrations in *Neurospora*, Carnegie Inst. Washington Year Book 53:254-260.

McClintock, B., 1948, Mutable loci in maize, Carnegie Inst. Washington Year Book 47:155-169.

McWhirter, K.S., and R.A. Brink,. 1962, Continuous variation in level of parmutation at the *R* locus in maize, Genetics 47:1053-1074.

Muller-Neumann, M., J. Yonder, and P. Starlinger, 1985, The sequence of the *Ac* element of *Zea mays*, Mol. Gen. Genet. (In press).

O'Reilly, C.O., N.S. Shepard, A. Pereira, Zs. Schwarz-Sommer, I. Bertram, D.S. Robertson, P.A. Peterson, and H. Saedler, 1985, Molecular cloning of the *al*

locus of *Zea mays* using the transposable elements *En* and *Spm,* EMBO J. 4:887-882.
Paz-Ares, J., U. Wienand, P.A. Peterson, and H. Saedler, 1986, Molecular cloning of the *c* locus of *Zea mays*, EMBO J. 5:829-834.
Pohlman, R., N.V. Fedoroff, and J. Messing, 1984, The nucleotide sequence of the maize controlling element *Activator*, Cell 37:635-643.
Rigby, P.W.J., M. Dieckmann, C. Rhodes, and P. Berg, 1977, Labeling deozxyridonucleic acid to high specific activity *in vitro* by nick translation with DNA polymerase I., J. Mol. Biol. 113, 237-251.
Shure, M., S. Wessler, and N. Fedoroff, 1983, Molecular identification and isolation of the *Waxy* locus in maize, Cell 35:225-233.
Southern, E., 1975, Detection of specific sequences among DNA fragments separated by gel electrophoresis, J. Mol. Biol. 98:503.
Stadler, L.J., 1951, Spontaneous mutation in maize, Cold Spring Harbor Symp. Quant. Biol. 16:49-63.
Styles, E.D., O. Ceska, and -K-T.Seah, 1973, Developmental differences in action of *R* and *B* alleles in maize, Can. J. Genet. Cytol. 15:59-72.
Theres, K., and P. Starlinger, 1986, Molecular cloning of *bz2-m*, Maize Genet. Coop. Newsletter 60:40.
Van Schaik, N.W., and R.A. Brink, 1959, Transpositions of *Modulator*, a component of the variegated pericarp allele in maize, Genetics 44:725-738.
Wienand, U., U.Weydemann, U. Niesbach-Klosgen, P. Peterson, and H. Saedler, 1986 Molecular analysis of the *c2* locus, Maize Genet. Coop. Newsletter 60:42.

GENETIC ENGINEERING FOR CROP IMPROVEMENT

Robert T. Fraley, Stephen G. Rogers, Robert B. Horsch, Ganesh M. Kishore, Roger N. Beachy, Nilgun N. Tumer, David A. Fischhoff, Xavier Delannay, Harry J. Klee and Dilip M. Shah

Monsanto Company (BB3B)
700 Chesterfield Village Parkway
St. Louis, MO 63198

INTRODUCTION

Since their development in 1983, the availability of gene transfer systems has already led to several important insights into the regulation of gene expression and protein function in plants. With the increasing efforts to refine and develop vector systems and the progress that has been made in the identification and isolation of plant genes, the next several years will bring about a dramatic expansion in our understanding of gene structure and function at the molecular level.

In many cases, the technical ability to isolate and transfer genes has surpassed the level of biochemical and cell biology understanding required for their rational manipulation. However, the wealth of interesting genetic, physiological, environmental and agronomic problems available for study ensures that there will be a rapid transition from basic research to agricultural applications. In this chapter, the development of genetic transformation systems will be discussed. Special emphasis will be placed on areas where advances have permitted extension of transformation technology to agronomic crops. The progress that has been made in the identification of agronomic traits which confer herbicide, virus, disease and insect tolerance will be reviewed with an emphasis on Monsanto research programs.

PLANT TRANSFORMATION

The rapid progress that has been made in the development of gene transfer systems for higher plants has surprised even the most optimistic of researchers in the field. Today, nearly two dozen species of crop plants can be routinely manipulated using available Agrobacterium tumefaciens transformation systems. Within the next 2-3 years it is likely that all major dicotyledonous crop species will be accessible to improvement using transformation technology. The major technical advances which have permitted the extension of gene transfer technology to crop plants have been: i) construction of fascile vectors based on the Agrobacterium tumefaciens Ti plasmid system, ii) availability of a broad spectrum of selectable and scoreable transformation marker genes and iii) development of explant-based regeneration systems which efficiently couple the transformation and regeneration processes. Each of these areas is discussed in detail below.

Ti Plasmid Vectors--The fundamental features of the Agrobacterium tumefacients Ti plasmid system have been previously reviewed (Bevan and Chilton, 1982; Schell and van Montagu,1983;Fraley et al., 1986). To harness this natural gene transfer system, derivative Ti plasmids which lack the phytohormone genes responsible for tumor formation, have been constructed. Examples of such plasmids include pGV3850 (Zambryski et al., 1983) and pTiB6S3-SE (Fraley et al., 1985); these plasmids retain a functional vir region, Agrobacterium origin of replication and optionally, T-DNA border sequences and DNA segments which facilitate homologous recombination with intermediate vectors. There are basically two methods for transferring specific genes into Agrobacterium using intermediate vectors: the first method developed utilizes integration by recombination to combine the intermediate vector with a disarmed Ti plasmid to reconstruct a disarmed T-DNA segment. Although this method works satisfactorily, a disadvantage is the low efficiency of the recombination event.

A second method takes advantage of the fact that vir region functions can operate in trans relative to the T-DNA, allowing the disarmed Ti plasmids and an intermediate plasmid containing a border sequence to exist independently in Agrobacterium. Such binary or trans vectors contain an Agrobacterium origin of replication and are represented by Bin 19 (Bevan, 1984) or pMON505 (Fraley et al., 1986).

Selectable and Scoreable Markers-- Aminoglycoside resistance conferred by various neomycin phosphotransferases (NPT) was the first widely used selectable marker in plant transformation studies (Herrera-Estrella et al., 1983; Fraley et al., 1983; Bevan et al., 1983) and it remains a marker of choice for many plant systems. In the last 5 years, many additional selectable markers have been developed which complement NPT and extend the useful range of plants in which selection of transformants can be readily achieved:

Selectable Marker	Reference
neomycin phosphotransferase (Type I)	Bevan et al., 1983 Fraley et al., 1983 Herrera-Estrella et al., 1983
neomycin phosphotransferase (Type II)	Fraley et al., 1983
hygromycin phosphotransferase	Lloyd et al., 1986 Van den Elzen et al., 1985
bacterial dihydrofolate reductase	de Block et al., 1984
mammalian dihydrofolate reductase	Eichholtz et al., 1987
gentamicin acetyltransferase	Hayford et al., in press
streptomycin phosphotransferase	Jones et al., 1987
EPSP synthase	Shah et al., 1986
acetolactate synthase	Haughn et al., in press
bromoxynil nitrilase	Stalker, et al., (pers. comm.)
phosphinothricin acetyl transferase	de Block et al., 1984

The availability of additional selectable marker genes allows for multiple transformation of selected lines and for optimization of the selection process for different species. For example, while NPT genes function efficiently in Solanaceous plants, it has been found that hygromycin resistance is superior in Arabidopsis (Lloyd et al., 1986) and that gentamicin resistance functions better in certain legumes including alfalfa (J. Fry, unpublished observation).

The nopaline and octopine synthase genes represented the first generation of useful scoreable marker genes, however these genes suffered from lack of sensitivity, flexibility and quantitation. Significant progress has been made in the identification and construction of other scoreable marker genes:

Scorable Markers	Reference
nopaline synthase	Depicker et al., 1982
octopine synthase	De Greve et al., 1983
β-galactosidase	Helmer et al., 1984
chloramphenicol acetyltransferase	De Block et al., 1984
firefly luciferase	Ow et al., 1986
β-glucuronidase	Jefferson et al., 1988

The new markers are not only useful in analyzing gene expression and inheritance of foreign DNA inserts, but also are valuable in demonstrating the targeting of transformation to particular cell types in tissue explants.

Regeneration Systems--Early transformation studies utilized protoplast-based methods; however the technical difficulties and species limitations associated with the use of protoplasts represent significant problems. The use of tissue explants such as leaves (Horsch et al., 1985), stems (Lloyd et al., 1986) or cotyledons (McCormick et al., 1986) effectively combined ease of regeneration with the unique ability of Agrobacterium to transform intact plant tissues. Since most plants can be regenerated from tissue explants, this powerful approach should extend transformation technology to all dicotyledonous crop species. A key area of current research is the optimization of strains, vectors and selectable markers with particular regeneration systems. The success of this approach is evident from the rapidly growing list of transgenic plant species:

Species in Which Transgenic Plants have been Produced

tomato	alfalfa	white clover	sugarbeet
potato	peas	flax	lotus
petunia	lettuce	cotton	cucumber
tobacco	sunflower	cabbage	asparagus
carrot	oil seed rape	pear	celery
poplar	arabidopsis		

CROP IMPROVEMENT

Plant breeding has played a major role in modern agriculture by providing plants with superior disease resistance and increased yields. Genetic engineering promises to have a significant impact on crop improvement by accelerating breeding efforts and by dramatically expanding the available gene pool. While efforts to identify and transfer genes of agronomic importance into plants are just beginning, some degree of success has already been achieved in engineering selective tolerance to herbicides, viral diseases and insect pests. A summary is presented of the results of these studies with an emphasis on Monsanto research programs.

Herbicide Tolerance--Advances have been particularly dramatic in the engineering of selective herbicide tolerance because existing knowledge of herbicide mode-of-action and metabolism has permitted rapid identification of key target genes. It seems clear now, that within the period of the next 5-10 years, that commercial level, slective tolerance will be available for many major existing herbicides, including Roundup , sulfonylureas and imidazolinones, as well as several newly introduced products. The availability of selective crop resistance to many of today's potent, broad spectrum herbicides will exert a dramatic effect on weed control practices by accelerating the trend towards commercializing fewer, more effective, less costly and environmentally more acceptable weed control products.

Glyphosate, the active ingredient in Roundup herbicide, is a broad spectrum non-selective herbicide used to control annual and biennial sedges, and broadleafed weed species. It is rapidly absorbed by foliar tissue and is quickly translocated to various plant organs.

Glyphosate is rapidly metabolized in the soil and has no residual activity. The shikimate pathway enzyme, 5-enolpyruvylshikimate-3-phosphate synthase (EPSPS), involved in aromatic amino acid biosynthase has been identified to be the specific target of glyphosate in bacteria (Steinrucken and Amrhein, 1980). Subsequent studies have shown that glyphosate also inhibits this enzyme in higher plants (Mousdale and Coggins, 1984; Rubin et al., 1984; Nafziger et al., 1984).

Monsanto researchers (Shash et al., 1986) utilized high-level expression of a plant EPSPS gene to engineer glyphosate-resistant plants. A full-length cDNA clone for EPSPS was isolated from a glyphosate-resistant suspension cell line of *Petunia hybrida*. The amino acid sequence predicted from the nucleotide sequence indicated that the enzyme was synthesized as a precursor polypeptide with an amino-terminal "transit peptide" sequence of 72 amino acids. The transit peptide is responsible for post-translational targeting of the precursor enzyme to the chloroplast (della-Cioppa et al., 1986). The wild-type petunia EPSPS cDNA was placed under control of the promoter for the 35S transcript of cauliflower mosaic virus (CaMV) and the resulting chimeric gene was transferred to a binary vector. Introduction of the chimeric EPSPS genes in petunia cells led to the growth of callus at concentrations of glyphosate sufficient to inhibit completely the proliferation of wild-type callus. Transformed petunia plants were regenerated and these plants were tolerant to application of formulated glyphosate at 0.9 kg/ha, approximately four times the quantity necessary to kill the control plants. However the growth of these plants was reduced relative to unsprayed controls.

More recently, experiments have been carried out with a mutant EPSPS gene which encodes an enzyme that is 6000-fold less sensitive to glyphosate inhibition than the wild-type petunia enzyme. Transformed tobacco plants expressing the glyphosate-resistant EPSPS gene were found to be significantly more tolerant to glyphosate than plants overexpressing the wild-type EPSPS gene. Tobacco plants expressing the mutant gene displayed no visible injury when sprayed with 0.9 kg/ha of glyphosate; the treated plants flowered normally and set seed at the levels identical to unsprayed controls.

In 1987, Monsanto Compnay carried out field evaluations of genetically engineered tomato plants containing the mutant EPSPS gene. The site chosen for the field test was a farm located 3 miles outside the farming community of Jerseyville, Illinois. The test plot was well-drained and was not located near a flood plain or tomato production areas. The site was fenced to protect the plants from damage by animals. The local community and state officials were briefed in advance; as expected, there was strong local support for the tests. Over 9000 tomato transplants were planted between the first and third weeks of June; agronomic conditions were chosen which closely paralleled commercial practices.

The Roundup tolerant tomatoes containing the mutant petunia EPSPS gene yielded comparably to controls in the absence of herbicide treatment indicating that alteration of EPSPS levels has no effect on plant growth. Twenty-two independently-derived transgenic tomato lines were evaluated in the test; the best plants survived Roundup applications of over 1 lb/acre, in contrast, control plants were rapidly killed at this rate. The treated transgenic and delayed flowering relative to unsprayed controls. Given the extreme sensitivity of tomatoes damaged by Roundup exposure, the dramatic increase in tolerance observed in the field results is extremely encouraging and indicates that tolerance to commercial rates (0.5-1 lb/acre) should be achievable.

Virus tolerant plants--An important application of genetic engineering technology is in the area of producing resistance to pathogens. At present, very little is known about genetic mechanisms of disease resistance in plants. Observations made 40 to 50 years ago pointed to a possible mechanism for increasing host resistance to viral pathogens. In 1929 McKinney observed that tobacco infected with one strain of a virus resists infection by a second related strain (McKinney, 1929). This phenomenon named "cross-protection" could be demonstrated with a number of viruses for which distinct strains could be found (reviewed by Hamilton, 1980; Sequira, 1984; Palukaitis and Zaitlin, 1984; Fulton, 1986). Although the mechanisms of cross-protection are not fully understood, several investigators have suggested that protection from viral disease could be achieved with the help of gene transfer methods (Hamilton, 1980; Sequira, 1984; Palukaitis and Zaitlin, 1984; Beachy et al., 1985; Bevan et al., 1985; Sanford and Johnston, 1985).

Powellabel et al. (1986) constructed an expression vector containing the 35S promotor from cauliflower mosaic virus (CaMV), a cDNA encoding the coat protein gene of the U1 strain of tobacco mosaic virus (TMB) and polyadenylation signal from the nopaline synthase gene. This vector was introduced into tobacco and tomato cells and plants were regenerated. Progeny of self-fertilized transgenic plants expressing high levels of the coat protein gene (0.05 to 0.1% of total soluble cell protein) were inoculated with TMV. These plants either did not develop an infection or disease symptoms developed more slowly than in the control plants after inoculation with the U1 strain of TMV. The plants expressing the coat protein (+CP) were also delayed in symptom development after inoculation with a severe TMB strain, PV230, which is immunologically related to the U1 strain (Nelson et al., 1987). The numbers of chlorotic and necrotic local lesions on the inoculated leaves of transgenic Xanthi and Xanthi 'nc' plants infected with PV230 or U1 were 70% and 90% lower, respectively, than the numbers observed on the control plants. As observed in classical cross-protection, the decrease in the numbers of chlorotic and necrotic lesions could at least be partially overcome by inoculation with viral RNA. Thus the presence of CP on the challenge virus was necessary for maximum protection to occur (Nelson et al., 1987).

The phenomenon of genetically engineered cross-protection was extended to another virus, alfalfa mosaic virus (AIMV) (Tumer et al., 1987). AIMV differs from TMV in many respects. TMV is a rigid rod shaped virus which encapsidates a single RNA molecule. It encodes at least four proteins in three open reading frames (Hirth and Richards, 1981). In contrast, AIMV is a bacilliform shaped virus which has a tripartite genome consisting of RNA molecules 1, 2 and 3. RNA 4, a subgenomic RNA of RNA 3 encodes the CP (Jaspars, 1985). Unlike infection with TMV RNA, which does not require CP, infection by AIMV RNA requires addition of RNA 4 or coat protein (Bol et al., 1971; Houwing and Jaspars, 1978). Thus, TMV and AIMV are clearly distinguished by their morphology, genome structure, strategies of viral gene expression, early steps in replication and modes of transmission (AIMV is transmitted mechanically and by aphids, TMV is primarily mechanically transmitted).

Tumer et al. (1987) have cloned RNA 4 encoding the CP of AIMV and engineered it into an expression vector downstream from the CaMV 35S promoter. The cDNA was flanked at the 3' end by the nopaline synthase polyadenylation site. Leaf disks of tobacco and tomato were transformed with _A. tumefaciens_ containing this construct, selected for kanamycin resistance and regenerated into plants Horsch et al., 1985; McCormick et al., 1986). The amount of AIMV CP expressed in different transgenic tobacco plants varied between 0.1 to 0.4% of total extractable leaf protein. The amount of AIMV CP expressed in transgenic tomato plants was between 0.1 to 0.8% of the total extractable leaf protein. The segregation ratio of the CP gene was determined in the seedling progeny of the self-fertilized transgenic plants. Progeny from a single transgenic tobacco plant were then inoculated with different concentrations of AIMV (20ug/ml, 10ug/ml and 5ug/ml) and symptom production was monitored in a growth chamber. Typical symptoms of AIMV infection appeared in 3-4 days on the inoculated leaves of the control seedlings. Of the seedlings that expressed the AIMV CP, 80% of the plants infected with 20 ug/ml of AIMV and 90% of the plants infected with 5 and 10 ug/ml of AIMV did not develop symptoms by the end of two weeks. As expected, decreasing the inoculum concentration decreased the number of +CP and -CP plants that showed symptoms on the inoculated leaves and increased the symptom production by one to twelve days. Similar results were obtained for transgenic tomato plants. Recently, cross protection against PVX and CMV has also been obtained in our laboratory using an identical strategy.

An important question which remains to be answered is if the protection observed in transgenic plants is the same as that found in plants that are cross-protected against virus infection. Although the answer is not known, the two types of protection show similarities. Both types of protection are due to a delay in disease development. In the case of the transgenic plants, the delay is the result of a 90 to 100% reduction on the numbers of lesions on the plant. Some plants escape infection completely and if infection is established, the spread of virus is slowed down in the transgenic plants. In the case of TMV, protection could be partially overcome both in the transgenic and cross-protected plants by increasing the concentration of the virus in the inoculum. In the case of AlMV, the protection was not overcome with the highest concentration of inoculum used (50 ug/ml). The protection observed against TMV could also be overcome in both tomato and tobacco plants if the challenge inoculum contains the virus RNA rather than virions (Nelson et al., 1987). Regardless of the precise mechanism involved in the virus disease resistance in transgenic plants, genetically engineered cross-protection provides a generally applicable way of producing virus resistant plants.

Transgenic tomatoes carrying the TMV coat protein gene were evaluated in field tests this summer. The TMV tolerant plants were hand inoculated with 104- ug/ml of TMV (U1 strain) at different intervals after transplanting. In all cases, 100% of the control plants exhibited virus symptoms and tested positive for the presence of TMV. In contrast, less than 5% of the transgenic tomatoes containing the chimeric coat protein gene displayed symptoms. Where symptoms did appear on the transgenic plants they were quite mild and were restricted to the inoculated leaves. Biochemical tests confirmed the lack of systemic spread of the virus in the transgenic plants. In the absence of virus pressure, the transgenic tomato plants expressing TMV coat protein yielded comparably to controls, indicating the presence of the coat protein does not affect plant growth. Interestingly, the transgenic plants showed no yield reduction after virus infection whereas the control plants suffered 23-33% yield losses. This result was somewhat surprising since the U1 TMV strain caused only mild symptoms on the control plants. These results may indicate, as many plant pathologists suspect, that "subclinical" levels of viral diseases in plants may cause significant but generally unobserved yield reductions.

Insect tolerant plants--Another application of genetic engineering with important implications for crop improvment has been the production of insect resistant plants. Progress in engineering insect resistance in transgenic plants has been achieved through the expression in plants of the insect toxin gene of Bacillus thuringiensis (B.t.). B.t. is an entomocidal bacterium which produces a parasporal protein crystal. Most strains of B.t. are toxic to lepidopteran larvae (Dulmage, 1981) although some strains with toxicity to dipteran (Goldberg and Margalit, 1977) or coleopteran (Krieg et al., 1983; Herrnstadt et al., 1986) larvae have also been described. The insect toxicity of B.t. resides in the parasporal protein crystal which in the case of lepidopteran-active strains is composed of toxin protein subunits of approximately 130,000 kDa (Bulla et al., 1981). Genes encoding the lepidopteran-specific toxins from several strains of B.t. have been cloned and sequenced (Adang et al., 1985; Schnepf et al., 1985; Shibano et al., 1985; Thorne et al., 1985; Hofte et al., 1986; Wabiko et al., 1986; Fischhoff et al., in preparation). These genes are largely similar; however, there are significant regions of variability at both the nucleotide sequence and amino acid sequence level. Deletion of variants of several of these genes have been constructed in vitro and tested for toxicity after expression in E. coli (Adang et al., 1985; Schnepf and Whiteley, 1985; Shibano et al., 1985; Hofte et al., 1986; Wabiko et al., 1986; Fischhoff et al., 1987). Taken together these experiments indicate that the region essential for toxicity resides in the N-terminal portion of the protein extending approximately from amino acid 29 to amino acid 610.

Fischhoff et al. (1987) have engineered and expressed in tomato plants a lepidopteran-specific toxin gene from B.t. subsp. kurstaki HD-1 (Watrud et al., 1985). DNA sequence analysis has shown that this gene is very similar to genes reported from B.t. subsp. berliner 1715 (Hofte et al., 1986; Wabiko et al., 1986). A fragment of this gene encoding amino acids 1 through 725 was isolated and shown to encode a functional toxin by expression in E. coli. A chimaeric truncated toxin gene was constructed in which the B.t. toxin gene is flanked by the CaMV 35S promoter and the 3'-end of the nopaline synthase gene.

Tomato plants transformed with this chimaeric gene were recovered by the method of McCormick et al., 1986. Plants were assayed for expression of the B.t. toxin gene by Northern analysis, and a polyadenylated RNA of the size expected of a full-length transcript was detected. The plants were also analyzed for toxicity to Manduca sexta (tobacco hornworm), an insect which is sensitive to the B.t. toxin and known to feed on tomato leaves. All of the larvae applied to the transgenic plants were killed within a few days, and the plants showed very little evidence of feeding damage. Larvae applied to control plants survived and eventually consumed the plants completely. These experiments demonstrate the feasibility of producing transgenic plants resistant to some lepidopteran insects through the expression of the B.t. toxin gene.

Transgenic tomato plants containing the B.t. gene were field tested this summer. The plants were infested with egg masses of tobacco hornworm and Heliothis zea (tomato fruitworm). The level of insect control observed in the field tests was generally superior to that observed in growth chamber experiments. Under conditions where control plants were totally defoliated by the hornworm infestation, tomato plants containing the B.t. gene suffered no agronomic damage. Similar results were obtained with control and B.t. containing tomato plants following natural or inoculated infestations by tomato fruit-worm. Fruit damage on control plots was 17-23%; whereas fruit damage on B.t. plants was only 4-9%. The excellent insect control observed under field conditions may be a result of using egg mass inoculations instead of young larvae as in the growth chamber experiments; this may have allowed more effective control at early stages in insect development. It is also possible the feeding deterent effect associated with the B.t. insecticidal protein are more pronounced under natural field conditions.

Future prospects for genetically engineering insect tolerant plants will likely continue to focus on the protein insect toxins from B.t. for several reasons. The B.t. toxins show a high degree of specificity. Individual toxins are only active against a single order of insects (e.g., lepidoptera or coleoptera), and within a given order not all insects are equally sensitive. In spite of this high degree of specificity, the B.t. toxins are active against some insect pests of major agronomic importance such as *Heliothis zea* (corn earworm/cotton bollworm) in the case of lepidopteran-specific toxins and *Leptinotarsa decemlineata* (Colorado potato beetle) in the case of the coleopteran-specific toxins. The leopidopteran-specific toxins, which have been intensively studied, are considered very safe; they have no known activity against mammals, fish or nontarget invertebrates.

CONCLUSIONS

The results described above indicate the potential for using gene-transfer technology for crop improvement. However with all the attention focused on agricultural applications, it is important to emphasize the impact this powerful tool will have on advancing basic research in all aspects of the plant sciences. Precise localization of the regulatory sequences for tissue-specific and environmentally modulated expression of genes have just begun for a small number of plant genes. The availability of specific molecular probes is already extending research in plant physiology, biochemistry and cell biology.

The field testing of genetically engineered plants containing herbicide, virus and insect tolerance genes represents an important step in the commercialization of plant biotechnology research. In view of recent breakthroughs, increased public awareness and the existence of new regulatory policies, such tests have undoubtedly taken on greater significance than deserved. It is important to remember that many of the first tests that have been conducted are not with commercial cultivars and that several years of plant breeding efforts, field evaluations and scale-up lie ahead before improved crops will be marketed. At the same time though, the technology is developing faster than most realize and already issues such as regulatory costs and registration timelines are becoming key concerns to companies attempting to develop improved genetically engineered crops. While the commercial significance of this year's field tests will only be determined after much additional evaluation, several important observations are noted below:

- o field experiments were carried out with local community support
- o no adverse impact of engineered plants on test site environment
- o field performance was as good or better than greenhouse tests

It will be important that the process for evaluating field testing of genetically engineered plants recognizes and responds quickly to the need for testing additional plants at multiple locations. It will also be important in subsequent evaluations that normal agronomic practices be employed in field tests, including completion of crop reproduction cycles and testing in normal production areas. Finally it is critical that regulatory requirements be formulated in such a fashion that recognizes the inherent low risk of the technology and which does not draw undue attention to the particular biotechnology process used to improve plants.

REFERENCES

Adang, M. J., Staver, M. J., Rochleau, T. A., Leighton, J., Barker, R. F., and Thompson, D. V., 1985, Characterized full length and truncated clones of the crystal protein Bacillus thuringiensis subsp. kurstaki HD73 and their toxicity to Manduca sexta, Gene, 36:289-300.

Baulcombe, D. C., Saunders, G. R., Bevan, M. W., Mayo, M. A., and Harrison, B. D., 1986, Expression of biologically active viral satellite RNA from the nuclear genome of transformed plants, Nature, 321:446-449.

Beachy, R. N., Powellabel, P., Oliver, M. J., De, B., Fraley, R. T., Rogers, S. G., and Horsch, R. B., 1985, Potential for applying genetic transformation to studies of viral pathogenesis and cross protection, In: "Biotechnology in Plant Science," M. Zaitlin, P. Day and A. Hollaender, Eds., Academic Press, Orlando, Florida, pp. 265-275.

Bevan, M., and Chilton, M. D., 1982, T-DNA of the Agrobacterium Ti and Ri plasmids, Annual Review of Genetics, 16:357-384.

Bevan, M., 1984, Agrobacterium vectors for plant transformation, Nucleic Acids Research, 12:8711-8718.

Bevan, M., Flavell, R. B., and, Chilton, M.-D., 1983, A chimeric antibiotic resistance gene as a selectable marker for plant cell transformation, Nature, 304:184-187.

Bevan, M. W., Mason, S. E., and Goelet, P., 1985, Expression of tobacco mosaic virus coat protein by a cauliflower mosaic virus promoter in plants transformed by Agrobacterium, EMBO J., 4:1921-1926.

Bol, J. F., Van Vloten-Doting, L., and Jaspers, E. M. J., 1971, A functional equivalence of top comonent a RNA and coat protein in the initiation of infection by alfalfa mosaic virus, Virology, 46:73-85.

Bulla, L. A., Kramer, K. J., Cox, D. J., Jones, B. L., Davidson, L. I., and Lookhart, G. L., 1981, Purification and characterization of the entomocidal protoxin of Bacillus thuringiensis, Journal of Biological Chemistry, 256:3000-3004.

Comai, L., Facciotti, D. Hiatt, W. R., Thompson, G., Rose, R. E., and Stalker, D. M., 1985, Expression in plants of a mutant aroA gene from Salmonella typhimurium confers tolerance to glyphosate, Nature, 317:741-744.

De Block, M., Herrera-Estrella, L., Van Montagu, M., Schell, J., and Zambryski, P., 1984, Expression of foreign genes in regenerated plants and their progeny, EMBO J., 3:1681-1689.

De Block, M., Botterman, J., Vandewiele, M., Dockx, J., Thoen, C., 1987, Engineering herbicide resistance in plants by expression of a detoxifying enzyme, EMBO J., 6:2513-2518.

De Greve, H., Dhaese, P., Suerinck, J., Lemmers, M., Van Montagu, M., Schell, J., 1983, Nucleotide sequence and transcript map of the Agrobacterium tumefaciens Ti plasmid-encoded octopine synthase gene, J. Mol. Appl. Genet., 1:499-511.

Della-Cioppa, G., Bauer, S. C., Klein, B. K., Shah, D. M., Fraley, R. T., and Kishore, G. M., 1986, Translocation of the precursor of 5-enolpyruvylshikimate-3-phosphate synthase into chloroplasts of higher plants in vivo, Proc. of the Nat. Acad. Sci., USA, 83:6873-6877.

Depicker, A., Stachel, S., Dhaese, P., Zambryski, P., and Goodman, H., 1982, Nopaline synthase: transcript mapping and DNA sequence, J. Mol. Appl. Genet., 1:561-573.

Dulmage, H. T., 1981, Insecticidal activity of isolates of Bacillus thuringiensis and their potential for pest control. In: "Microbial Control of Pests and Plant Diseases 1970-1980," H. D. Burges, Ed., Academic Press, New York, pp. 193-222.

Eichholtz, D., Rogers, S., Horsch, R., Klee, H., Hayford, M., Hoffman, N., Bradford, S., Fink, C., Flick, J., O'Connell, K., and Fraley, R., 1987, Expression of mouse dihydrofolate reductase gene confers methotrexate resistance in transgenic petunia plants, Som. Cell and Mol. Genet., 13:67-76.

Fischhoff, D. A., Bowdish, K. S., Perlak, F. J., Marrone, P. G., McCormick, S. M., Niedermeyer, J. G., Dean, D. A., Kusano-Krettzmer, K., Mayer, E. J., Rochester, D. E., Rogers, S. G., and Fraley, R. T., 1987, Insect tolerant transgenic tomato plants, Bio/Technology, 5:807-813.

Fraley, R. T., Rogers, S. G., Horsch, R. B., Sanders, P., Flick, J., Adams, S., Bittner, M., Brand, L., Fink, C., Fry, J., Galluppi, J., Goldberg, S., Hoffman, N., and Woo, S., 1983, Expression of bacterial genes in plant cells, Proceedings of the National Academy of Sciences U.S.A., 80:4803-4807.

Fraley, R. T., Rogers, S. G., Horsch, R. B., Eichholtz, D. E., Flick, J., Fink, C., Hoffman, N., and Sanders, P., 1985, The SEV system: a new disarmed Ti plasmid for plant transformation., Bio/Technology, 3:629-637.

Fraley, R. T., Rogers, S. G., and Horsch, R. B., 1986, Genetic transformation in higher plants, CRC Critical Reviews in Plant Sciences, 4:1-46.

Fromm, M. E., Taylor, L. P., and Walbot, V., 1986, Stable transformation of maize after gene transfer by electroporation, Nature, 319:791-793.

Fulton, R. W., 1986, Practices and precautions in the use of cross protection for plant virus disease, Annual Review of Phytopathology, 24:67-81.

Goldberg, L. J., and Margalit, J., 1977, A bacterial spore demonstrating rapid larvicidal activity against Anopheles sergentii, Uranatenia unguiculata, Culex univittatus, Aedes aegypti and Culex pipiens., Mosquito News, 37:353-358.

Hamilton, R. I., 1980, Defenses triggered by previous invaders: Viruses. In: "Plant Disease," J. G. Horsfall and E. B. Cowling, Eds., Academic Press, volume 5, pp. 279-303.

Haughn, G., Smith, J., Mazur, B., and Somerville, C., 1988, Molec. Gen. Genet., (in Press).

Hayford, M., Medford, J., Hoffman, N., Rogers, S., and Klee, H., 1988, Plant Physiol., (in Press).

Helmer, G., Casadaban, M., Bevan, M., Kayes, L., and Chilton, M.-D., 1984, A new chimeric gene as a marker for plant transformation: the expression of Escherichia coli β-galactosidase in sunflower and tobacco cells, Bio/Technology, 2:520-527.

Herrera-Estrella, L., Deblock, M., Van Montagu, M., and Schell, J., 1983, Chimeric genes as dominant selectable markers in plant cells, EMBO J., 2:987-994.

Herrnstad, C., Soares, G. G., Wilcox, E. R., and Edwards, D. L., 1986, A new strain of Bacillus thuringiensis with activity against Coleopteran insects, Bio/Technology 4:305-308.

Hirth, L., and Richards, K. E., 1981, Tobacco mosaic virus: Model for structure and function of a simple virus, Advances in Virus Research, 26:145-199.

Hofte, H., De Greve, H., Suerinck, J., Jansens, S., Mahillon, J., Ampe, C., Vandekerckhove, J., Vanderbruggen, H., Van Montagu, M., and Vaeck, M., 1986, Structural and functional analysis of a cloned delta endotoxin of Bacillus thuringiensis berliner 1715, European Journal of Biochemistry, 161:273-280.

Horsch, R. B., Fry, J. B., Hoffman, N. L., Eichholtz, D., Rogers, S. G., and Fraley, R. T., 1985, A simple and general method for transferring genes into plants, Science, 227:1229-1231.
Houwing, C. J., and Jaspers, E. M. J., 1978, Coat protein binds to 3'-terminal part of RNA4 of alfalfa mosaic virus, Biochemistry. 17:2927-2933.
Jaspers, E. M. J., 1985, Interaction of Alfalfa Mosaic Virus Nucleic Acid and Protein. In: "Molecular Plant Virology," J. W. Davies, Ed., CRC Press, Boca Raton, volume 1, pp. 155-221.
Jefferson, R., Kavanaugh, T., and Bevan, M., 1988, GUS fusions: β glucuronidase as a sensitive and versatile gene fusion marker in higher plants, EMBO J., 6:3901-3908.
Jones, J., Svab, Z., Harper, E., Hurwitz, C., and Maliga, P., 1987, A dominant nuclear streptomycin resistance marker for plant cell transformation, Mol. Gen. Genet., 210:86-91.
Krieg, A., Huger, A. M., Langenbruch, G. A., and Schnetter, W., 1983, Bacillus thuringiensis var. tenebrionis, ein neuer, gegenuber Larven von Coleopteran wirksamer, Pathotyp. Z Angew Entomol., 96:500-508.
Lamppa, G. K., Morelli, G., and Chua, N.-H., 1985, Structure and developmental regulation of a Wheat gene encoding the major chlorophyll a/b-binding polypeptide, Molecular and Cellular Biology, pp. 1370-1378.
Lloyd, A. M., Barnason, A. R., Rogers, S. G., Byrne, M. C., Fraley, R. T., and Horsch, R. B., 1986, Transformation of Arabidopsis thaliana with Agrobacterium tumefaciens, Science, 234:464-466.
McCormick, S., Niedermeyer, J., Fray, J., Barnason, A., Horsch, R., and Fraley, R., 1986, Leaf disc transformation of cultivated tomato (L. esculentum) using Agrobacterium tumefaciens, Plant Cell Reports, 5:81-84.
McKinney, H. H., 1929, Mosaic diseases in the Canary islands, West Africa and Gibraltar, Journal of Agricultural Research, 39: 557-578.
Mousdale, D. M., and Coggins, J. R., 1984, Purification and properties of 5-enolpyruvylshikimate-3-phosphate synthase from seedlings of Pisum sativum, Planta, 160:78-83.
Nafziger, E. D., Widholm, J. M., Steinrucken, H. C., and Kilmer, J. L., 1984, Selection and characterization of a carrot cell line tolerant to glyphosate, Plant Physiology, 76:571-574.
Nelson, R. S., Powellabel, P., and Beachy, R., 1988, Fewer viral infection sites in transgenic tobacco plants expressing the coat protein gene of tobacco mosaic virus, Virology, (in press).
Ow, D., Wood, K., Deluca, M., De Wet, J., Helinski, D., and Howell, S., 1986, Transient and stable expression of the firefly luciferase gene in plant cells and transgenic plants, Science, 234:856-859.
Palukaitis, P., and Zaitlin, M., 1984, A model to explain the "cross-protection" phenomenon shown by plant viruses and viroids. In: "Plant-Microbe Interactions. Molecular and Genetic

Perspectives," T. Kosuge and E. N. Nester, Eds., Macmillan Press, New York, 1, pp. 420-429.
Powellabel, P., Nelson, R. S., De, B., Hoffman, N., Rogers, S. G., Fraley, R. T., and Beachy, R. N., 1986, Delay of disease development in transgenic plants that express the tobacco mosaic virus coat protein gene, Science, 232:738-743.
Rubin, J. L., Gaines, C., and Jansen, R. A., 1984, Glyphosate inhibition of 5-enolpyruvylshikimate-3-phosphate synthase from suspension-culture cells of Nicotiana sylvestris, Plant Physiology, 75:839-845.
Sanford, J. C., and Johnston, S. A., 1985, The concept of parasite-derived resistance-deriving resistance genes from the parasite's own genome, Journal of Theoretical Biology, 113:395-405.
Sequeira, L., 1984, Cross protection and induced resistance: their potential for plant disease control, Trends in Biotechnology, 2: 25-29.
Schell, J., and Van Montagu, M., 1983, The Ti plasmids as natural and as practical vectors for plants, Bio/Technology, 1:175-180.
Schnepf, H. E., and Whiteley, H. R., 1985, Delineation of toxin encoding segment of a Bacillus thuringiensis crystal protein gene Journal of Biological Chemistry, 260:6273-6280.
Shah, D. M., Horsch, R. B., Klee, H. J., Kishore, G. M., Winter, J. A., Tumer, N. E., Hironaka, C. M., Sanders, P. R., Gasser, C. S., Aykent, S., Siegel, N. R., Rogers, S. G., and Fraley, R. T., 1986, Engineering herbicide tolerance in transgenic plants, Science, 233:478-481.
Shibano, Y., Yamagata, A., Nakamura, N., Iizuka, T., Sugisaki, H., and Takanami, M., 1985, Nucleotide sequence coding for the insecticidal fragment of the Bacillus thuringiensis crystal protein, Gene, 34:243-251.
D. Stalker (personal communication)
Steinrucken, H. C., and Amrhein, N., 1984, The herbicide glyphosate is a potent inhibitor of 5-enolpyruvylshikimic acid-3-phosphate synthase, Biochemical and Biophysical Research Communications, 94:1207-1212.
Thorne, L., Garduno, F., Thompson, T., Decker, D., Zounes, M., Wild, M., Walfield, A. M., and Pollock, T. J., 1986, Structural similarity between the Lepidoptera and Diptera specific insecticidal endotoxin genes of Bacillus thuringiensis subsp. kurstaki and israelensis, Journal of Bacteriology, 166:801-811.
Tumer, N. E., O'Connell, K. M., Nelson, R. S., Sanders, P. R., Beachy, R. N., Fraley, R. T., and Shah, D. M., 1987, Expression of alfalfa mosaic virus coat protein gene confers cross-protection in transgenic tobacco and tomato plants, EMBO J., 6:1181-1188.
Wabiko, H., Raymond, K. C., and Bulla, L. A., 1986, Bacillus thuringiensis entomocidal protoxin gene sequence and gene product analysis, DNA, 5:305-314.
Waldron, C., Murphy, E., Roberts, J., Gustafson, G., Armour, S., and Malcom, S., 1985, Resistance of hygromycin-B, Plant Mol. Biol., 5:103-108.

Watrud, L. S., Perlak, F. J., Tran, M.-T., Kusano, K., Mayer, E. J., Miller-Wideman, M. A., Obukowicz, M. G., Nelson, D. R., Kreitinger, J. P., and Kaufman, R. J., 1985, Cloning of the Bacillus thuringiensis subsp. kurstaki delta endotoxin gene into Pseudomonas fluorescens: Molecular biology and ecology of an engineered microbiol pesticide. In: "Engineered Organisms in the Environment," H. O. Halvorson, D. Pramer, M. Rogul, Eds., American Society for Microbiology, Washington, pp. 40-46.

Van Den Elzen, P., Townsend, J. Lee, K., and Bedbrook, J., 1985, A chimeric hygromycin resistance gene as a selectable marker in plant cells, Plant Mol. Biol. 5:299-302.

Zambryski, P., Joos, H., Maes, M., Warren, G., Van Montagu, M., and Schell, J., 1983, Ti plasmid vector for the introduction of DNA into plant cells without alteration of their normal regeneration capacity, EMBO J., 2:2143-2150.

THE CURRENT STATUS OF CHROMOSOME ANALYSIS IN WHEAT

Bikram S. Gill
R. G. Sears

Department of Plant Pathology
and Department of Agronomy
Throckmorton Hall
Kansas State University
Manhattan, KS 66506 U.S.A.

I. INTRODUCTION

The first definitive observations on the morphology (size and arm-ratio) of the chromosomes of common wheat (2n=6x=42) were made on monosomic laggard chromosomes at anaphase I or telophase II of meiosis of the 21 different monosomics (Morrison, 1953; Sears, 1954). These observations allowed the cytogenetic identifications of chromosomes 1B and 6B (satellited but with different arm ratios) and chromosome 5B, being highly heterobrachial with the longest long arm of the whole chromosome complement. The remaining chromosomes were not distinctive enough to be identified in somatic cells from the morphological data.

It soon became apparent that monosomic laggard chromosomes tended to misdivide during meiosis and misdivision products, i.e., short arm and long arm telocentrics, could be recovered in the progenies of such monosomics with relative ease. Telocentric chromosomes can be identified in somatic cells, and in meiosis of interspecific and intergeneric hybrids as part of heteromorphic bivalents or univalents. With the aid of telocentrics, all 21 chromosomes of wheat could be identified. Thus, among many other applications, phenomena such as somatic association (or lack of it) of homologous chromosomes, centromere mapping and location, and amount of alien chromatin transfers could be monitored (Sears and Sears, 1978). Even at present, telocentric analysis provides the most accurate method of chromosome identification, especially if it can be combined with chromosome banding analysis as will be illustrated later.

Understandably, although the first modern methods of chromosome identification were developed in plants (Hsu, 1973), there was little enthusiasm to explore alternative methods of chromosome identification in wheat. While cytogenetic stocks and marker chromosomes provided reliable means of chromosome identification, little was known about the structure of the individual chromosomes of wheat. The advent of chromosome banding methods led to the recognition of biologically meaningful heterochromatic and euchromatic regions of somatic chromosomes of wheat (Gill, 1987, for review). Molecular cytogenetic techniques allowed the location of defined DNA sequences in those regions, (May and Appels, 1987, for review). Both can be considered the beginning of a second revolution in wheat cytogenetics. The present review will mainly be concerned with the recent work on these aspects of the structure of wheat chromosomes.

II. HETEROCHROMATIN STRUCTURE OF WHEAT CHROMOSOMES

Two techniques, C-banding and N-banding, have been useful in analyzing the heterochromatin structure of wheat chromosomes. The C-banding, which detects constitutive heterochromatin, is a general method for detecting all classes of heterochromatin in plant and animal chromosomes. The N-banding procedure, which originally was developed to stain nucleolus organizer regions of animal and plant chromosomes, also detects specialized heterochromatic regions in animal and plant chromosomes (Funaki et al., 1975; Pimpinelli et al., 1976; Jewell, 1981). The variable results with the N-banding technique may be ascribed to modifications of the original procedure by different workers. Most of the other methods for differentially staining of various chromosomal regions have been inadequately studied in wheat (Vosa, 1981; Bianchi et al., 1985). Various fluorescent dyes have yielded little longitudinal differentiation in wheat chromosomes (Appels, 1982; Schlegel and Gill, unpublished results). Apparent G-bands have recently been reported in wheat chromosomes (Zhu et al., 1986), but reproducible cytogenetic identification of wheat chromosomes with these new methods remains to be demonstrated.

The relationship between C-banded and N-banded heterochromatin in cereal chromosomes has been clarified during the last few years. When Gerlach (1977) originally reported N-bands in nine chromosomes of Chinese Spring wheat (n = 21), he also noted that the location of N-bands coincided with sites of hybridization with a polypyrimidine tract sequence DNA. Endo and Gill (1984a) reported an improved N-banding technique that revealed N-bands on 16 chromosomes (Gill, 1987) (Fig. 1), and these also coincided with observed sites of hybridization with polypyrimidine tract sequence following longer autoradiograph exposure (Appels et al., 1982).

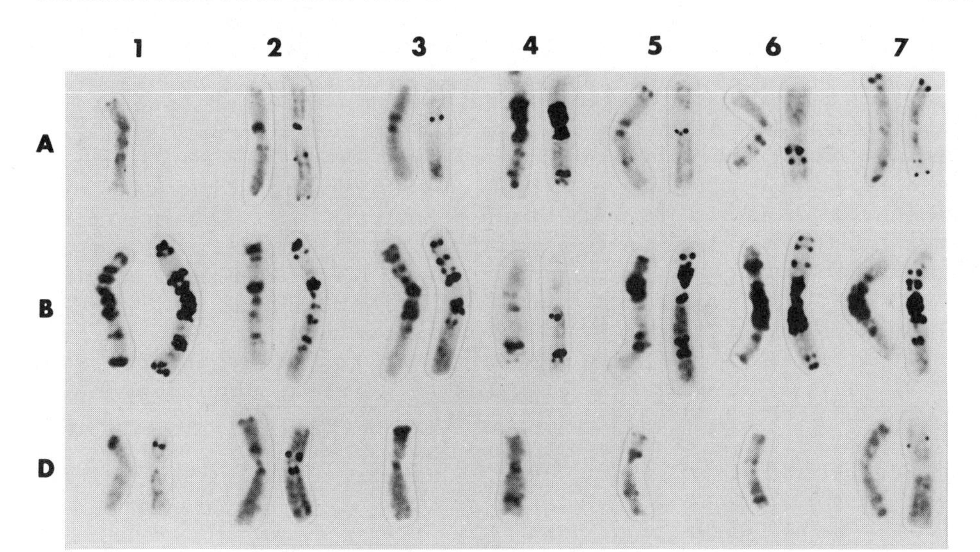

Fig. 1. The heterochromatin structure of Chinese Spring wheat chromosomes. Each chromosome pair was stained by C-banding (left) and N-banding (right). Chromosomes 1A, 3D, 4D, 5D, and 6D lack N-bands and are not shown. Note that certain bands are exclusively revealed by the C-banding technique (C^+N^-) and most are revealed by both techniques (C^+N^+) thereby differentiating two types of heterochromatin.

A more conclusive relationship between N-banded and C-banded heterochromatin has been demonstrated in rye. Schlegel and Gill (1984) sequentially stained the same metaphase rye chromosomes in a cell with acetocarmine/C-banding/N-banding techniques. The C-banded rye chromosomes showed the large terminal and a few interstitial C-bands on all rye chromosomes. The N-banding procedure showed one interstitial band on each of three pairs of rye chromosomes in close proximity to the centromeres. Appels et al. (1978) had earlier located polypyrimidine tract DNA sequences at the position of the observed N-bands. Thus, in both wheat and rye, heterochromatin can be described as C^+N^+ or C^+N^- and only the C^+N^+ heterochromatin contains polypyrimidine tract DNA sequences. Similarly, N-bands appear to be also confined to the location of polypyrimidine-polypurine DNA sequences in <u>Aegilops</u> and barley (Gerlach et al., 1978; Dennis et al., 1980; Chen and Gill, 1983; Singh and Tsuchiya, 1982). It should not be construed from these data that polypyrimidine sequence DNA per se is stained by the N-banding method; rather, it appears that polypyrimidine tract sequences and associated phosphoproteins are resistant to N-banding acid hydrolysis and stain as dark bands with Giemsa (Jewell, 1981; Buys and Osinga, 1982).

Another characterization of heterochromatin, the so-called 'cold sensitive' heterochromatin, was observed in the form of a tertiary constriction in the long arm of chromosome 6B following 24 h exposure of root tips at 0°C. A faint heterochromatic band was observed at the same location after banding treatment (Endo and Gill, 1984a). Cold sensitive heterochromatin is more dramatically observed in rye chromosomes after treatment of root tips for 16-24 h at 0°C (Bhattacharyya and Jenkins, 1960). Apparent satellites and tertiary constrictions are observed in several pairs of rye chromosomes (Endo and Gill, unpublished results; Fig. 2). After C-banding, these areas show prominent heterochromatic bands in the terminal regions and faint bands in the interstitial tertiary constrictions. Cold sensitive heterochromatin, which apparently may result from reversible uncoiling of specific chromatin after cold treatment (Rocchi, 1982), was not observed in chromosome preparations from root tips

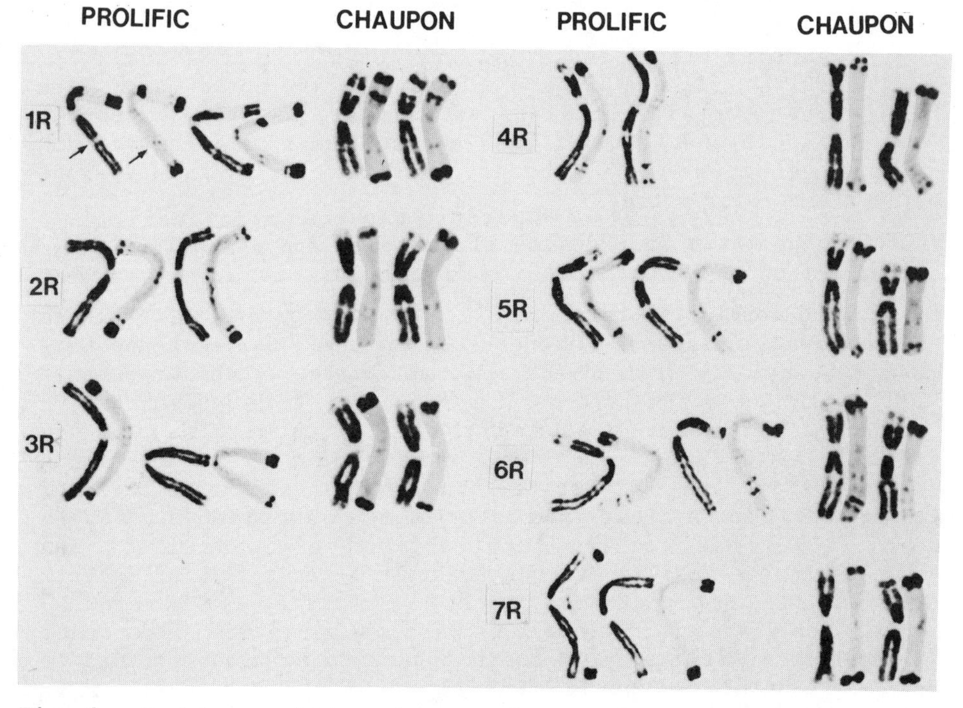

Fig. 2. Cold sensitive heterochromatin seen as tertiary constrictions and apparent satellites in most acetocarmine-stained chromosomes in two varieties of rye (<u>Secale</u> <u>cereale</u>). Sequential C-banding showed C-bands at the site of tertiary constrictions (arrows in 1R as an example).

pretreated at room temperature (22 ± 1°C) with 1-bromonaphthalene. There are many additional techniques that can be used to study the heterogeneity of heterochromatin (Rocchi, 1982; Miller et al, 1983); but these have been little explored to study the heterochromatin structure of wheat chromosomes.

The usefulness of characterization of heterochromatin in tracing evolutionary affinities of individual chromosomes in polyploid species was recently demonstrated by Morris and Gill (1987). They analyzed polyploid _Elymus trachycaulus_ (Link) Gould ex Shinners (2n=28, SSHH) chromosomes by C-banding and N-banding and found that seven chromosomes mostly contained C^+N^+ heterochromatin, and the remaining seven chromosomes mostly contained C^+N^- heterochromatin. Analysis of the S genome and H genome diploid progenitor species indicated H-genome heterochromatin to be mostly C^+N^+ and S-genome heterochromatin exclusively C^+N^-. Thus, the C^+N^+ heterochromatin-containing chromosomes in _E. trachycaulus_ were allocated to the H genome and C^+N^- to the S genome (Fig. 3). Similarly, the genomic affinities of individual _E. trachycaulus_ chromosome addition lines in Chinese Spring wheat were also determined. This allocation of addition lines was further confirmed by the detection of S-genome and H-genome specific spacer 18S-28S rDNA and 5S DNA sequences under different stringencies of molecular hybridization (Gill et al., 1988; for a more detailed discussion see part VII).

Another aspect of chromosome evolution in the polyploid genome of _E. trachycaulus_ is apparent from the amplification of C^+N^+ heterochromatin in H-genome chromosomes and deamplification or deletion of terminal C^+N^- heterochromatin in S-genome chromosomes. Moreover, most of the S-genome chromosomes contain traces of C^+N^+ heterochromatin, whereas S-genome progenitor species lacked this class of heterochromatin entirely. The highly heterochromatic H7 chromosome was novel and was not observed in the H-genome progenitor species (Fig. 3). One explanation of these results may be that S-genome and H-genome progenitor species existed with the heterochromatin structure observed in modern _E. trachycaulus_ species. However, a more likely explanation may be that polyploid genome evolution has been more dynamic, involving extensive amplification, deamplification, and horizontal spread of DNA sequences contained in various heterochromatin bands.

The practical utility of banding techniques in karyotypic analysis of various _Triticeae_ species was emphasized by Endo and Gill (1984b). Due to differential contraction even among homologous chromosomes in a cell, the value of karyotypes constructed from conventional preparations (on the basis of size and arm-ratio alone) is questionable. Therefore, sequential acetocarmine/banding analysis has not only been helpful in the

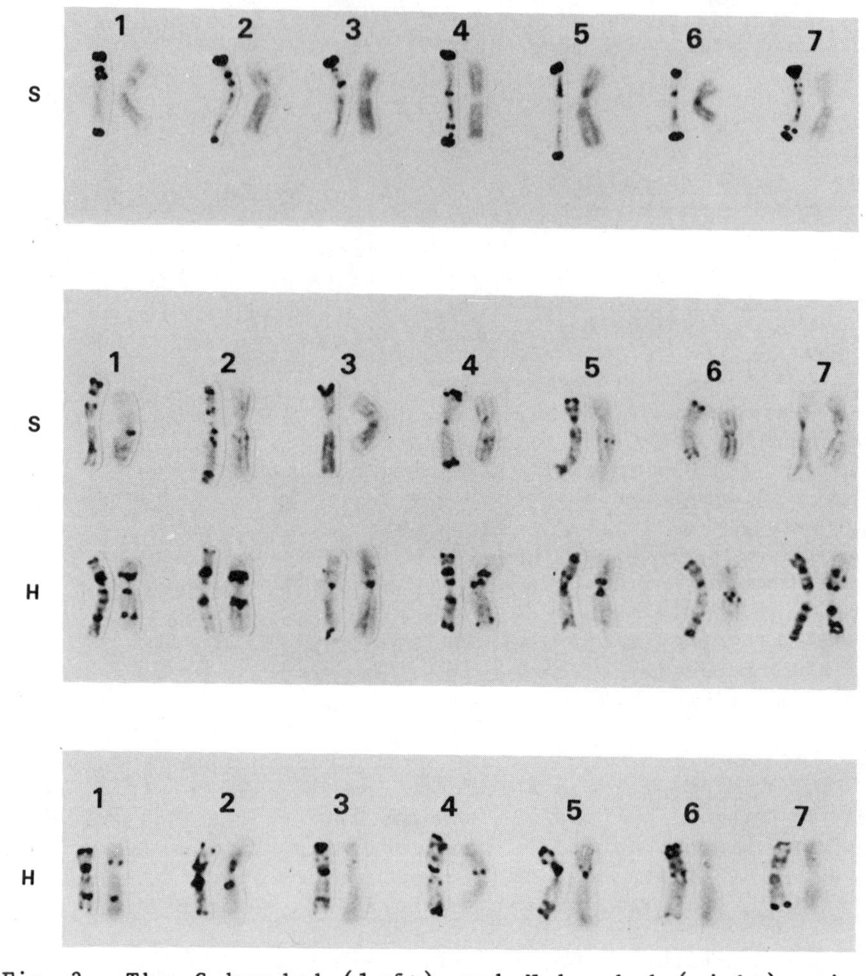

Fig. 3. The C-banded (left) and N-banded (right) pairs of chromosomes of diploid S-genome progenitor species *Pseudoroegneria spicata* (top), diploid H-genome progenitor species *Critesion bogdanii* (bottom) and allopolyploid SH-genome species *Elymus trachycaulus* (middle). Note similar chromosomes S1, H1, H2, and others in progenitor and polyploid species. See text for further discussion.

correct identification of different chromosomes but has also provided additional information on the heterochromatin content of the chromosomes.

As discussed earlier, the heterochromatic bands provide physical landmarks for the identification of individual chromosomes and subregions of chromosomes. Thus, genetic stocks containing polymorphic bands or deletions can be used in physical mapping of wheat chromosomes (Snape et al., 1985; Dvorak et al., 1984; Jampates and Dvorak, 1986; Kota and Dvorak, 1986). This aspect of cytogenetic mapping in wheat will eventually reveal more information on the genetic content of chromosomal subregions including heterochromatic bands.

III. MEIOTIC CHROMOSOME PAIRING ANALYSIS

The genome-analyzer method, pioneered by Kihara, and its many refinements expounded by Kimber (1982) is used to deduce evolutionary relationships based on the degree of chromosome pairing at metaphase I of meiosis in interspecific F_1 hybrids. Although "the genome designations tend to be oversimplifications" as pointed out by Sears (1948), they "serve as useful guides to further research." The banding techniques can be used to determine the identity of individual chromosomes involved in various pairing configurations, and are more useful than conventional chromosome pairing analysis, especially in the resolution of individual chromosome homologies (Fig. 4).

The usefulness of this method can be illustrated by a comparison of metaphase I pairing between *Triticum turgidum* L.

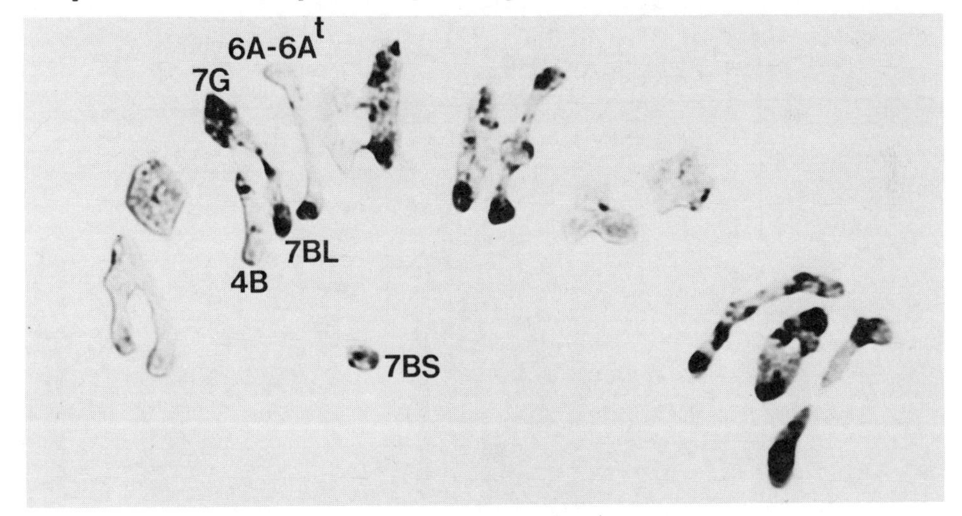

Fig. 4. The N-banded metaphase I pollen mother cell of F_1 hybrid *Triticum turgidum*/*T. timopheevii* where paired configurations involving B-G genome chromosomes (highly banded) can be easily distinguished from A-A^t genome chromosomes (lightly banded). The *T. turgidum* telosomic chromosome 7BL is paired in a trivalent association with chromosomes 7G and 4B. Telosome 7BS is univalent.

(2n=28, AABB) and *T. timopheevii* Zhuk. chromosomes (2n=28, A^tA^tGG) by conventional staining and by Giemsa staining methods (Table 1). The conventional staining identifies five kinds of configurations, that is univalents, bivalents (rods and rings), trivalents, and quadrivalents. However, 17 types of configurations can be identified based on the identity of chromosomes involved in those configurations after Giemsa staining. Thus, it can be determined that most univalents and rod bivalents involve B/G genome chromosomes indicating poor pairing between those two genomes. Conversely, A/A^t genome chromosomes are poorly represented in the univalent category and mostly pair as ring bivalents, indicating greater homology between those two genomes. Translocations involving chromosomes of various genomes can also be identified.

The accuracy of this method can be further enhanced by the use of telocentric marker chromosomes to identify individual chromosome homologies (Fig. 4). This kind of analysis led Gill and Chen (1987) to conclude that polyploid *T. timopheevii* and *T. turgidum* harbor chromosomes $4A^t$ (A-genome chromosome) and 4B (genome origin unknown), respectively, that are unique to each lineage. Furthermore, in *T. timopheevii*, it was found that $4A^t$ is translocated with chromosome $3A^t$. The chromosomes $6A^t$ and 6G were found to be satellited, $6A^t$ probably translocated from 1G as $6A^tS$ arm showed homoeologous pairing with 1BS satellite arm. These two

Table 1. High resolution N-banding analysis of metaphase I of meiosis of F_1 hybrids: *Triticum turgidum*(AABB)/ *T. timopheevii*(A^tA^tGG)[a]

Average pairing per pollen mother cell													
I		II						III				IV	
		Rod			Ring								
7.81		5.46			3.43			0.97				0.03	
0	1	0	1	2	0	1	2	0	1	2	3	0	3
1.80	6.01	1.87	0.06	3.53	2.89	0.04	0.50	0.75	0.04	0.06	0.12	0.004	0.023

[a]For N-banded PMCs, each chromosome association indicates number of banded (B,G genome) chromosomes involved (0, 1, 2, 3). The A, A^t genome chromosomes were considered unbanded (0). For IV association, five types are possible (0, 1, 2, 3, 4), but only two types involving all unbanded chromosomes (0.004) and three banded (+1 unbanded) chromosomes (0.023) were observed.

translocations are apparently ubiquitous in T. timopheevii (and there may be others), and probably arose at the time of the origin of timopheevii wheats.

Gill and Chen (1987) also found that 4B is a translocated chromosome with its long arm pairing with the 7G chromosome of T. timopheevii (Fig. 4) and the 7AS arm of T. turgidum in T. turgidum/Aegilops speltoides Tausch. hybrids. Recently, Naranjo et al. (1987) have provided cytogenetic evidence (again from banding analysis of chromosome pairing) of a cyclical translocation involving chromosomes 4B, 5A, and 7B as illustrated in Fig. 5. They speculated that the 4B chromosome originated from Triticum monococcum L. and a 4B-5A translocation was present in T. monococcum that hybridized with a B-genome species. At the tetraploid level, among the translocated chromosomes 4B-5A and 5A-4B of the A-genome donor, 4B-5A underwent a translocation with 7B of the B-genome donor to produce translocated chromosomes 4B-7B and 7B-5A. Thus, the modern day T. turgidum carries a cyclical translocation with chromosome constitution of translocated chromosomes as follows: 4BS.4BL-7BS + 5AS.5AL-4BL + 7BL.7BS-5A.

Gill and Chen (1987) have proposed that 4B is not an A-genome chromosome and that the cyclical translocation may have coincided with the introgression of chromosome 4B and the origin of T. turgidum species. Irrespective of the manner of origin of chromosome 4B and the cyclical translocation, there is additional evidence for the existence of a cyclical translocation involving chromosomes 4B, 5A, and 7B. Riley and Chapman (1966) analyzed

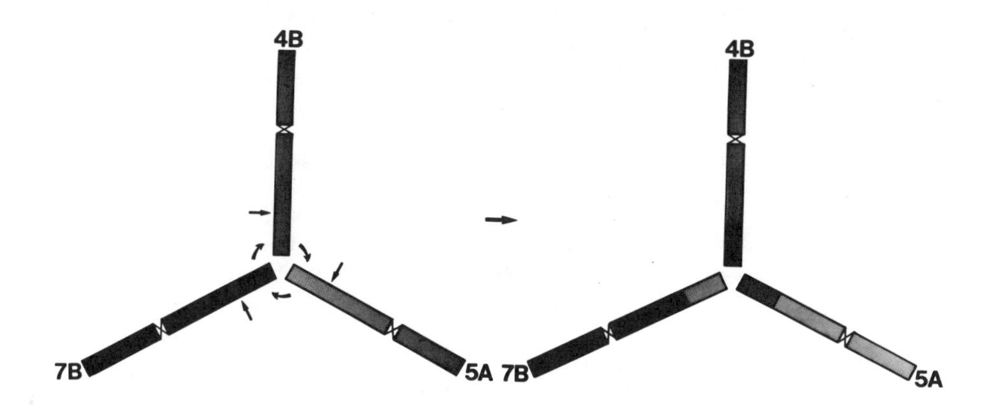

Fig. 5. Apparent cyclical translocation showing putative normal chromosomes 4B, 5A, and 7B (left) and translocated chromosomes 4B-7B, 5A-4B, and 7B-5A (right) that accompanied the evolution of modern day wheats. Figure based on paper by Naranjo et al. (in press) and our own research.

homoeologous pairing among the 5AL, 5BL, and 5DL arms and observed that 5AL paired with 5BL and 5DL in 7% of the cells as compared to 44% pairing between the 5BL and 5DL arms. The reduced pairing of the 5AL arm may be caused by a segment of 4BL which is translocated distally to the 5AL arm. In addition, Ainsworth et al. (1983) located β-amylase loci on arms 4AL, 4DL, and 5AL, the last presumably on the 4BL segment translocated to 5A as 5AS.5AL-4BL. Additional evidence for the translocated nature of the 4B chromosome (4BS.4BL-7BS) has been presented by Gill and Chen (1987). It is noteworthy that whereas *T. turgidum* has a translocated 7B (7BL.7BS-5AL), *T. timopheevii* chromosome 7G is intact, as observed from pairing of the 7BS segment of 4BL (4BS.4BL-7BS) telosomic with a 7G arm and complete lack of pairing of 7BS telo (.7BS-5AL) with 7G of *T. timopheevii* (Gill and Chen, 1987; see also Fig. 4).

Apart from analysis of evolutionary affinities and ancient translocations, the role of heterochromatin on various facets of meiotic pairing became amenable to analysis by banding techniques. Thomas and Kaltsikes (1974) were the first to observe that in triticale meiosis, most univalents belonged to rye chromosomes with large heterochromatic knobs. Other facets of meiotic pairing analysis, including the relative roles of arm length and heterochromatin content on chiasma frequency, coorientation of multivalents and univalents, chiasma terminalization, and achiasmate associations, have been investigated (Ferrer et al., 1984; Fominaya and Jouve, 1985; Naranjo and Lacedena, 1982; Jones, 1978; Orellana, 1985). In fact, the ease with which heterochromatin differentiation and chromosome identification can be achieved in wheat and its many hybrids, now makes this group of plants the most suitable material for many kinds of theoretical and applied cytogenetic studies.

IV. DETECTION OF AMOUNT AND LOCATION OF ALIEN CHROMATIN IN WHEAT

Wheat, being a polyploid crop, can tolerate alien chromosome segments transferred from related genera. The general procedures for alien chromatin transfers have been reviewed (Sears, 1972; Sears, 1981). There are several interesting properties of alien chromatin segments introduced into wheat. The genes on the alien chromosome segment form a single linkage block, because they do not pair or cross over with the homoeologous wheat chromosome segments (Fig. 6). Thus, they can be handled as a single 'super gene' in breeding populations. Such a concentration of desirable genes on a small alien chromatin segment in the wheat background also makes them useful targets for eventual molecular cloning of the desirable genes. Moreover, the alien segments may serve as 'home' to genetically engineered genes, and permit gene pyramiding for durable resistance breeding.

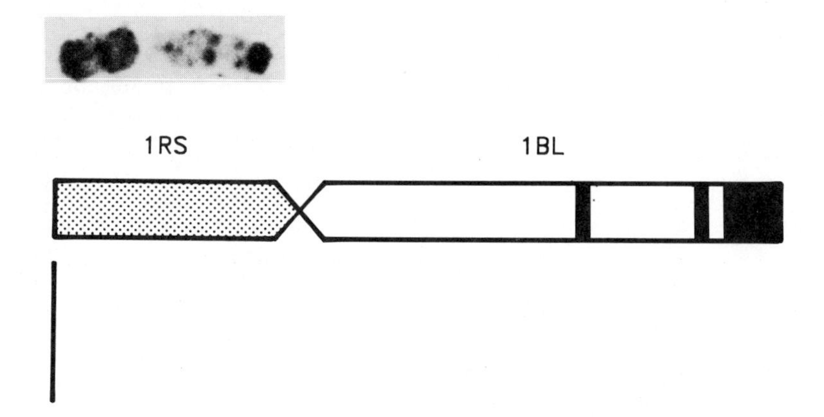

Lr26 Sr27 Yr9 Pm7 Wcm
Gb Sec1 Gpi-1 Per-1
5S DNA 18S·26S rDNA pSC74 pSC119

Fig. 6. The *in situ* hybridization pattern of translocated chromosome 1RS.1BL in Bobwhite wheat labeled with pSC119 rye repeated DNA probe. The rye arm is completely labeled. The 1BL arm shows three distinct sites of hybridization. The idiogram shows the composition of the translocated chromosome as determined from *in situ* hybridization analysis. Below are listed a compilation of genes and DNA sequences that have been found in different 1RS arms of rye transferred into wheat (Bobwhite, Amigo, and others). *Wcm* indicates resistance to wheat curl mite, other gene symbols are standard designations.

Often alien chromatin segments may also carry deleterious genes, however. Wheats containing the 1RS arm of rye have a sticky-dough problem, and may be unsuitable for bread making in some situations. Further cytogenetic manipulation of the alien segment may be needed and has been attempted for the 1RS rye segment (Koebner and Shepherd, 1986).

In such cytogenetic manipulations, the alien chromatin may be detected by chromosome pairing analysis and a variety of genetic markers including the use of appropriate DNA probes in Southern blots (Gill et al., 1988; Sears, 1972; Sears, 1981; Koebner et al., 1986). These techniques generally do not provide an accurate estimate of the breakage point and the amount of alien chromatin. The chromosome bands also do not mark the alien or the wheat chromosome in their entirety. Instead, a combination of genome-specific repeated DNA probes, in conjunction with in situ hybridization, may provide an accurate measure of the amount and location of alien chromatin introduced into wheat (Appels and Moran, 1984).

Recently, in our laboratory, we have used a biotin-labelling technique that has some distinct advantages over radiolabelled probes. The procedure is rapid, requiring only 6 h for hybridization and detection steps. The sites of hybridization lie on the chromosomes. As a result, translocation junctions involving two types of chromatin can be easily distinguished. Moreover, DNA sequence deletion or amplification changes involving highly repeated DNA sequences can be clearly observed. These and other advantages and applications of this technique have been described (Rayburn and Gill, 1985a; Lapitan et al., 1986; Rayburn and Gill, 1986; Rayburn and Gill, 1987a,b,c).

Lapitan et al. (1986) used the biotin-labelling technique to observe translocation breakpoints of wheat and rye arms in 1BL.1RS chromosome in 'Bobwhite' (Fig. 6) and 1AL.1RS chromosome in 'Amigo.' A rye repeated DNA probe pSCl19 was used. Originally, Bedbrook et al. (1980) described pScl19 to be part of a 120 bp tandem repeated DNA family. Recent work (Appels and Frankel, personal communication) has shown that pSCl19 actually consists of three fragments (after digestion with HindIII): one is dispersed, one heterochromatic, and one (a small fragment) in too low a copy number to be detected by in situ hybridization. Lapitan et al. (1986) found pSCl19 to be dispersed on all rye chromosomes, producing a brown precipitate following in situ hybridization. In wheat, the pSCl19 probe gave 1-4 distinct and major sites of hybridization on 10 chromosomes; the remaining chromosomes and chromosome segments had a blue appearance indicating lack of hybridization. In the translocation chromosomes 1AL.1RS and 1BL.1RS, the short arms were labelled all over, indicating them to be rye chromosome arms. The 1AL arm was unlabelled, and the 1BL arm showed typical sites of hybridization - a faint interstitial and a distinct terminal site of hybridization (Fig. 6). This analysis identified these to be wheat chromosome arms. In both cases, the translocation breakpoint was at the centromere. Thus, biotin-labelling provided an efficient and sensitive method for detection of the amount, location, and breakage-point of the introduced alien chromatin. The 1B/1R translocation first arose in German wheats from spontaneous centromeric misdivision (Mettin et al., 1973). The fact that Amigo wheat also contained a centromeric breakage-point may indicate its origin by spontaneous centromeric misdivision also, although x-ray irradiation was used in the original experiment (Sebesta and Wood, 1978).

With respect to the pSCl19 probe, it is noteworthy that Jones and Flavell (1982a,b) found it to be primarily distributed in the heterochromatic bands of rye following the application of 65,000 cpm (^{3}H) and 21-28 day autoradiographic exposure in one experiment and 110,000 cpm/slide and exposure of 14 days in another experiment. In the study by Lapitan et al. (1986) on chromosomes of Chaupon rye, in addition to dark brown labelling of heterochromatic regions, all chromosomes also gave an overall light brown appearance indicating the dispersed nature of the sequence. This roughly provides a relative comparison of the sensitivities of the two techniques. Under these conditions biotin-labelling proved to be more sensitive. However, it should be pointed out that with radiolabelled probes, there is room for improving the efficiency of the technique by use of high specific activity probes and longer autoradiographic exposures.

V. DNA SEQUENCE CHANGES DETECTED AT THE CHROMOSOME LEVEL IN TISSUE CULTURE

Lapitan et al. (1984) observed enlarged C-bands on certain rye chromosomes (2RL, 7RS) in tissue culture-regenerated wheat-rye hybrids. Similar enlargement of C-bands on chromosomes 1R, 2R, 3R, 4R, and 5R has been observed in natural and cross-breeding populations of rye and triticale (Lapitan, 1984; Gustafson et al., 1983). Lapitan et al. (1984) speculated that enlarged C-bands may arise from translocations (with other C-banded material), although it seemed unlikely since the remaining chromosomes had normal C-bands. Alternatively, it was argued that enlarged C-bands may be due to the amplification of repeated DNA sequences present in them. With the availability of the biotin-labelling technique, the molecular nature of the enlarged C-bands on chromosome arm 7RS was analyzed (Lapitan et al., in press) and the results are summarized in Fig. 7.

a

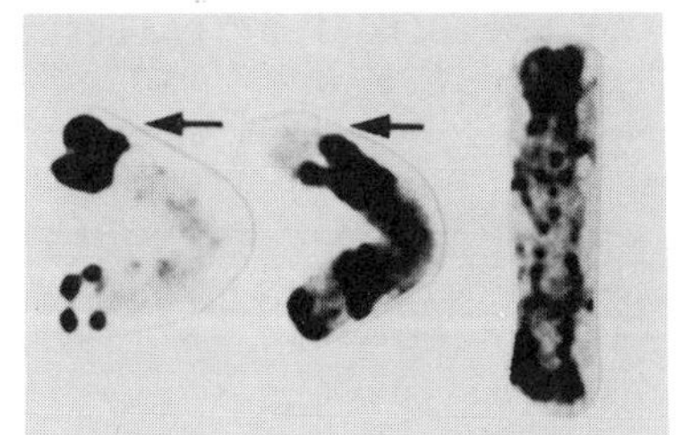

b

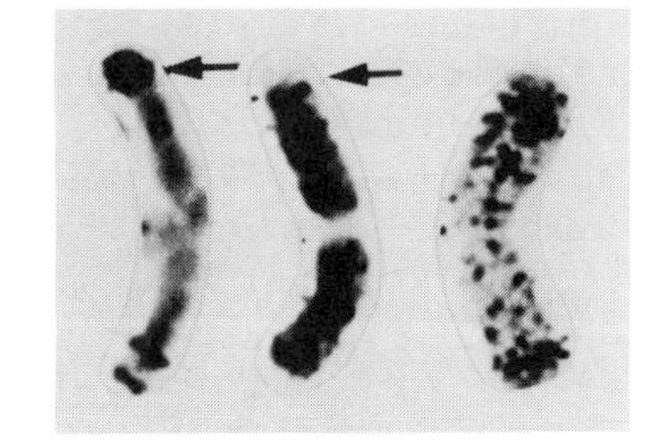

Fig. 7. Chromosome 7R with increased hybridization with pSC74 (left arrow), unlabelled with pSC119 (middle arrow), and doubly labelled with probes pSC74 and pSC119. b. Another set of rye chromosomes showing amplified pSC74 sequence (left arrow) that is missing the pSC119 sequence (middle arrow), and is labelled all over with probes pSC74 and pSC119 (right).

The exclusive amplification of 480 bp sequences (assayed with pSC74 probe) was observed in enlarged C-bands on the 7RS arm. The amplified 480-bp sequences were detected by enlarged hybridization sites with the pSC74 probe; the same 7RS sites were unlabelled with the 120-bp pSC119 probe which otherwise labels normal 7RS telomeric bands (Fig. 7). This means that a substantial portion of the 7RS telomeric band suffered a terminal deletion of 120 bp containing sequences, followed by amplification of the remaining and now distally located 480-bp sequences. This arrangement of the two repeated DNA families in the rye C-bands was reported previously (Jones and Flavell, 1982a,b). Several deletion and amplification events must have taken place, as a range of deletion and amplification events were observed in several hybrids of independent origin.

The reasons for the instability of the 7RS telomeric C-band in tissue culture and the mechanism of deletion and amplification events are not known. Gustafson et al. (1983) speculated that C-band deletion and reduplication (giving rise to enlarged C-bands) may occur spontaneously in somatic cells by chromosome breakage and redistribution of telomeric C-bands. The C-band deletions may occur by chromosome breakage; enlarged C-bands are not simple additions of C-band material from other chromosome ends but involve amplification of specific DNA sequences, as shown here.

VI. GENOME-SPECIFIC DNA SEQUENCES

Most of the genome-specific repeated DNA sequences discussed in the previous sections have been isolated in rye. Recently, repeated DNA sequences have been described in *Aegilops squarrosa* L. (syn. *T. tauschii*), the D genome donor of wheat (Rayburn and Gill, 1987a,b; Henry, 1988). The D genome is a pivotal genome in several polyploid *Aegilops* species. This species exhibits tremendous diversity in both genetic and physiological traits that are useful in wheat improvement (Gill and Raupp, 1987). The diverse races of the species may vary in DNA content by 20% (Furuta et al., 1975) and a study of the mechanisms of the genome plasticity are of particular interest. *Aegilops squarrosa* has been analyzed for DNA restriction fragment length polymorphisms and for genome-specific repeated DNA sequences as an approach towards mapping the genome of wheat as well as the analysis of genome plasticity in *A. squarrosa* (Henry 1988; Kam-Morgan et al., 1987; Rayburn and Gill, 1987a).

Rayburn and Gill (1987a) reported a pAS1 clone that preferentially labelled D-genome chromosomes following *in situ* hybridization (Fig. 8). The sequence is predominantly found in the telomeric regions of D-genome chromosomes with only a few interstitial sites. Trace sites of hybridizations can also be

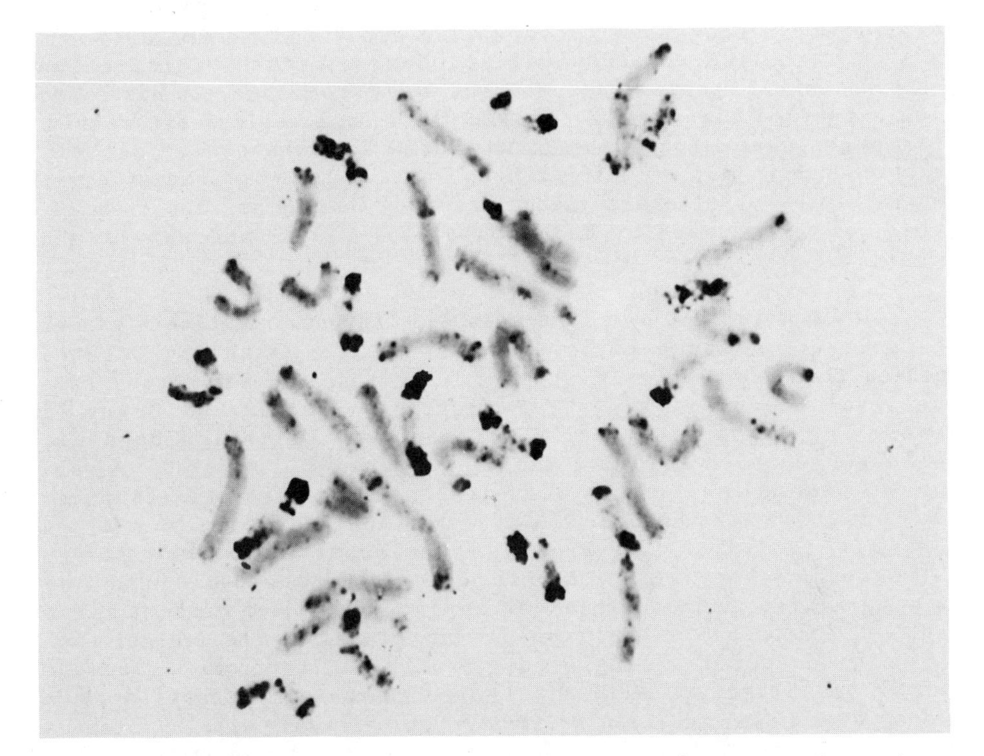

Fig. 8. *In situ* hybridization sites on Chinese Spring wheat chromosomes probed with biotin-labelled pAS1 clone, a D-genome repeated DNA sequence. The heavily labelled chromosomes belong to the D-genome, smaller sites of pAS1 sequence can be seen on A and B genome chromosomes.

observed in some A- and B-genome chromosomes in Fig. 8, a result further confirmed by the presence of pAS1 sequences in AB-genome DNA in Southern blots (Gill, unpublished results). The pAS1 sequence was also distributed in the other D-genome containing polyploid species. Most of the pAS1 related sequences appear to be located at the site of C-bands indicating that this sequence comprises a significant component of constitutive heterochromatin of D-genome chromosomes (Rayburn and Gill, 1986). The pAS1 probe may be useful in analyzing the arrangement of D-genome chromosomes in the wheat nucleus (Rayburn and Gill, 1987c). Two other repeated DNA clones were dispersed on A, B, and D genome chromosomes (Rayburn and Gill 1986).

Henry (1988) further analyzed a large number of D-genome clones (in pUC8) and selected those that appeared to be specific to the D-genome based on dot blot assays. Further analysis of these clones by *in situ* hybridization and Southern blot analyses revealed that they were present in A and B genome DNA as well. In situ hybridization resulted in three general hybridization patterns: 1) dispersed over most of the chromosome, 2) dispersed with a few localized areas, and 3) distinct localized sites only. It appears that either the A, B, and D genomes are so closely related that it may be difficult to isolate truly genome-specific DNA sequences or alternate techniques are needed as, for example, those that were used by Bedbrook et al. (1980) and Appels and Moran (1984).

As the work with rye indicates, it has been relatively easier to isolate genome-specific DNA probes from distantly related species that are absent in wheat. Thus, repeated DNA clones have been isolated from Thinopyrum elongata (Host.) D. R. Dewey (E genome), Haynaldia villosa L. (V genome), Critesion bogdanii (Wilensky) Love (H genome), and Pseudoroegneria spicata (Pursh) Love (S genome) that are widely dispersed in these taxa but are not found in wheat (Appels and Moran, 1984; McIntyre et al., in press, a,b; Gill et al., 1988). These repeated DNA probes are valuable for assaying alien chromatin in Southern blots and by in situ hybridization. Another application that remains to be explored is the use of the repeated DNA clones in the isolation of chromosome-specific single copy probes for restriction fragment length polymorphism (RFLP) analysis. In this approach, cosmid clones from alien-addition or translocation lines (with homologous DNA sequences to a known wheat chromosome) may be identified with genome-specific repeated DNA probes. From these cosmid clones, unique sequence clones may be isolated for RFLP analysis of a specific wheat chromosome (unique sequences have been conserved throughout the Triticeae tribe).

VII. GENOMIC AFFINITIES OF INDIVIDUAL LOCI AND CHROMOSOMES IN POLYPLOID GENOMES

There are very few methods that can be reliably used for assigning genomic affinities and evolutionary relationships at the chromosomal level among species and genera. The chromosome pairing method, as discussed earlier, oversimplifies the complex nature of genome evolution. It is only in combination with marker chromosomes and other techniques such as chromosome banding analysis, and additional supporting evidence, that a clearer picture of phylogenetic relationships may be deduced (Gill and Chen, 1987). There are instances, however, where a lack of meiotic pairing may be an inherent property of a chromosome(s) or an entire genome when transferred into a different genetic background, and meiotic analysis actually may lead to misleading conclusions (Kota et al., 1986). Introgressive hybridization may also alter the complexion of a genome in a similar fashion, leading to misleading conclusions from pairing analysis. Finally,

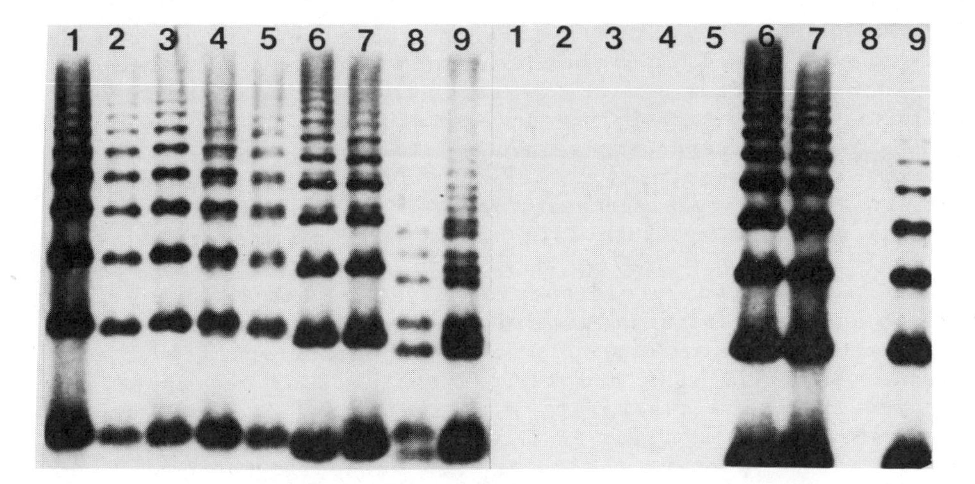

Fig. 9. BamHl digested DNA samples from H-genome species (lanes 1-6), *Critesion bogdanii* (1), *C. roshevitzii* (2), *C. brachyantherum* (3), *C. chilense* (4), *C. californicum* (5), S-genome species *Pseudoroegneria spicata* (6), SH-genome species *Elymus trachycaulus* (7) and derived wheat-*E. trachycaulus* disomic addition lines $5H^t$ (8), and $4S^t$ (9) probed with pPST2 (S-genome 5S DNA clone) at 37°C (left) and 55°C (right). The S-genome specificity of the probe (lanes 6, 7) and the 5S DNA locus on $4S^t$ chromosome (lane 9) can be seen at 55°C.

individual chromosomes may undergo structural alterations in a manner that may affect their pairing ability, and hence pairing analysis may offer no clues to their evolutionary affinity.

Among the alternative approaches, genome-specific repeated DNA sequences, although invaluable in cytogenetic analysis, have limited value in phylogenetic analysis. The genome specificity of repeated DNA sequences is not absolute and certain repeated sequences previously thought to be limited to a single genome have been found distributed in diverse genomes (Lapitan et al., 1987; Xin and Appels, in press). Moreover, in certain cases the repeated DNA sequences that are restricted to one of the progenitor species may be spread across two or more genomes in a polyploid species due, in part, to the nature of polyploid-genome evolution (Gill et al., 1988). In certain cases, the distinctive pattern of repeated DNA sequence distribution may be useful in phylogenetic analysis (Rayburn and Gill, 1985b). However, in such cases, one must assume constancy of distribution of the repeated DNA sequence in a chromosome or genome. In most cases, this constancy is unlikely, due to the rapid rate of evolutionary change in nucleotide sequence, location, and amount of repeated DNA (McIntyre et al., a,b, in press).

Perhaps defined DNA sequences whose homologies can be determined based on relative stringencies of molecular hybridization offer a more promising tool for assaying genomic affinities of individual loci and chromosomes. In ribosomal RNA (rRNA) loci, unlike coding sequences which are highly conserved, the spacer DNA sequences in 18S.28S, and 5S genes can diverge rapidly and have been successfully used in phylogenetic analysis (Dvorak and Appels, 1982; Gill and Appels, in press; McIntyre et al.,b, in press; Scoles et al., in press). Apparently, in these instances polyploid species may have had a more recent origin and genomic affinities of their rRNA loci could still be assayed by spacer rDNA probes from progenitor species or vice versa. Using 5S rDNA and 18S.28S rDNA spacer sequences, we assayed the genomic affinities of individual E. trachycaulus chromosomes added to wheat as illustrated in Fig. 9 (see also Gill et al., 1988). The limitation of this method is the paucity of the loci that are amenable to this kind of molecular analysis.

VIII. CONCLUSIONS

The N-banding and C-banding techniques provide a rapid means of detection and characterization of the heterochromatin structure of wheat chromosomes. It can be seen that B-genome heterochromatin is distinctive (C^+N^+) from heterochromatin of A- and D-genome chromosomes (variable C^+N^+ or C^+N^-). Molecular analysis further reveals the heterogeneity of heterochromatin, as the heterochromatin bands within chromosomes differ in repeated DNA composition. Some repeated DNA sequences may be present in the heterochromatin of many different genomes; species-specific repeated DNA sequences are also present, however. Molecular mechanisms apparently exist for rapid turnover of repeated DNA sequences. This characterization of heterochromatin is of great value in analysis of individual chromosomal homologies in evolutionary studies and in practical plant breeding. Furthermore, heterochromatic bands provide valuable landmarks for the physical mapping of the genome of wheat.

ACKNOWLEDGEMENTS

The research reviewed in this contribution was carried out by a large number of graduate students, postdoctoral fellows, and visiting scientists over several years and we express our grateful thanks to T. R. Endo, Fan Lu, R. Schlegel, K. L. D. Morris, J. Henry, P. D. Chen, N. L. V. Lapitan, and A. L. Rayburn. One of us (B.S.G.) thanks Rudi Appels for his hospitality during his sabbatical stay at CSIRO, Canberra, where part of the work was done. W. J. Raupp deserves our thanks for skillful technical assistance.

REFERENCES

Ainsworth, C. C., Gale, M. D., and Baird, S., 1983, The genetics of β-amylase isozymes in wheat. I. Allelic variation among hexaploid varieties and intrachromosomal gene locations, Theor. Appl. Genet., 66:39-50.

Appels, R., 1982, The molecular cytology of wheat-rye hybrids. Int. Rev. Cytol., 80:83-132.

Appels, R., Driscoll, C., and Peacock, W. J., 1978, Heterochromatin and highly repeated DNA sequences in rye (Secale cereale), Chromosoma, 70:67-89.

Appels, R., Gustafson, J. P., and May, C. E., 1982, Structural variation in the heterochromatin of rye chromosomes in triticales, Theor. Appl. Genet., 63:235-244.

Appels, R., and L. Moran, 1984, Molecular analysis of alien chromatin introduced into wheat, in: "Gene Manipulation in Plant Improvement," 16th Stadler Genet. Symp., J. P. Gustafson, ed., Plenum Press, NY. pp. 529-558.

Bedbrook, J. R., Jones, J., O'Dell, M., Thompson, R. D., and Flavell, R. B., 1980, A molecular description of telomeric heterochromatin in Secale species, Cell, 18:545-561.

Bhattacharyya, N. K., and B. C. Jenkins, 1960, Karyotype analysis and chromosome designations for Secale cereale L. 'Dakold,' Can. J. Genet. Cytol., 2:268-277.

Bianchi, M. S., Bianchi, N. O., Pantelias, G. E., and Wolff, S., 1985, The mechanism and pattern of banding induced by restriction endonucleases in human chromosomes, Chromosoma, 91:131-136.

Buys, Ch. H. C. M., and Osinga, J., 1982, A relation between G-, C-, and N-band patterns as revealed by progressive oxidation of chromosomes and a note on the nature of N-bands, Genetica, 58:3-9.

Chen, P. D., and B. S. Gill, 1983, The origin of chromosome 4A, and genomes B and G of tetraploid wheats, Proc. 6th Int. Wheat Genet. Symp., Kyoto, Japan, 39-48.

Dennis, E. S., Gerlach, W. L. and Peacock, W. J., 1980, Identical polypyrimidine-polypurine satellite DNAs in wheat and barley, Heredity, 44:349-366.

Dvorak, J., and Appels, R., 1982, Chromosome and nucleotide sequence differentiation in genomes of polyploid Triticum species, Theor. Appl. Genet., 73:349-360.

Dvorak, J., Chen, K. C., and Giorgi, B., 1984, The C-band pattern of a Ph^{-} mutant of a durum wheat, Can. J. Genet. Cytol., 26:360-363.

Endo, T. R., and Gill, B. S., 1984a, Somatic karyotype, heterochromatin distribution, and nature of chromosome differentiation in common wheat, Triticum aestivum L. em Thell, Chromosoma 89:361-369.

Endo, T. R., and B. S. Gillb, 1984, The heterochromatin distribution and genome evolution in diploid species of Elymus and

Agropyron, Can. J. Genet. Cytol., 26:669-678.

Ferrer, E., Gonzalez, J. M., and Jouve, N., 1984, The meiotic pairing of nine wheat chromosomes, Theor. Appl. Genet., 69:193-198.

Fominaya, A., and Jouve, N., 1985, Metaphase I centromere coorientation in interchange heterozygotes of Triticum aestivum L., J. Hered., 76:191-193.

Funaki, K., Matsui, S., and Sasaki, M., 1975, Location of nucleolar organizers in animal and plant chromosomes by means of an improved N-banding technique, Chromosoma, 49:357-370.

Furuta, Y., Nishikawa, K., and Makino, T., 1975, Intraspecific variation of nuclear DNA content in Aegilops squarrosa L., Jpn. J. Genet., 50:257-263.

Gerlach, W. L., 1977, N-banded karyotypes of wheat species, Chromosoma 62:49-56.

Gerlach, W. L., Appels, R., Dennis, E. S., and Peacock, W. J., 1978, Evolution and analysis of wheat genomes using highly repeated DNA sequences, Proc. 5th Int. Wheat Genet. Symp., 1:81-91.

Gill, B. S., 1987, Chromosome banding methods, standard chromosome band nomenclature, and applications in cytogenetic analysis, in: "Wheat and Wheat Improvement," 2nd ed., E. G. Heyne, ed., Am. Soc. Agron., Madison, WI. pp. 243-254.

Gill, B. S., and R. Appels, Evolutionary change in grasses of the Triticeae, III, Relationships between Nor-loci of different species, Plant Syst. Evol. (in press).

Gill, B. S., and P. D. Chen, 1987, Role of cytoplasm-specific introgression in the evolution of polyploid wheats, Proc. Nat. Acad. Sci. USA, 84:6800-6804.

Gill, B. S., Morris, K. L. D., and Appels, R., 1988, Assignment of the genomic affinities of chromosomes from polyploid Elymus species added to wheat, Genome (in press).

Gill, B. S., and Raupp, W. J., 1987, Direct genetic transfers from Aegilops squarrosa L. to hexaploid wheat, Crop Sci. 27: 445-450.

Gustafson, J. P., Lukaszewski, A. J., and Bennett, M. D., 1983, Somatic deletion and redistribution of telomeric heterochromatin in the genus Secale and in triticale, Chromosoma, 88:293-298.

Henry, J., 1988, Characterization of D-genome repeated DNA clones from Aegilops squarrosa, M.S. thesis, Kans. St. Univ., Manhattan.

Hsu, T. C., 1973, Longitudinal differentiation of chromosomes, Annu. Rev. Genet., 7:153-176.

Jampates, R., and J. Dvorak, 1986, Location of the Phl locus in the metaphase chromosome map and the linkage map of the 5Bq arm of wheat, Can. J. Genet. Cytol., 28:511-519.

Jewell, D. C., 1981, Recognition of two types of positive staining chromosomal material by manipulation of critical steps in the N-banding technique, Stain Technol., 56:227-234.

Jones, G. H., 1978, Giemsa C-banding of rye meiotic chromosomes and the nature of 'terminal' chiasmata, Chromosoma, 66:45-57.

Jones, J. D. G., and R. B. Flavell, 1982a, The mapping of highly repeated DNA families and their relationship to C-bands in chromosomes of Secale cereale, Chromosoma, 86:595-612.

Jones, J. D. G., and R. B. Flavell, 1982b, The structure, amount and chromosome localisation of defined repeated DNA sequences in species of the genus Secale, Chromosoma, 86:613-641.

Kam-Morgan, L. N. W., 1987, DNA restriction fragment length polymorphisms as genetic markers in wheat, Ph.D. thesis, Kans. St. Univ., Manhattan.

Kimber, G., 1982, Evolutionary relationships and their influence on plant breeding, in: "Gene Manipulation in Plant Improvement," 16th Stadler Genet. Symp.:281-293.

Koebner, R. M. D., and Shepherd, K. W., 1986, Controlled introgression of genes from rye chromosome arm 1RS by induction of allosyndesis, I. Isolation of recombinants, Theor. Appl. Genet., 73:197-208.

Koebner, R. M. D., Shepherd, K. W., and Appels, R., 1986, Controlled introgression to wheat of genes from rye chromosome arm 1RS by induction of allosyndesis, II, Characterization of recombinants, Theor. Appl. Genet. 73:209-217.

Kota, R. S., and J. Dvorak, 1986, Mapping of a chromosome pairing gene and 5S rRNA genes in Triticum aestivum L. by a spontaneous deletion in chromosome arm 5Bp, Can. J. Genet. Cytol., 28:266-271.

Kota, R. S., McGuire, P. E., and Dvorak, J., 1986, Latent nonstructural differentiation among homoeologous chromosomes at the diploid level, Chromosome $6B^l$ of Aegilops longissima, Genetics 114:579-592.

Lapitan, N. L. V., Gill, B. S., and Sears, R. G., 1987, Genomic and phylogenetic relationships among rye and perennial species in the Triticeae, Crop Sci., 27:682-687.

Lapitan, N. L. V., Sears, R. G., and Gill, B. S., 198 , Amplification of repeated DNA sequences in wheat x rye hybrids regenerated from tissue culture, Theor. Appl. Genet., (in press).

Lapitan, N. L. V., Sears, R. G., and Gill, B. S., 1984, Translocations and other karyotypic structural changes in wheat x rye hybrids regenerated from tissue culture, Theor. Appl. Genet. 68:547-554.

Lapitan, N. L. V. Sears, R. G. Rayburn, A. L., and Gill, B.S., 1986, Wheat-rye translocations: Detection of chromosome breakpoints by in situ hybridization with a biotin-labeled DNA probe, J. Hered. 77:415-419.

May, C. E., and Appels, R., 1987, The molecular genetics of wheat: Toward an understanding of 16 billion base pairs of DNA, in: "Wheat and Wheat Improvement," 22nd ed., E. G. Heyne, ed., Am. Soc. Agron., Madison, WI. pp. 166-168.

McIntyre, C. L., Clarke, B. S., and Appels, R., 198 a. Evolutionary change in grasses of the Triticeae, I. Amplification and dispersion of repetitive DNA sequences, Plant Syst. Evol. (in press).

McIntyre, C. L., Clarke, B. C., and Appels, R., 198 b., IV. Evolutionary change in grasses of the Triticeae. IV. DNA sequence analyses of the ribsomal DNA spacer regions, Plant Syst. Evol. (in press).

Mettin, D., Bluthner, W. D., and Schegel, G., 1973, Additional evidence on spontaneous 1B/1R wheat-rye substitutions and translocations, Proc. 4th Int. Wheat Genet. Symp., Mo. Agric. Exp. Stn., Columbia, MO. pp. 179-184.

Miller, D. A., Choi, Y-C., and Miller, O. J., 1983, Chromosome localization of highly repetitive human DNAs and amplified ribosomal DNA with restriction enzymes, Science, 219:395-397.

Morris, K. L. D., and B. S. Gill, 1987, Genomic affinities of individual chromosomes based on C- and N-banding analyses of tetraploid Elymus species and their diploid progenitor species, Genome, 29:247-252.

Morrison, J. W., 1953, Chromosome behavior in wheat monosomics, Heredity, 7:207-207.

Naranjo, T., and Lacedena, J. R., 1982, Wheat univalent orientation at anaphase I in wheat-rye derivatives, Chromosoma, 84: 653-661.

Naranjo, T., Roca, A., Goicoechea, P. G., and Giraldez, R., 198 , Arm homoeology of wheat and rye chromosomes, Genome (in press).

Orellana, J., 1985, Most of the homoeologous pairing at metaphase I in wheat-rye hybrids is not chiasmatic, Genetics, 111: 917-931.

Pimpinelli, S., Santini, G., and Gatti, M., 1976, Characterization of Drosophila heterochromatin, Chromosoma, 57:377-386.

Rayburn, A. L., and B. S. Gill, 1985a, Use of biotin-labeled probes to map specific DNA sequences on wheat chromosomes, J. Hered., 76:78-81.

Rayburn, A. L., and B. S. Gill, 1985b, Molecular evidence for the origin and evolution of chromosome 4A in polyploid wheats, Can. J. Genet. Cytol. 27:246-250.

Rayburn, A. L., and B. S. Gill, 1986, Molecular identification of D-genome chromosomes of wheat, J. Hered., 77:253-255.

Rayburn, A. L., and B. S. Gill, 1987a, Isolation of a D-genome specific repeated DNA sequence from Aegilops squarrosa, Plant Mol. Biol. Reptr., 4:102-109.

Rayburn, A. L., and B. S. Gill, 1987b, Molecular analysis of the D-genome of the Triticeae, Theor. Appl. Genet., 73:385-388.

Rayburn, A. L., and B. S. Gill, 1987c, Use of repeated DNA sequences as cytological markers, Am. J. Bot., 74:574-580.

Riley, R., and Chapman, V., 1966, Estimates of the homoeology of wheat chromosomes by measurements of differential affinity at meiosis, in: "Chromosome Manipulation and Plant

Genetics," R. Riley and K. R. Lewis, eds., Oliver and Boyd, Edinburgh. pp. 46-58.

Rocchi, A., 1982, On the heterogeneity of heterochromatin, Caryologia, 35:169-189.

Schlegel, R., and B. S. Gill, 1984, N-banding analysis of rye chromosomes and the relationship between N-banded and C-banded heterochromatin, Can. J. Genet. Cytol., 22:765-769.

Scoles, G., Gill, B. S., Xin, Z-Y., Clarke, B. C., McIntyre, L. L., and Appels, R., 198 , Evolutionary change in grasses of the Triticeae, V. Fragment duplication and deletion events in 5S DNA spacer regions, Plant Syst. Evol. (in press).

Sears, E. R., 1948, The cytology and genetics of the wheats and their relatives, Adv. Genet. II:239-270.

Sears, E. R., 1954, The aneuploids of common wheat, Mo. Agric. Exp. Stn. Res. Bull. 572. 58 pp.

Sears, E. R., 1972, Chromosome engineering in wheat, Stadler Genet. Symp., Mo. Agric. Exp. Stn. 4:23-38.

Sears, E. R., 1981, Transfer of alien genetic material to wheat, in: "Wheat Science - Today and Tomorrow," L. T. Evans and W. J. Peacock, eds., Cambridge University Press, London. pp. 75-89.

Sears, E. R., and L. M. S. Sears, 1978, The telocentric chromosomes of common wheat, Proc. 5th Int. Wheat Genet. Symp., New Delhi:389-407.

Sebesta, E. E., and Wood, E. A. Jr. 1978, Transfer of greenbug resistance from rye to wheat with x-rays, Agron. Abstrs. pp. 61-62.

Singh, R. J., and Tsuchiya, T., 1982, Identification and designation of telocentric chromosomes in barley by means of Giemsa N-banding technique, Theor. Appl. Genet., 64:13-24.

Snape, J. W., Flavell, R. B., O'Dell, W. G., Hughes, W. G., and Payne, P. I., 1985, Intrachromosomal mapping of nucleolar organizing region relative to 3 marker loci of 1B of T. aestivum., Theor. Appl. Genet., 69:263-270.

Thomas, J. B., and P. J. Kaltsikes, 1974, A possible effect of heterochromatin on chromosome pairing, Proc. Natl. Acad. Sci. USA, 71:2787-2790.

Vosa, C. G., 1981, A new feulgen method for SCE-detection in plant chromosomes, Caryologia, 34:15-361.

Xin, Z. Y., and Appels, R., 198 , Evolutionary change in grasses of the Triticeae, II. Distribution of rye (Secale cereale) 350-family DNA sequences, Plant Syst. Evol., (in press).

Zhu, F., Wei, J., Fu, J., and Liancheng, L., 1986, High resolution banding in plant chromosomes, First Int. Symp. Chromosome Engg. Plants, Xian, China (Abstr.).

Contribution no. 88-407-A, Department of Plant Pathology and the Wheat Genetics Resource Center, Kansas Agricultural Experiment Station, Kansas State University, Manhattan.

INDEX